*Stochastic Mechanics*
*Random Media*
*Signal Processing*
*and Image Synthesis*
*Mathematical Economics*
*Stochastic Optimization*
*Stochastic Control*

# Applications of Mathematics

*Stochastic Modelling
and Applied Probability*

# 41

*Edited by* I. Karatzas
M. Yor

*Advisory Board* P. Brémaud
E. Carlen
W. Fleming
D. Geman
G. Grimmett
G. Papanicolaou
J. Scheinkman

# Springer
*New York
Berlin
Heidelberg
Barcelona
Hong Kong
London
Milan
Paris
Singapore
Tokyo*

# Applications of Mathematics

*(continued after index)*

Terje Aven    Uwe Jensen

# Stochastic Models
# in Reliability

*With 12 Figures*

 Springer

Terje Aven
Department of Mathematics
University of Oslo
N-0316 Oslo, Norway

Uwe Jensen
Department of Stochastics
University of Ulm
D-89069 Ulm, Germany

*Managing Editors*

I. Karatzas
Departments of Mathematics and Statistics
Columbia University
New York, NY 10027, USA

M. Yor
CNRS, Laboratoire de Probabilités
Université Pierre et Marie Curie
4, Place Jussieu, Tour 56
F-75252 Paris Cedex 05, France

Mathematics Subject Classification (1991): 62L20, 93E10, 93E23, 65C05, 93E35, 93-02, 90C15

Library of Congress Cataloging-in-Publication Data
Aven, T. (Terje)
    Stochastic models in reliability / Terje Aven, Uwe Jensen.
        p.    cm. — (Applications of mathematics ; 41)
    Includes bibliographical references and index.
    ISBN 0-387-98633-2 (alk. paper)
    1. Reliability (Engineering)—Mathematical models.   2. Stochastic
processes.   I. Jensen, Uwe, 1931–  .   II. Title.   III. Series.
TA169.A95   1998
620′.00452′015118—dc21                                    98-31039

Printed on acid-free paper.

Production managed by Allan Abrams; manufacturing supervised by Joe Quatela.
Photocomposed copy prepared from the authors' Scientific Word files.
Printed and bound by Edwards Brothers, Inc., Ann Arbor, MI.
Printed in the United States of America.

9 8 7 6 5 4 3 2 1

ISBN 0-387-98633-2 Springer-Verlag New York Berlin Heidelberg    SPIN 10695247

# Preface

As can be seen from the files of the databases of *Zentralblatt/Mathematical Abstracts* and *Mathematical Reviews,* about 1% of all mathematical publications are connected to the keyword *reliability.* This gives an impression of the importance of this field and makes it clear that it is impossible to include all the topics connected to reliability in one book. The existing literature on reliability covers *inter alia* lifetime analysis, complex systems and maintenance models, and the books by Barlow and Proschan [33, 34] can be viewed as first milestones in this area. Since then the models and tools have been developed further. The aim of *Stochastic Models in Reliability* is to give a comprehensive up-to-date presentation of some of the classical areas of reliability, based on a more advanced probabilistic framework using the modern theory of stochastic processes. This framework allows the analyst to formulate general failure models, establish formulas for computing various performance measures, as well as to determine how to identify optimal replacement policies in complex situations. A number of special cases analyzed previously can be included in this framework. Our book presents a unifying approach to some of the key research areas of reliability theory, summarizing and extending results obtained in recent years. Having future work in this area in mind, it will be useful to have at hand a general set-up where the conditions and assumptions are formulated independently of particular models.

This book comprises five chapters in addition to two appendices.

Chapter 1 gives a short introduction to stochastic models of reliability, linking existing theory and the topics treated in this book. It also contains an overview of some questions and problems to be treated in the book.

In addition Section 1.1.6 explains why martingale theory is a useful tool for describing and analyzing the structure of complex reliability models. In the final section of the chapter we briefly discuss some important aspects of reliability modeling and analysis, and present two real-life examples. To apply reliability models in practice successfully, there are many challenges related to modeling and analysis that need to be faced. However, it is not within the scope of this book to discuss these challenges in detail. Our text is an introduction to the topic and of motivational character.

Chapter 2 presents an overview of some parts of basic reliability theory: the theory of complex (monotone) systems, both binary and multistate systems, as well as lifetime distributions and nonparametric classes of lifetime distributions. The aim of this chapter has not been to give a complete overview of the existing theory, but to highlight important areas and give a basis for the coming chapters.

Chapter 3 presents a general set-up for analyzing failure-prone systems. A (semi-) martingale approach is adopted. This general approach makes it possible to formulate a unifying theory of both nonrepairable and repairable systems, and it includes point processes, counting processes, and Markov processes as special cases. The time evolution of the system can also be analyzed on different information levels, which is one of the main attractions of the (semi-) martingale approach. Attention is drawn to the failure rate process, which is a key parameter of the model. Several examples of application of the set-up are given, including a monotone (coherent) system of possibly dependent components, and failure time and (minimal) repair models. A model for analyzing the time to failure based on risk reserves (the difference between total income and accumulated costs of repairs) is also covered.

In the next two chapters we look more closely at types of models for analyzing situations where the system and its components could be repaired or replaced in the case of failures, and where we model the downtime or costs associated with downtimes.

Chapter 4 gives an overview of availability theory of complex systems, having components that are repaired upon failure. Emphasis is placed on monotone systems comprising independent components, each generating an alternating renewal process. Multistate systems are also covered, as well as systems comprising cold standby components. Different performance measures are studied, including the distributions of the number of system failures in a time interval and the downtime of the system in a time interval. The chapter gives a rather comprehensive asymptotic analysis, providing a theoretical basis for approximation formulae used in cases where the time interval considered is long or the components are highly available.

Chapter 5 presents a framework for models of maintenance optimization, using the set-up described in Chapter 3. The framework includes a number of interesting special cases dealt with by other authors.

By allowing different information levels, it is possible to extend, for example, the classical age replacement model and minimal repair/replacement model to situations where information is available about the underlying condition of the system and the replacement time is based on this information. Again we illustrate the applicability of the model by considering monotone systems.

Chapters 3 to 5 are based on stochastic process theory, including theory of martingales and point, counting, and renewal processes. For the sake of completeness and to help the reader who is not familiar with this theory, two appendices have been included summarizing the mathematical basis and some key results. Appendix A gives a general introduction to probability and stochastic process theory, whereas Appendix B gives a presentation of results from renewal theory. Appendix A also summarizes basic notation and symbols.

Although conceived mainly as a research monograph, this book can also be used for graduate courses and seminars. It primarily addresses probabilists and statisticians with research interests in reliability. But at least parts of it should be accessible to a broader group of readers, including operations researchers and engineers. A solid basis in probability and stochastic processes is required, however. In some countries many operations researchers and reliability engineers now have a rather comprehensive theoretical background in these topics, so that it should be possible to benefit from reading the more sophisticated theory presented in this book. To bring the reliability field forward, we believe that more operations researchers and engineers should be familiar with the probabilistic framework of modern reliability theory. Chapters 1 and 2 and the first part of Chapters 4 and 5 are more elementary and do not require the more advanced theory of stochastic processes.

References are kept to a minimum throughout, but readers are referred to the bibliographic notes following each chapter, which give a brief review of the material covered and related references.

*Acknowledgments.* We express our gratitude to our institutions, the Stavanger University College, the University of Oslo, and the University of Ulm, for providing a rich intellectual environment, and facilities indispensable for the writing of this book. The authors are grateful for the financial support provided by the Norwegian Research Council and Deutscher Akademischer Austauschdienst. We would also like to acknowledge our indebtedness to Jelte Beimers, Jørund Gåsemyr, Harald Haukås, Tina Herberts, Karl Hinderer, Günter Last, Volker Schmidt, Richard Serfozo, Marcel Smith, Fabio Spizzichino and Rune Winther for making helpful comments and suggestions on the manuscript. Thanks for TeXnical support go to Jürgen Wiedmann.

We especially thank Bent Natvig, University of Oslo, for the great deal of time and effort he spent reading and preparing comments. Thanks also go to the three reviewers for providing advice on the content and organization of the book. Their informed criticism motivated several refinements and improvements. Of course, we take full responsibility for any errors that remain.

We also acknowledge the editing and production staff at Springer for their careful work. In particular, we appreciate the smooth cooperation of John Kimmel.

Stavanger                                                        TERJE AVEN
Ulm                                                              UWE JENSEN

December 1998

# Contents

# 1
# Introduction

This chapter gives an introduction to the topics covered in this book: failure time models, complex systems, different information levels, maintenance and optimal replacement. We also include a section on reliability modeling, where we draw attention to some important factors to be considered in the modeling process. Two real life examples are presented: a reliability study of a system in a power plant and an availability analysis of a gas compression system.

## 1.1 Lifetime Models

In reliability we are mainly concerned with devices or systems that fail at an unforeseen or unpredictable (this term is defined precisely later) random age of $T > 0$. This random variable is assumed to have a distribution $F$, $F(t) = P(T \leq t), t \in \mathbb{R}$, with a density $f$. The hazard or failure rate $\lambda$ is defined on the support of the distribution by

$$\lambda(t) = \frac{f(t)}{\bar{F}(t)},$$

with the survival function $\bar{F}(t) = 1 - F(t)$. The failure rate $\lambda(t)$ measures the proneness to failure at time $t$ in that $\lambda(t) \triangle t \approx P(T \leq t + \triangle t | T > t)$ for small $\triangle t$. The (cumulative) hazard function is denoted by $\Lambda$,

$$\Lambda(t) = \int_0^t \lambda(s)\, ds = -\ln\{\bar{F}(t)\}.$$

The well-known relation

$$\bar{F}(t) = P(T > t) = \exp\{-\Lambda(t)\} \tag{1.1}$$

establishes the link between the cumulative hazard and the survival function. Modeling in reliability theory is mainly concerned with additional information about the state of a system, which is gathered during the operating time of the system. This additional information leads to updated predictions about proneness to system failure. There are many ways to introduce such additional information into the model. In the following sections some examples of how to introduce additional information and how to model the lifetime $T$ are given.

### 1.1.1 Complex Systems

As will be introduced in detail in Chapter 2, a complex system comprises $n$ components with positive random lifetimes $T_i, i = 1, 2, ..., n, n \in \mathbb{N}$. Let $\Phi : \{0,1\}^n \to \{0,1\}$ be the structure function of the system, which is assumed to be monotone. The possible states of the components and of the system, "intact" and "failed," are indicated by "1" and "0," respectively. Then $\Phi_t = \Phi(\mathbf{X}_t)$ describes the state of the system at time $t$, where $\mathbf{X}_t = (X_t(1), ..., X_t(n))$ and $X_t(i)$ denotes the indicator function

$$X_t(i) = I(T_i > t) = \begin{cases} 1 & \text{if } T_i > t \\ 0 & \text{if } T_i \leq t, \end{cases}$$

which is 1, if component $i$ is intact at time $t$, and 0 otherwise. The lifetime $T$ of the system is then given by $T = \inf\{t \in \mathbb{R}_+ : \Phi_t = 0\}$.

**Example 1** As a simple example the following system with three components is considered, which is intact if component 1 and at least one of the components 2 or 3 are intact:

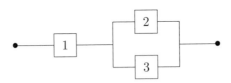

In this example $\Phi_t = X_t(1)\{1 - (1 - X_t(2))(1 - X_t(3))\}$ is easily obtained with $T = \inf\{t \in \mathbb{R}_+ : \Phi_t = 0\} = T_1 \wedge (T_2 \vee T_3)$, where as usual $a \wedge b$ and $a \vee b$ denote $\min\{a, b\}$ and $\max\{a, b\}$, respectively. The additional information about the lifetime $T$ is given by the observation of the state of the single components. As long as all components are intact, only a failure of component 1 leads to system failure. If one of the components 2 or 3 fails first, then the next component failure is a system failure.

Under the classical assumption that all components work independently, i.e., the random variables $T_i$, $i = 1, ..., n$, are independent, certain characteristics of the system lifetime are of interest:

- Determining the system lifetime distribution from the known component lifetime distributions or at least finding bounds for this distribution (see Sections 2.1 and 2.2).

- Are certain properties of the component lifetime distributions like IFR (increasing failure rate) or IFRA (increasing failure rate average) preserved by forming monotone systems? One of these closure theorems states, for example, that the distribution of the system lifetime is IFRA if all component lifetimes have IFRA distributions (see Section 2.2).

- In what way does a certain component contribute to the functioning of the whole system? The answer to this question leads to the definition of several importance measures (see Section 2.1).

### 1.1.2 Damage Models

Additional information about the lifetime $T$ can also be introduced into the model in a quite different way. If the state or damage of the system at time $t \in \mathbb{R}_+$ can be observed and this damage is described by a random variable $X_t$, then the lifetime of the system may be defined as

$$T = \inf\{t \in \mathbb{R}_+ : X_t \geq S\},$$

i.e., as the first time the damage hits a given level $S$. Here $S$ can be a constant or, more general, a random variable independent of the damage process. Some examples of damage processes $X = (X_t)$ of this kind are described in the following subsections.

### Wiener Process

The damage process is a Wiener process with positive drift starting at 0 and the failure threshold $S$ is a positive constant. The lifetime of the system is then known to have an inverse Gaussian distribution. Models of this kind are especially of interest if one considers different environmental conditions under which the system is working, as, for example, in so-called burn-in models. An accelerated aging caused by additional stress or different environmental conditions can be described by a change of time. Let $\tau : \mathbb{R}_+ \to \mathbb{R}_+$ be an increasing function. Then $Z_t = X_{\tau(t)}$ denotes the actual observed damage. The time transformation $\tau$ drives the speed of the deterioration. One possible way to express different stress levels in time intervals $[t_i, t_{i+1}), 0 = t_0 < t_1 < ... < t_k, i = 0, 1, ..., k - 1, k \in \mathbb{N}$, is the choice

$$\tau(t) = \sum_{j=0}^{i-1} \beta_j(t_{j+1} - t_j) + \beta_i(t - t_i), \ t \in [t_i, t_{i+1}), \beta_v > 0.$$

In this case it is seen that if $F_0$ is the inverse Gaussian distribution function of $T = \inf\{t \in \mathbb{R}_+ : X_t \geq S\}$, and $F$ is the distribution function of the lifetime $T_a = \inf\{t \in \mathbb{R}_+ : Z_t \geq S\}$ under accelerated aging, then $F(t) = F_0(\tau(t))$. A generalization in another direction is to consider a random time change, which means that $\tau$ is a stochastic process. By this, randomly varying environmental conditions can be modeled.

## Compound Point Processes

Processes of this kind describe so-called shock processes where the system is subject to shocks that occur from time to time and add a random amount to the damage. The successive times of occurrence of shocks, $T_n$, are given by an increasing sequence $0 < T_1 \leq T_2 \leq ...$ of random variables, where the inequality is strict unless $T_n = \infty$. Each time point $T_n$ is associated with a real-valued random mark $V_n$, which describes the additional damage caused by the $n$th shock. The marked point process is denoted $(T, V) = (T_n, V_n), n \in \mathbb{N}$. From this marked point process the corresponding compound point process $X$ with

$$X_t = \sum_{n=1}^{\infty} I(T_n \leq t)V_n \qquad (1.2)$$

is derived, which describes the accumulated damage up to time $t$. The simplest example is a compound Poisson process in which the shock arrival process is Poisson and the shock amounts $(V_n)$ are i.i.d. random variables. As before, the lifetime $T$ is the first time the damage process $(X_t)$ hits the level $S$. If we go one step further and assume that $S$ is not deterministic and fixed, but a random failure level, then we can describe a situation in which the observed damage process does not carry complete information about the (failure) state of the system; the failure can occur at different damage levels $S$.

Another way to describe the failure mechanism is the following. Let the accumulated damage up to time $t$ be given by the shock process $X_t$ as in (1.2). If the system is up at $t-$ just before $t$, the accumulated damage equals $X_{t-} = x$ and a shock of magnitude $y$ occurs at $t$, then the probability of failure at $t$ is $p(x + y)$, where $p(x)$ is a given $[0, 1]$-valued function. In this model failures can only occur at shock times and the accumulated damage determines the failure probability.

### 1.1.3   Different Information Levels

It was pointed out above in what way additional information can lead to a reliability model. But it is also important to note that in one and the same model different observation levels are possible, i.e., the amount of actual available information about the state of a system may vary. The following examples will show the effect of different degrees of information.

### 1.1.4   Simpson's Paradox

This paradox says that if one compares the death rates in two countries, say A and B, then it is possible that the crude overall death rate in country A is higher than in B although all age-specific death rates in B are higher than in A. This can be transferred to reliability in the following way. Considering a two-component parallel system, the failure rate of the system lifetime may increase although the component lifetimes have decreasing failure rates. The following proposition, which can be proved by some elementary calculations, yields an example of this.

**Proposition 1** *Let $T = T_1 \vee T_2$ with i.i.d. random variables $T_i$, $i = 1, 2$, following the common distribution $F$,*

$$F(t) = 1 - e^{-u(t)}, \ t \geq 0, u(t) = \gamma t + \alpha(1 - e^{-\beta t}), \ \alpha, \beta, \gamma > 0.$$

*If $2\alpha e^{\alpha} < \left(\frac{\gamma}{\beta}\right)^2 < 1$, then the failure rate $\lambda$ of the lifetime $T$ increases, whereas the component lifetimes $T_i$ have decreasing failure rates.*

This example shows that it makes a great difference whether only the system lifetime can be observed (aging property: IFR) or additional information about the component lifetimes is available (aging property: DFR). The aging property of the system lifetime of a complex system does not only depend on the joint distribution of the component lifetimes but also, of course, on the structure function. Instead of a two-component parallel system, consider a series system where the component lifetimes have the same distributions as in Proposition 1. Then the failure rate of $T_{ser} = T_1 \wedge T_2$ decreases, whereas $T_{par} = T_1 \vee T_2$ has an increasing failure rate.

### 1.1.5   Predictable Lifetime

The Wiener process $X = (X_t), t \in \mathbb{R}_+$, with positive drift $\mu$ and variance scaling parameter $\sigma$, is a popular damage threshold model. The process $X$ can be represented as $X_t = \sigma B_t + \mu t$, where $B$ is standard Brownian motion. If one assumes that the failure level $S$ is a fixed known constant, then the lifetime $T = \inf\{t \in \mathbb{R}_+ : X_t \geq S\}$ follows an inverse Gaussian distribution with a finite mean $ET = S/\mu$. One criticism of this model is that the paths of $X$ are not monotone. As a partial answer, one can respond that maintenance actions also lead to improvements and thus $X$ could be decreasing at some time points. A more severe criticism from the point of view of the available information is the following. It is often assumed that in this model the paths of the damage process can be observed continuously. But this would make the lifetime $T$ a predictable random time (a precise definition follows in Chapter 3), i.e., there is an increasing sequence $\tau_n, n \in \mathbb{N}$, of random time points that announces the failure. In this model one could choose $\tau_n = \inf\{t \in \mathbb{R}_+ : X_t \geq S - 1/n\}$, and take $n$ large enough and stop operating the system at $\tau_n$ "just" before failure, to carry out some preventive maintenance, cf. Figure 1.1. This does not usually apply in practical situations. This example shows that one has to distinguish carefully between the different information levels for the model formulation (complete information) and for the actual observation (partial information).

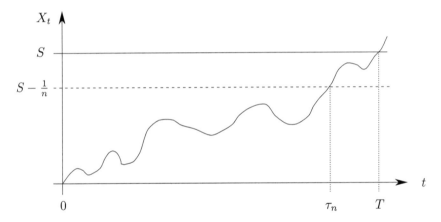

FIGURE 1.1. Predictable Stopping Time

## 1.1.6    A General Failure Model

The general failure model considered in Chapter 3 uses elements of the theory of stochastic processes and particularly some martingale theory. Some of the readers might wonder whether sophisticated theory like this is necessary and suitable in reliability, a domain with engineering applications. Instead of a comprehensive justification we give a motivating example.

**Example 2** We consider a simple two-component parallel system with independent $\mathrm{Exp}(\alpha_i)$ distributed component lifetimes $T_i, i = 1, 2$. The system lifetime $T = T_1 \vee T_2$ has distribution function

$$F(t) = P(T_1 \leq t, T_2 \leq t) = (1 - e^{-\alpha_1 t})(1 - e^{-\alpha_2 t})$$

with an ordinary failure rate

$$\lambda(t) = \frac{\alpha_1 e^{-\alpha_1 t} + \alpha_2 e^{-\alpha_2 t} - (\alpha_1 + \alpha_2) e^{-(\alpha_1 + \alpha_2)t}}{e^{-\alpha_1 t} + e^{-\alpha_2 t} - e^{-(\alpha_1 + \alpha_2)t}}.$$

This formula is rather complicated for such a simple system and reveals nothing about the structure of the system. Using elementary calculus it can be shown that for $\alpha_1 \neq \alpha_2$ the failure rate is increasing on $(0, t^*)$ and decreasing on $(t^*, \infty)$ for some $t^* > 0$. This property of the failure rate, however, is neither obvious nor immediate to see. We also know that $F$ is of IFRA type.

But is it not more natural and simpler to say that a failure rate (process) should be 0 as long as both components work (no system failure can occur) and, when the first component failure occurs, then the rate switches to $\alpha_1$ or $\alpha_2$ depending on which component survives? We want to derive a model that allows such a simple failure rate process and also includes the ordinary failure rate. Of course, this simple failure rate process, which can be expressed as

$$\lambda_t = \alpha_1 I(T_2 \leq t < T_1) + \alpha_2 I(T_1 \leq t < T_2),$$

needs knowledge about the random component lifetimes $T_i$. Now the failure rate $\lambda_t$ is a stochastic process and the information about the status of the components at time $t$ is represented by a filtration. The model allows for changing the information level and the ordinary failure rate can be derived from $\lambda_t$ on the lowest level possible, namely no information about the component lifetimes.

The modern theory of stochastic processes allows for the development of a general failure model that incorporates the above aspects: time dynamics and different information levels. Chapter 3 presents this model. The failure rate process $\lambda_t$ is one of the basic parameters of this set-up. If we consider the lifetime $T$, under some mild conditions we obtain the failure rate process on $\{T > t\}$ as the limit of conditional expectations with respect to the pre-$t$-history ($\sigma$-algebra) $\mathcal{F}_t$,

$$\lambda_t = \lim_{h \to 0+} \frac{1}{h} P(T \leq t + h | \mathcal{F}_t),$$

extending the classical failure rate $\lambda(t)$ of the system. To apply the set-up, focus should be placed on the failure rate process $(\lambda_t)$. When this process has been determined, the model has basically been established. Using the above interpretation of the failure rate process, it is in most cases rather straightforward to determine its form. The formal proofs are, however, often quite difficult.

If we go one step further and consider a model in which the system can be repaired or replaced at failure, then attention is paid to the number $N_t$

of system failures in $[0, t]$. Given certain conditions, the counting process $N = (N_t), t \in \mathbb{R}_+$, has an "intensity" that as an extension of the failure rate process can be derived as the limit of conditional expectations

$$\lambda_t = \lim_{h \to 0+} \frac{1}{h} E[N_{t+h} - N_t | \mathcal{F}_t],$$

where $\mathcal{F}_t$ denotes the history of the system up to time $t$. Hence we can interpret $\lambda_t$ as the (conditional) expected number of system failures per unit of time at time $t$ given the available information at that time. Chapter 3 includes several special cases that demonstrate the broad spectrum of potential applications.

## 1.2   Maintenance

To prolong the lifetime, to increase the availability, and to reduce the probability of an unpredictable failure, various types of maintenance actions are being implemented. The most important maintenance actions include:

- Preventive replacements of parts of the system or of the whole system

- Repairs of failed units

- Providing spare parts

- Inspections to check the state of the system if not observed continuously

Taking maintenance actions into account leads, depending on the specific model, to one of the following subject areas: Availability Analysis and Optimization Models.

### 1.2.1   Availability Analysis

If the system or parts of it are repaired or replaced when failures occur, the problem is to characterize the performance of the system. Different measures of performance can be defined as, for example,

- The probability that the system is functioning at a certain point in time (point availability)

- The mean time to the first failure of the system

- The probability distribution of the downtime of the system in a given time interval.

Traditionally, focus has been placed on analyzing the point availability and its limit (the steady-state availability). For a single component, the steady-state formula is given by $MTTF/(MTTF+MTTR)$, where $MTTF$ and $MTTR$ represent the mean time to failure and the mean time to repair (mean repair time), respectively. The steady-state probability of a system comprising several components can then be calculated using the theory of complex (monotone) systems.

Often, performance measures related to a time interval are used. Such measures include the distribution of the number of system failures, and the distribution of the downtime of the system, or at least the mean of these distributions. Measures related to the number of system failures are important from an operational and safety point of view, whereas measures related to the downtime are more interesting from a productional point of view. Information about the probability of having a long downtime in a time interval is important for assessing the economic risk related to the operation of the system. For production systems, it is sometimes necessary to use a multistate representation of the system and some of its components, to reflect different production levels.

Compared to the steady-state availability, it is of course more complicated to compute the performance measures related to a time interval, in particular the probability distributions of the number of system failures and of the downtime. Using simplifications and approximations, it is however possible to establish formulas that can be used in practice. For highly available systems, a Poisson approximation for the number of system failures and a compound Poisson approximation for the downtime distribution are useful in many cases.

These topics are addressed in Chapter 4, which gives a detailed analysis of the availability of monotone systems. Emphasis is placed on performance measures related to a time interval. Sufficient conditions are given for when the Poisson and the compound Poisson distributions are asymptotic limits.

## 1.2.2 Optimization Models

If a valuation structure is given, i.e., costs of replacements, repairs, downtime, etc., and gains, then one is naturally led to the problem of planning the maintenance action so as to minimize (maximize) the costs (gains) with respect to a given criterion. Examples of such criteria are expected costs per unit time and total expected discounted costs.

**Example 3** We resume Example 2, p. 6, and consider the simple two-component parallel system with independent $\text{Exp}(\alpha_i)$ distributed component lifetimes $T_i, i = 1, 2$, with the system lifetime $T = T_1 \vee T_2$. We now allow preventive replacements at costs of $c$ units to be carried out before failure, and a replacement upon system failure at cost $c + k$. It seems intuitive that $T_1 \wedge T_2$, the time of the first component failure, should be a candidate for an optimal replacement time with respect to some cost criterion, at least if $c$ is "small" compared to $k$. How can we prove that this random time $T_1 \wedge T_2$ is optimal among all possible replacement times? How can we characterize the set of all possible replacement times?

These questions can only be answered in the framework of martingale theory and are addressed in Chapter 5.

One can imagine that thousands of models (and papers) can be created by combining the different types of lifetime models with different maintenance actions. The general optimization framework formulated in Chapter 5 incorporates a number of such models. Here the emphasis is placed on determining the optimal replacement time of a deteriorating system. The framework is based on the failure model of Chapter 3, which means that rather complex and very different situations can be studied. Special cases include monotone systems, (minimal) repair models, and damage processes, with different information levels.

## 1.3   Reliability Modeling

Models analyzed in this book are general, in the sense that they do not refer to any *specific* real life situation but are applicable in a number of cases. This is the academic and theoretical approach of mathematicians (probabilists, statisticians) who provide tools that can be used in applications.

The reliability engineer, on the other hand, has a somewhat different starting point. He or she is faced with a real problem and has to analyze this problem using a mathematical model that describes the situation appropriately. Sometimes it is rather straightforward to identify a suitable model, but often the problem is complex and it is difficult to see how to solve it. In many cases, a model needs to be developed. The modeling process requires both experience on the part of the practitioner and knowledge on the part of the theorist.

However, it is not within the scope of this book to discuss in detail the many practical aspects related to reliability modeling and analysis. Only a few issues will be addressed. In this introductory section we will highlight important factors to be considered in the modeling process and two real life examples will be presented.

The objectives of the reliability study can affect modeling in many ways, for example, by specifying which performance measures and which factors (parameters) are to be analyzed. Different objectives will require different approaches and methods for modeling and analysis. Is the study to provide decision support in a design process of a system where the problem is to choose between alternative solutions; is the problem to give a basis for specifying reliability requirements; or is the aim to search for an optimal preventive maintenance strategy? Clearly, these situations call for different models.

The objectives of the study may also influence the choice of the computational approach. If it is possible to use analytical calculation methods, these would normally be preferred. For complex situations, Monte Carlo simulation often represents a useful alternative, cf., e.g., [16, 70].

The modeling process starts by clarifying the characteristics of the situation to be analyzed. Some of the key points to address are:

Can the system be decomposed into a set of independent subsystems (components)? Are all components operating normally or are some on stand-by? What is the state of the component after a repair? Is it "as good as new"? What are the resources available for carrying out the repairs? Are some types of preventive maintenance being employed? Is the state of the components and the system continuously monitored, or is it necessary to carry out inspections to reveal their condition? Is information available about the underlying condition of the system and components, such as wear, stress, and damage?

Having identified important features of the system, we then have to look more specifically at the various elements of the model and resolve questions like the following:

- How should the deterioration process of the components and system be modeled? Is it sufficient to use a standard lifetime model where the age of the unit is the only information available? How should the repair/replacement times be modeled?

- How are the preventive maintenance activities to be reflected in the model? Are these activities to be considered fixed in the model or is it possible to plan preventive maintenance action so that costs (rewards) are minimized (maximized)?

- Is a binary (two-state) approach for components and system sufficiently accurate, or is multistate modeling required?

- How are the system and components to be represented? Is a reliability block diagram appropriate?

- Are time dynamics to be included or is a time stationary model sufficient?

- How are the parameters of the model to be determined? What kind of input data are required for using the model? How is uncertainty to be dealt with?

Depending on the answers to these questions, relevant models can be identified. It is a truism that no model can cover all aspects, and it is recommended that one starts with a simple model describing the main features of the system.

The following application examples give further insight into the situations that can be modeled using the theory presented in this book.

## 1.3.1  Nuclear Power Station

In this example we consider a small part of a very complex technical system, in which safety aspects are of great importance. The nuclear power station under consideration consists of two identical boiling water reactors in commercial operation, each with an electrical power of 1344 MW. They started in 1984 and 1985, respectively, working with an efficiency of 35%.

Nuclear power plants have to shut down from time to time to exchange the nuclear fuel. This is usually performed annually. During the shutdown phase a lot of maintenance tasks and surveillance tests are carried out. One problem during such phases is that decay heat is still produced and thus has to be removed. Therefore, residual heat removal (RHR) systems are in operation. At the particular site, three identical systems are available, each with a capacity of 100%. They are designed to remove decay heat during accident conditions occurring at full power as well as for operational purposes in cooldown phases.

One of these RHR systems is schematically shown in Figure 1.2. It consists of three different trains including the closed cooling water system. Several pumps and valves are part of the RHR system. The primary cooling system can be modeled as a complex system comprising the following main components:

- Closed cooling water system pump (CCWS)

- Service water system pump (SWS)

- Low-pressure pump with a pre-stage (LP)

- High-pressure pump (HP)

- Nuclear heat exchanger (RHR)

- Valves $(V_1, V_2, V_3)$

For the analysis we have to distinguish between two cases:

1. The RHR system is not in operation.

   Then the functioning of the system can be viewed as a binary structure of the main components as is shown in the reliability block diagram in Figure 1.3. When the system is needed, it is possible that single components or the whole system fails to start on demand. In this case, to calculate the probability of a failure on demand, we have

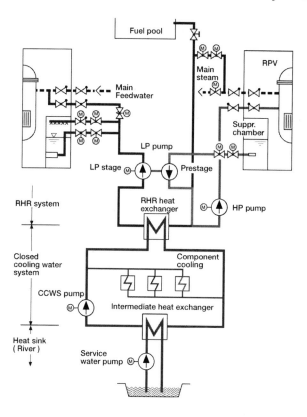

FIGURE 1.2. Cooling System of a Power Plant

to take all components in the reliability block diagram into consideration. Two of the valves, $V_1$ and $V_2$, are in parallel. Therefore, the RHR system fails on demand if either $V_1$ *and* $V_2$ fail or at least one of the remaining components LP,..., HP, $V_3$ fails. We assume that the time from a check of a component until a failure in the idle state is exponentially distributed. The failure rates are $\lambda_{v_1}, \lambda_{v_2}, \lambda_{v_3}$ for the valves and $\lambda_{p_1}, \lambda_{p_2}, \lambda_{p_3}, \lambda_{p_4}, \lambda_h$ for the other components. If the check (inspection or operating period) dates $t$ time units back, then the probability of a failure on demand is given by

$$1 - \{1 - (1 - e^{-\lambda_{v_1} t})(1 - e^{-\lambda_{v_2} t})\} e^{-(\lambda_{p_1} + \lambda_{p_2} + \lambda_{p_3} + \lambda_{p_4} + \lambda_h + \lambda_{v_3}) t}.$$

2. The RHR system is in operation.

During an operation phase, only the pumps and the nuclear heat exchanger can fail to operate. If the valves have once opened on demand when the operation phase starts, these valves cannot fail during operation. Therefore, in this operation case, we can either ignore the

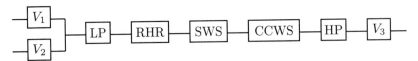

FIGURE 1.3. Reliability Block Diagram

valves in the block diagram or assign failure probability 0 to $V_1, V_2, V_3$.
The structure reduces to a simple series system. If we assume that
the failure-free operating times of the pumps and the heat exchanger
are independent and have distributions $F_{p_1}, F_{p_2}, F_{p_3}, F_{p_4}$, and $F_h$, respectively, then the probability that the system fails before a fixed
operating time $t$ is just

$$1 - \bar{F}_{p_1}(t)\bar{F}_{p_2}(t)\bar{F}_{p_3}(t)\bar{F}_{p_4}(t)\bar{F}_h(t),$$

where $\bar{F}(t)$ denotes the survival probability.

In both cases the failure time distributions and the failure rates have to
be estimated. One essential condition for the derivation of the above formulae is that all components have stochastically independent failure times
or lifetimes. In some cases such an independence condition does not apply. In Chapter 3 a general theory is developed that also includes the case
of complex systems with dependent component lifetimes. The framework
presented covers different information levels, which allow updating of reliability predictions using observations of the condition of the components
of the system, for example.

### 1.3.2   Gas Compression System

This example outlines various aspects of the modeling process related to
the design of a gas compression system.

A gas producer was designing a gas production system, and one of the
most critical decisions was related to the design of the gas compression
system.

At a certain stage of the development, two alternatives for the compression system were considered:

i) One gas train with a maximum throughput capacity of 100%

ii) Two trains in parallel, each with a maximum throughput capacity of
50%.

Normal production is 100%. For case i) this means that the train is operating normally and a failure stops production completely. For case ii) both trains are operating normally. If one train fails, production is reduced to 50%. If both trains are down, production is 0.

Each train comprises compressor–turbine, cooler, and scrubber. A failure of one of these "components" results in the shutdown of the train. Thus a train is represented by a series structure of the three components compressor–turbine, cooler, and scrubber.

The following failure and repair time data were assumed:

| Component | Failure rate (unit of time: 1 year) | Mean repair time (unit of time: 1 hour) |
|---|---|---|
| Compressor–turbine | 10 | 12 |
| Cooler | 2 | 50 |
| Scrubber | 1 | 20 |

To compare the two alternatives, a number of performance measures were considered. Particular interest was shown in performance measures related to the number of system shutdowns, the time the system has a reduced production level, and the total production loss due to failures of the system. The gas sales agreement states that the gas demand is to be met with a very high reliability, and failures could lead to considerable penalties and loss of goodwill, as well as worse sales perspectives for the future.

Using models as will be described in Chapter 4, it was possible to compute these performance measures, given certain assumptions.

It was assumed that each component generates an alternating renewal process, which means that the repair brings the component to a condition that is as good as new. The uptimes were assumed to be distributed exponentially, so that the component in the operating state has a constant failure rate. The failure rate used was based on experience data for similar equipment. Such a component model was considered to be sufficiently accurate for the purpose of the analysis. The exponential model represents a "first-order approximation," which makes it rather easy to gain insight into the performance of the system. For a complex "component" with many parts to be maintained, it is known that the overall failure rate exhibits approximately exponential nature. Clearly, if all relevant information is utilized, the exponential model is rather crude. But again we have to draw attention to the purpose of the analysis: provide decision support concerning the choice of design alternatives. Only the essential features should be included in the model.

A similar type of reasoning applies to the problem of dependency between components. In this application all uptimes and downtimes of the components were assumed to be independent. In practice there are, of course, some dependencies present, but by looking into the failure causes and the way the components were defined, the assumption of independence was not considered to be a serious weakness of the model, undermining the results of the analysis.

To determine the repair time distribution, expert opinions were used. The repair times, which also include fault diagnosis, repair preparation, test and restart, were assessed for different failure modes. As for the uptimes, it was assumed that no major changes over time take place concerning component design, operational procedures, etc.

Uncertainty related to the input quantities used was not considered. Instead, sensitivity studies were performed with the purpose of identifying how sensitive the results were with respect to variations in input parameters.

Of the results obtained, we include the following examples:

- The gas train is down 2.7% of the time in the long run.

- For alternative i), the average system failure rate, i.e., the average number of system failures per year, equals 13. For alternative ii) it is distinguished between failures resulting in production below 100% and below 50%. The average system failure rates for these levels are approximately 26 and 0.7, respectively. Alternative ii) has a probability of about 50% of having one or more complete shutdowns during a year.

- The mean lost production equals 2.7% for both alternatives. The probability that the lost production during one year is more than 4% of demand is approximately equal to 0.16 for alternative i) and 0.08 for alternative ii).

This last result is based on assumptions concerning the variation of the repair times. Refer to Section 4.7.1, p. 155, where the models and methods used to compute these measures are summarized.

The results obtained, together with an economic analysis, gave the management a good basis for choosing the best alternative.

**Bibliographic Notes.** There are now many journals strongly devoted to reliability, for example, the *IEEE Transactions on Reliability* and *Reliability Engineering and System Safety*. In addition, there are many journals in Probability and Operations Research that publish papers in this field.

As mentioned before, there is an extensive literature covering a variety of stochastic models of reliability. Instead of providing a long and, inevitably, almost certainly incomplete list of references, some of the surveys and review articles are quoted, as well as some of the reliability books.

From time to time, the *Naval Research Logistics Quarterly* journal publishes survey articles in this field, among them the renowned article by

Pierskalla and Voelker [140], which appeared with 259 references in 1976, updated by Sherif and Smith [155] with an extensive bibliography of 524 references in 1981, followed by Valdez-Flores and Feldman [170] with 129 references in 1989. Bergman's review [41] reflects the author's experience in industry and emphasizes the usefulness of reliability methods in applications. Gertsbakh's paper [83] reviews asymptotic methods in reliability and especially investigates under what conditions the lifetime of a complex system with many components is approximately exponentially distributed. Natvig [136] gives a concise overview of importance measures for monotone systems. The surveys of Arjas [4] and Koch [119] consider reliability models using more advanced mathematical tools as marked point processes and martingales. A guided tour for the non-expert through point process and intensity-based models in reliability is presented in the article of Hokstad [98]. The book of Thompson [167] gives a more elementary presentation of point processes in reliability. Other reliability books that we would like to draw attention to are Aven [16], Barlow and Proschan [33, 34], Beichelt and Franken [38], Bergman and Klefsjö [42], Gaede [79], Gertsbakh [82], Høyland and Rausand [99], and Kovalenko, Kuznetsov, and Pegg [121]. Some of the models addressed in this introduction are treated in the overview of Jensen [105] where related references can also be found.

# 2
# Basic Reliability Theory

This chapter presents some basic theory of reliability, including complex system theory and properties of lifetime distributions. Basic availability theory and models for maintenance optimization are included in Chapters 4 and 5, respectively.

The purpose of this chapter is not to give a complete overview of the existing theory, but to introduce the reader to common reliability concepts, models, and methods. The exposition highlights basic ideas and results, and it provides a starting point for the more advanced theory presented in Chapters 3–5.

## 2.1   Complex Systems

This section gives an overview of some basic theory of complex systems. Binary monotone (coherent) systems are covered, as well as multistate monotone systems.

### 2.1.1   Binary Monotone Systems

In this section we give an introduction to the classical theory of monotone (coherent) systems. First we study the structural relations between a system and its components. Then methods for calculation of system reliability are reviewed when the component reliabilities are known. When not stated

FIGURE 2.1. Series Structure

otherwise, the random variables representing the state of the components are assumed to be independent.

## Structural properties

We consider a system comprising $n$ components, which are numbered consecutively from 1 to $n$. In this section we distinguish between two states: a functioning state and a failure state. This dichotomy applies to the system as well as to each component. To indicate the state of the $i$th component, we assign a binary variable $x_i$ to component $i$:

$$x_i = \begin{cases} 1 & \text{if component } i \text{ is in the functioning state} \\ 0 & \text{if component } i \text{ is in the failure state.} \end{cases}$$

(The term *binary variable* refers to a variable taking on the values 0 or 1.) Similarly, the binary variable $\Phi$ indicates the state of the system:

$$\Phi = \begin{cases} 1 & \text{if the system is in the functioning state} \\ 0 & \text{if the system is in the failure state.} \end{cases}$$

We assume that

$$\Phi = \Phi(\mathbf{x}),$$

where $\mathbf{x} = (x_1, x_2, \ldots, x_n)$, i.e., the state of the system is determined completely by the states of the components. We refer to the function $\Phi(\mathbf{x})$ as the *structure function* of the system, or simply the structure. In the following we will often use the phrase structure in place of system.

**Example 4** A system that is functioning if and only if each component is functioning is called a series system. The structure function for this system is given by

$$\Phi(\mathbf{x}) = x_1 \cdot x_2 \cdot \ldots \cdot x_n = \prod_{i=1}^{n} x_i.$$

A series structure can be illustrated by the reliability block diagram in Figure 2.1. "Connection between $a$ and $b$" means that the system functions.

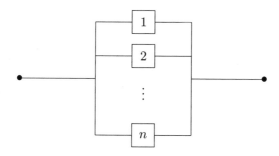

FIGURE 2.2. Parallel Structure

**Example 5** A system that is functioning if and only if at least one component is functioning is called a parallel system. The corresponding reliability block diagram is shown in Figure 2.2.

The structure function is given by

$$\Phi(\mathbf{x}) = 1 - (1 - x_1)(1 - x_2) \cdots (1 - x_n) = 1 - \prod_{i=1}^{n}(1 - x_i). \qquad (2.1)$$

The expression on the right-hand side in (2.1) is often written $\coprod x_i$. Thus, a parallel system with two components has structure function

$$\Phi(\mathbf{x}) = 1 - (1 - x_1)(1 - x_2) = \coprod_{i=1}^{2} x_i,$$

which we also write as $\Phi(\mathbf{x}) = x_1 \coprod x_2$.

**Example 6** A system that is functioning if and only if at least $k$ out of $n$ components are functioning is called a $k$-out-of-$n$ system. A series system is an $n$-out-of-$n$ system, and a parallel system is a 1-out-of-$n$ system. The structure function for a $k$-out-of-$n$ system is given by

$$\Phi(\mathbf{x}) = \begin{cases} 1 & \text{if} \quad \sum_{i=1}^{n} x_i \geq k \\ 0 & \text{if} \quad \sum_{i=1}^{n} x_i < k. \end{cases}$$

As an example, we will look at a 2-out-of-3 system. This system can be illustrated by the reliability block diagram shown in Figure 2.3. An airplane that is capable of functioning if and only if at least two of its three engines are functioning is an example of a 2-out-of-3 system.

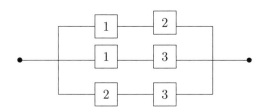

FIGURE 2.3.  2-out-of-3 Structure

**Definition 1 (*Monotone system*).** *A system is said to be monotone if*
1. *its structure function $\Phi$ is nondecreasing in each argument, and*
2. *$\Phi(\mathbf{0}) = 0$ and $\Phi(\mathbf{1}) = 1$.*

Condition 1 says that the system cannot deteriorate (that is, change from the functioning state to the failed state) by improving the performance of a component (that is, replacing a failed component by a functioning component). Condition 2 says that if all the components are in the failure state, then the system is in the failure state, and if all the components are in the functioning state, then the system is in the functioning state.

All the systems we consider are monotone. In the reliability literature, much attention has be devoted to *coherent* systems, which is a subclass of monotone systems. Before we define a coherent system we need some notation.

The vector $(\cdot_i, \mathbf{x})$ denotes a state vector where the state of the $i$th component is equal to 1 or 0; $(1_i, \mathbf{x})$ denotes a state vector where the state of the $i$th component is equal to 1, and $(0_i, \mathbf{x})$ denotes a state vector where the state of the $i$th component is equal to 0; the state of component $j$, $j \neq i$, equals $x_j$. If we want to specify the state of some components, say $i \in J$ $(J \subset \{1, 2, \ldots, n\})$, we use the notation $(\cdot_J, \mathbf{x})$. For example, $(\mathbf{0}_J, \mathbf{x})$ denotes the state vector where the states of the components in $J$ are all 0 and the state of component $i$, $i \notin J$, equals $x_i$.

**Definition 2 (*Coherent system*).** *A system is said to be coherent if*
1. *its structure function $\Phi$ is nondecreasing in each argument, and*
2. *each component is relevant, i.e., there exists at least one vector $(\cdot_i, \mathbf{x})$ such that $\Phi(1_i, \mathbf{x}) = 1$ and $\Phi(0_i, \mathbf{x}) = 0$.*

It is seen that if $\Phi$ is coherent, then $\Phi$ is also monotone. We also need the following terminology.

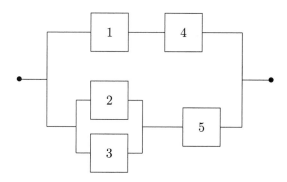

FIGURE 2.4. Example of a Reliability Block Diagram

**Definition 3** (*Minimal cut set*). *A cut set $K$ is a set of components that by failing causes the system to fail, i.e., $\Phi(\mathbf{0}_K, \mathbf{1}) = 0$. A cut set is minimal if it cannot be reduced without losing its status as a cut set.*

**Definition 4** (*Minimal path set*). *A path set $S$ is a set of components that by functioning ensures that the system is functioning, i.e., $\Phi(\mathbf{1}_S, \mathbf{0}) = 1$. A path set is minimal if it cannot be reduced without losing its status as a path set.*

**Example 7** Consider the reliability block diagram presented in Figure 2.4. The minimal cut sets of the system are: $\{1, 5\}, \{4, 5\}, \{1, 2, 3\}$, and $\{2, 3, 4\}$. Note that, for example, $\{1, 4, 5\}$ is a cut set, but it is not minimal. The minimal path sets are $\{1, 4\}, \{2, 5\}$, and $\{3, 5\}$. In the following we will refer to this example as the "5-components example."

## Computing System Reliability

Let $X_i$ be a binary random variable representing the state of the $i$th component at a given point in time, $i = 1, 2, \ldots, n$. Let

$$
\begin{aligned}
p_i &= P(X_i = 1) \\
q_i &= P(X_i = 0) \\
h &= h(\mathbf{p}) = P(\Phi(\mathbf{X}) = 1) \\
g &= g(\mathbf{q}) = P(\Phi(\mathbf{X}) = 0),
\end{aligned}
\tag{2.2}
$$

where $\mathbf{p} = (p_1, p_2, \ldots, p_n)$, $\mathbf{q} = (q_1, q_2, \ldots, q_n)$, and $\mathbf{X} = (X_1, X_2, \ldots, X_n)$. The probabilities $p_i$ and $q_i$ are referred to as the reliability and unreliability of component $i$, respectively, and $h$ and $g$ the corresponding reliability and unreliability of the system.

The problem is to compute the system reliability $h$ given the component reliabilities $p_i$. Often it will be more efficient to let the starting point of the calculation be the unreliabilities. Note that $h + g = 1$ and $p_i + q_i = 1$.

Before we present methods for computation of system reliability for a general structure, we will look closer into some special cases. We start with the series structure.

**Example 8 (Reliability of a series structure).** For a series structure the system functioning means that all the components function, hence

$$
\begin{aligned}
h &= P(\Phi(\mathbf{X}) = 1) = P(\prod_{i=1}^{n} X_i = 1) \\
&= P(X_1 = 1, X_2 = 1, \ldots, X_n = 1) \\
&= \prod_{i=1}^{n} P(X_i = 1) = \prod_{i=1}^{n} p_i.
\end{aligned}
\tag{2.3}
$$

**Example 9 (Reliability of a parallel structure).** The reliability of a parallel structure is given by

$$
h = 1 - \prod_{i=1}^{n}(1 - p_i) = \coprod_{i=1}^{n} p_i.
\tag{2.4}
$$

The proof of (2.4) is analogous with the proof of (2.3).

**Example 10 (Reliability of a $k$-out-of-$n$ structure).** The reliability of a $k$-out-of-$n$ structure of independent components, which all have the same reliability $p$, equals

$$
h = \sum_{i=k}^{n} \binom{n}{i} p^i (1 - p)^{n-i}.
$$

This formula holds since $\sum_{i=1}^{n} X_i$ has a binomial distribution with parameters $n$ and $p$ under the given assumptions. The case that the component reliabilities are not equal is treated later.

Next we look at an arbitrary series–parallel structure. By using the calculation formulae for a series structure and a parallel structure it is relatively straightforward to calculate the reliability of combinations of series and parallel structures, provided that each component is included in just one such structure. Let us consider an example.

**Example 11** Consider again the reliability block diagram in Figure 2.4. The system can be viewed as a parallel structure of two independent modules: the structure comprising the components 1 and 4, and the structure comprising the components $2, 3$, and 5. The reliability of the former structure equals $p_1 p_4$, whereas the reliability of the latter equals $(1 - (1 - p_2)(1 - p_3))p_5$. Thus the system reliability is given by

$$h = 1 - \{1 - p_1 p_4\}\{1 - (1 - (1 - p_2)(1 - p_3))p_5\}.$$

Assuming that $q_1 = q_2 = q_3 = 0.02$ and $q_4 = q_5 = 0.01$, this formula gives $h = 0.9997$, i.e., $g = 3 \cdot 10^{-4}$.

If, for example, a 2-out-of-3 structure of independent components with the same reliability $p$ is in series with the above system, the total system reliability will be as above multiplied by the reliability of the 2-out-of-3 structure, which equals

$$\binom{3}{2} p^2(1 - p) + \binom{3}{3} p^3(1 - p)^0 = 3p^2(1 - p) + p^3.$$

Now consider a general monotone structure. Computation of system reliability for complex systems might be a formidable task (in fact, impracticable in some cases) unless an efficient method (algorithm) is used. Developing such methods is therefore an important area of research within reliability theory.

There exist a number of methods for reliability computation of a general structure. Many of these methods are based on the minimal cut (path) sets. For smaller systems the so-called inclusion–exclusion method may be applied, but this method is primarily a method for approximate calculations for systems that are either very reliable or unreliable.

**Inclusion–Exclusion Method.** Let $A_j$ be the event that minimal cut set $K_j$ is not functioning, $j = 1, 2, \ldots, k$. Then clearly,

$$P(A_j) = \prod_{i \in K_j} q_i$$

and

$$g = P(\bigcup_{j=1}^{k} A_j).$$

Furthermore, let

$$
\begin{aligned}
w_1 &= \sum_{j=1}^{k} P(A_j) \\
w_2 &= \sum_{i<j} P(A_i \cap A_j) \\
&\vdots \\
w_r &= \sum_{1 \le i_1 < i_2 < \cdots < i_r \le k} P(\bigcap_{j=1}^{r} A_{i_j}).
\end{aligned}
$$

Then the well-known inclusion–exclusion formula states that

$$g = w_1 - w_2 + w_3 - \cdots + (-1)^{k+1} w_k \qquad (2.5)$$

and for $r \leq k$

$$g \ \leq \ w_1 - w_2 + w_3 - \cdots + w_r, \quad r \text{ odd}$$
$$g \ \geq \ w_1 - w_2 + w_3 - \cdots - w_r, \quad r \text{ even.}$$

Although in general it is not true that the upper bounds decrease and the lower bounds increase, in practice it may be necessary to calculate only a few $w_r$ terms to obtain a close approximation. If the component unreliabilities $q_i$ are small, i.e., the reliabilities $p_i$ are large, then the $w_2$ term will usually be negligible compared to $w_1$, such that $g \approx w_1$. Note that $w_1$ is an upper bound for $g$. By using $w_1$ as an estimate for the system unreliability, we will overestimate the system unreliability. In most cases, such an underestimation of reliability is preferable compared to an overestimation of reliability.

With a large number of minimal cut sets, the exact calculation using (2.5) will be extensive. The number of terms in the sum in $w_r$ equals $\binom{k}{r}$. Thus the total number of terms is

$$\sum_{r=1}^{k} \binom{k}{r} = (1+1)^k - 1 = 2^k - 1.$$

**Example 12** (Continuation of Examples 7 and 11). The problem is to calculate the unreliability of the 5-components system of Figure 2.4 by means of the approximation method described above. We assume that $q_1 = q_2 = q_3 = 0.02$ and $q_4 = q_5 = 0.01$. We find that $w_1 = 3 \cdot 10^{-4}$, which means that $g \approx 3 \cdot 10^{-4}$. It is intuitively clear that the error term by using this approximation will not be significant. Calculating $w_2$ confirms this:

$$
\begin{aligned}
w_2 \ &= \ q_1 q_4 q_5 + q_1 q_2 q_3 q_5 + q_1 q_2 q_3 q_4 q_5 + q_1 q_2 q_3 q_4 q_5 + q_2 q_3 q_4 q_5 + q_1 q_2 q_3 q_4 \\
&= \ 2.2 \cdot 10^{-6}.
\end{aligned}
$$

There exist also other bounds and approximations for the system reliability. For example, it can be shown that

$$1 - \prod_{j=1}^{k}\left(1 - \prod_{i \in K_j} q_i\right) = 1 - \prod_{j=1}^{k} \prod_{i \in K_j} p_i$$

is an upper bound for $g$, and a good approximation for small values of the component unreliabilities $q_i$; see Barlow and Proschan [34], p. 35. This bound is always as good as or better than $w_1$. In the following we sketch some alternative methods for reliability computation.

**Method Using the Minimal Cut Set Representation of the Structure Function**. Using

$$\Phi(\mathbf{X}) = \prod_{j=1}^{k} \coprod_{i \in K_j} X_i,$$

and by multiplying out the right-hand side of this expression, we can find an exact expression of $h$ (or $g$). As an illustration consider a 2-out-of-3 system. Then

$$\Phi = (X_1 \coprod X_2) \cdot (X_1 \coprod X_3) \cdot (X_2 \coprod X_3)$$

and by multiplication we obtain

$$\Phi = X_1 \cdot X_2 + X_1 \cdot X_3 + X_2 \cdot X_3 - 2 \cdot X_1 \cdot X_2 \cdot X_3.$$

We have used $X_i^r = X_i$ for $r = 1, 2, \dots$. It follows by taking expectations that

$$h = p_1 p_2 + p_1 p_3 + p_2 p_3 - 2 p_1 p_2 p_3.$$

For systems with low reliabilities, it is possible to establish similar results based on the minimal path sets.

**State Enumeration Method**. Of the direct methods that do not use the minimal cut (path) sets, the state enumeration method is conceptually the simplest. With this method reliability is calculated using

$$h = E\Phi(\mathbf{X}) = \sum_{\mathbf{x}} \Phi(\mathbf{x}) P(\mathbf{X} = \mathbf{x}) = \sum_{\mathbf{x}:\Phi(\mathbf{x})=1} \prod_{i=1}^{n} p_i^{x_i} (1 - p_i)^{1-x_i}.$$

This method, however, is not suitable for larger systems, since the number of terms in the sum can be extremely large, up to $2^n - 1$.

**Factoring Method**. Of other methods we will confine ourselves to describing the so-called *factoring algorithm* (pivot-decomposition method). The basic idea of this method is to make a conditional probability argument using the relation

$$h(\mathbf{p}) = p_i h(1_i, \mathbf{p}) + (1 - p_i) h(0_i, \mathbf{p}), \tag{2.6}$$

where $h(x_i, \mathbf{p})$ equals the reliability of the system given that the state of component $i$ is $x_i$. Formula (2.6) follows from the law of total probability. This process repeats until the system comprises only series–parallel structures. To illustrate the method we will give an example.

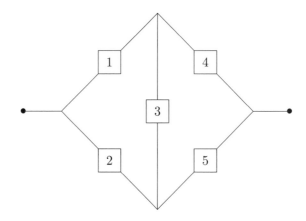

FIGURE 2.5. Bridge Structure

**Example 13** Consider a bridge structure as given by the diagram shown in Figure 2.5. If we first choose to pivot on component 3, formula (2.6) holds with $i = 3$. It is not difficult to see that given $x_3 = 1$, the system structure has the form

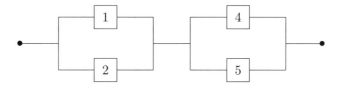

and that given $x_3 = 0$, the system structure has the form

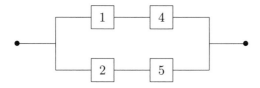

These two structures are both of series–parallel form, and we see that

$$\begin{aligned} h(1_3, \mathbf{p}) &= (p_1 \amalg p_2)(p_4 \amalg p_5) \\ h(0_3, \mathbf{p}) &= p_1 p_4 \amalg p_2 p_5. \end{aligned}$$

Thus a formula for the exact computation of $h(\mathbf{p})$ is established. Note that it was sufficient to perform only one pivotal decomposition in this case. If the structure given $x_3 = 1$ had not been in a series–parallel form, we would have had to perform another pivotal decomposition, and so on.

For a monotone structure $\Phi$ we have

$$\Phi(\mathbf{x} \textstyle\coprod \mathbf{y}) \geq \Phi(\mathbf{x}) \textstyle\coprod \Phi(\mathbf{y}), \tag{2.7}$$

where $\mathbf{x} \coprod \mathbf{y} = (x_1 \coprod y_1, \ldots, x_n \coprod y_n)$. This is seen by noting that $\Phi(\mathbf{x} \coprod \mathbf{y})$ is greater than or equal to both $\Phi(\mathbf{x})$ and $\Phi(\mathbf{y})$. It follows from (2.7) that

$$h(\mathbf{p} \textstyle\coprod \mathbf{p}') \geq h(\mathbf{p}) \textstyle\coprod h(\mathbf{p}')$$

for all $\mathbf{0} \leq \mathbf{p} \leq \mathbf{1}$ and $\mathbf{0} \leq \mathbf{p}' \leq \mathbf{1}$. These results state that redundancy at the component level is more effective than redundancy at system level. This principle is well known among design engineers. Note that if the system is a parallel system, then equality holds in the above inequalities. If the system is coherent, then equality holds if and only if the system is a parallel system.

## Time Dynamics

The above theory can be applied to different situations, covering both repairable and nonrepairable systems. As an example, consider a monotone system in a time interval $[0, t_0]$, and assume that the components of the system are "new" at time $t = 0$ and that a failed component stays in the failure state for the rest of the time interval. Thus the component is not repaired or replaced. This situation, for example, can describe a system with component failure states that can only be discovered by testing or inspection. We assume that the lifetime of component $i$ is determined by a lifetime distribution $F_i(t)$ having failure rate function $\lambda_i(t)$. To calculate system reliability at a fixed point in time, i.e., the reliability function at this point, we can proceed as above with $q_i = F_i(t)$ and $p_i = \bar{F}_i(t)$. Thus, for a series system the reliability at time $t$ takes the form

$$h = \prod_{i=1}^{n} \bar{F}_i(t). \tag{2.8}$$

But $\bar{F}_i(t)$ can be expressed by means of the failure rate $\lambda_i(t)$:

$$\bar{F}_i(t) = e^{-\int_0^t \lambda_i(u)\,du}. \tag{2.9}$$

By putting (2.9) into formula (2.8) we obtain

$$h = e^{-\int_0^t [\sum_{i=1}^n \lambda_i(u)]\,du}. \tag{2.10}$$

From (2.10) we can conclude that the failure rate of a series structure of independent components equals the sum of the failure rates of the components of the structure. In particular this means that if the components have constant failure rates $\lambda_i$, $i = 1, 2, \ldots, n$, then the series structure has constant failure rate $\sum \lambda_i$.

For a parallel structure we do not have a similar result. With constant failure rates of the components, the system will have a time-dependent failure rate; cf. Example 2, p. 6.

## Reliability Importance Measures

An important objective of many reliability and risk analyses is to identify those components or events that are most important (critical) from a reliability/safety point of view and that should be given priority with respect to improvements. Thus, we need an *importance measure*. A large number of such measures have been suggested (see Bibliographic Notes, p. 45). Here we briefly describe two measures, Improvement Potential and Birnbaum's measure.

Consider again the 5-components example (cf. pp. 23, 24, and 26). The unreliability of the system equals

$$
\begin{aligned}
g &= \{1 - p_1 p_4\}\{1 - p_5(p_2 + p_3 - p_2 p_3)\} \approx w_1 \\
w_1 &= q_1 q_5 + q_4 q_5 + q_1 q_2 q_3 + q_2 q_3 q_4 \\
&= 0.02 \cdot 0.01 + 0.01 \cdot 0.01 \\
&\quad + 0.02 \cdot 0.02 \cdot 0.02 + 0.02 \cdot 0.02 \cdot 0.01 \\
&= 3 \cdot 10^{-4}.
\end{aligned}
$$

If we look at the subsystems comprising the minimal cut sets, it is clear from the above expression that subsystems $\{1,5\}$ and $\{4,5\}$ are most important in the sense that they are contributing most to unreliability. To decide which components are most important, we must define more precisely what is meant by important. For example, we might decide to let the component with the highest potential for increasing the system reliability be most important (measure for reliability improvement potential) or the component that has the largest effect on system reliability by a small improvement of the component reliability (Birnbaum's measure).

**Improvement Potential**. The following reliability importance measure for component $i$, $I_i^A$, is appropriate in a large number of situations, in particular during design:

$$
I_i^A = h(1_i, \mathbf{p}) - h(\mathbf{p}),
$$

where $h(\mathbf{p})$ is the reliability of the system and $h(1_i, \mathbf{p})$ is the reliability assuming that component $i$ is in the best state 1. The measure $I_i^A$ expresses the system reliability improvement potential of the component, in other words, the unreliability that is caused by imperfect performance of component $i$. This measure can be used for all types of reliability definitions, and it can be used for repairable or nonrepairable systems.

For a highly reliable monotone system the measure $I_i^A$ is equivalent to the well-known Vesely–Fussell importance measure [95]. In fact, in this case $I_i^A$ is approximately equal to the sum of the unreliabilities of the minimal cut sets that include component $i$, i.e.,

$$
I_i^A \approx \sum_{j:\, i \in K_j} \prod_{l \in K_j} q_l. \tag{2.11}
$$

This is seen by applying the inclusion–exclusion formula. This formula states that $1 - h(\mathbf{p}) \approx \sum_{j=1}^{k} \prod_{l \in K_j} q_l$. Putting $q_i = 0$ in this formula and subtracting, we obtain the desired approximation formula for $I_i^A$. Note that, like the Vesely–Fussell measure, the measure $I_i^A$ gives the same importance to all the components of a parallel system, irrespective of component reliabilities, namely, $I_i^A = \prod_{j=1}^{n} q_j$. This is as it should be because each one of the components has the *potential* of making the system unreliability negligible, for example, by introducing redundancy.

**Example 14** Computation of $I_i^A$ for the 5-components example gives

$$
\begin{aligned}
I_1^A &= 2 \cdot 10^{-4}, & I_2^A &= 1 \cdot 10^{-5}, \\
I_3^A &= 1 \cdot 10^{-5}, & I_4^A &= 1 \cdot 10^{-4}, \\
I_5^A &= 3 \cdot 10^{-4}.
\end{aligned}
$$

Thus component 5 is the most important component based on this measure. Components 1 and 4 follow in second and third place, respectively.

**Birnbaum's Measure.** Birnbaum's measure for the reliability importance of component $i$, $I_i^B$, is defined by

$$
I_i^B = \frac{\partial h}{\partial p_i}.
$$

Thus Birnbaum's measure equals the partial derivative of the system reliability with respect to $p_i$. The approach is well known from classical sensitivity analyses. We see that if $I_i^B$ is large, a small change in the reliability of component $i$ will give a relatively large change in system reliability.

Birnbaum's measure might be appropriate, for example, in the operation phase where possible improvement actions are related to operation and maintenance parameters. Before looking closer into specific improvement actions of the components, it will be informative to measure the sensitivity of the system reliability with respect to small changes in the reliability of the components.

To compute $I_i^B$ the following formula is often used:

$$
I_i^B = h(1_i, \mathbf{p}) - h(0_i, \mathbf{p}). \tag{2.12}
$$

This formula is established using (2.6), p. 27.

**Example 15** Using (2.12) we find that

$$
\begin{aligned}
I_1^B &= 1.03 \cdot 10^{-2} = 1 \cdot 10^{-2}, & I_2^B &= I_3^B = 6 \cdot 10^{-4}, \\
I_4^B &= 1.02 \cdot 10^{-2} = 1 \cdot 10^{-2}, & I_5^B &= 3 \cdot 10^{-2}.
\end{aligned}
$$

We see that for this example the Birnbaum measure gives the same ranking of the components as the measure $I_i^A$. However, this is not true in general.

It is not difficult to see that

$$
\begin{aligned}
I_i^B &= E[\Phi(1_i, \mathbf{X}) - \Phi(0_i, \mathbf{X})] = P(\Phi(1_i, \mathbf{X}) - \Phi(0_i, \mathbf{X}) = 1) \\
&= P(\Phi(1_i, \mathbf{X}) = 1, \ \Phi(0_i, \mathbf{X}) = 0).
\end{aligned}
$$

If $\Phi(1_i, \mathbf{x}) - \Phi(0_i, \mathbf{x}) = 1$, we call $(1_i, \mathbf{x})$ a *critical path vector* and $(0_i, \mathbf{x})$ a *critical cut vector* for component $i$. For simplicity, we often say that component $i$ is *critical* for the system.

Thus we have shown that $I_i^B$ equals the probability that the system is in a state so that component $i$ is critical for the system. If the components are dependent, this probability is often used as the definition of Birnbaum's measure. Now set $p_j = 1/2$ for all $j \neq i$. Then

$$
I_i^B = \frac{1}{2^{n-1}} \sum_{(\cdot_i, \mathbf{x})} [\Phi(1_i, \mathbf{x}) - \Phi(0_i, \mathbf{x})] = \frac{1}{2^n} \sum_{\mathbf{x}} [\Phi(1_i, \mathbf{x}) - \Phi(0_i, \mathbf{x})].
$$

This quantity is used as a measure of the *structural* importance of component $i$.

**Some Comments on the Use of Importance Measures**. The two importance measures presented in this section can be useful tools in the system optimization process/system improvement process. This process can be described as follows:

1. Identify the most important units by means of the chosen importance measure

2. Identify possible improvement actions/measures for these units

3. Estimate the effect on reliability by implementing the measure

4. Perform cost evaluations

5. Make an overall evaluation and take a decision.

The importance measure to be used in a particular case depends on the characteristics we want the measure to reflect. Undoubtedly, different situations call for different importance measures. In a design phase the system reliability improvement potential $I_i^A$ might be the most informative measure, but for a system with frozen design, the Birnbaum measure might be more informative, since this measure reflects how small component reliability improvements affect system reliability.

## *Dependent Components*

One of the most difficult tasks in reliability engineering is to analyze dependent components (often referred to as common mode failures). It is difficult to formulate the dependency in a mathematically stringent way and at the same time obtain a realistic model and to provide data for the model. Whether we succeed in incorporating a "correct" contribution from common mode failures is very much dependent on the modeling ability of the analyst. By defining the components in a suitable way, it is often possible to preclude dependency. For example, common mode failures that are caused by a common external cause can be identified and separated out so that the components can be considered as independent components. Another useful method for "elimination" of dependency is to redefine components. For example, instead of including a parallel structure of dependent components in the system, this structure could be represented by one component. Of course, this does not remove the dependency, but it moves it to a lower level of the analysis. Special techniques, such as Markov modeling, can then be used to analyze the parallel structure itself, or we can try to estimate/assign reliability parameters directly for this new component.

Although it is often possible to "eliminate" dependency between components by proper modeling, it will in many cases be required to establish a model that explicitly takes into account the dependency. Refer to Chapter 3 for examples of such models.

Another way of taking into account dependency is to obtain bounds to the system reliability, assuming that the components are *associated* and not necessarily independent. Association is a type of positive dependency, for example, as a result of components supporting loads. The precise mathematical definition is as follows (cf. [34]):

**Definition 5** *Random variables* $T_1, T_2, \ldots, T_n$ *are associated if*

$$cov[f(\mathbf{T}), g(\mathbf{T})] \geq 0$$

*for all pairs of increasing binary functions f and g.*

A number of results are established for associated components, for example, the following inequalities:

$$\max_{1 \leq j \leq s} \prod_{i \in S_j} p_i \leq h \leq 1 - \max_{1 \leq j \leq k} \prod_{i \in K_j} q_i,$$

where $S_j$ equals the $j$th minimal path set, $j = 1, 2, \ldots, s$ and $K_j$ equals the $j$th minimal cut set, $j = 1, 2, \ldots, k$. This method usually leads to very wide intervals for the reliability.

### 2.1.2  Multistate Monotone Systems

In this section parts of the theory presented in Section 2.1.1 will be generalized to include multistate systems where components and system are allowed to have an arbitrary (finite) number of states/levels. Multistate monotone systems are used to model, e.g., production and transportation systems for oil and gas, and power transmission systems.

We consider a system comprising $n$ components, numbered consecutive from 1 to $n$. As in the binary case, $x_i$ represents the state of component $i$, $i = 1, 2, \ldots, n$, but now $x_i$ can be in one out of $M_i + 1$ states,

$$x_{i0}, x_{i1}, x_{i2}, \ldots, x_{iM_i} \quad (x_{i0} < x_{i1} < x_{i2} < \cdots < x_{iM_i}).$$

The set comprising these states is denoted $S_i$. The states $x_{ij}$ represent, for example, different levels of performance, from the worst, $x_{i0}$, to the best, $x_{iM_i}$. The states $x_{i0}, x_{i1}, \ldots, x_{i,M_i-1}$ are referred to as the failure states of the components.

Similarly, $\Phi = \Phi(\mathbf{x})$ denotes the state (level) of the system. The various values $\Phi$ can take are denoted

$$\Phi_0, \Phi_1, \ldots, \Phi_M \quad (\Phi_0 < \Phi_1 < \cdots < \Phi_M).$$

We see that if $M_i = 1$, $i = 1, 2, \ldots, n$, and $M = 1$, then the model is identical with the binary model of Section 2.1.1.

**Definition 6 (*Monotone system*).** *A system is said to be monotone if*
1. *its structure function $\Phi$ is nondecreasing in each argument, and*
2. $\Phi(x_{10}, x_{20}, \ldots, x_{n0}) = \Phi_0$ *and* $\Phi(x_{1M_1}, x_{2M_2}, \ldots, x_{nM_n}) = \Phi_M$.

In the following we will restrict attention to monotone systems. As usual, we use the convention that $(x_1, x_2, \ldots, x_n) > (z_1, z_2, \ldots, z_n)$ means that $x_i \geq z_i$, $i = 1, 2, \ldots, n$, and there exists at least one $i$ such that $x_i > z_i$.

**Definition 7 (*Minimal cut vector*).** *A vector $\mathbf{z}$ is a cut vector to level $c$ if $\Phi(\mathbf{z}) < c$. A cut vector to level $c$, $\mathbf{z}$, is minimal if $\Phi(\mathbf{x}) \geq c$ for all $\mathbf{x} > \mathbf{z}$.*

**Definition 8 (*Minimal path vector*).** *A vector $\mathbf{y}$ is a path vector to level $c$ if $\Phi(\mathbf{y}) \geq c$. A path vector to level $c$, $\mathbf{y}$, is minimal if $\Phi(\mathbf{x}) < c$ for all $\mathbf{x} < \mathbf{y}$.*

**Example 16** Figure 2.6 shows a simple example of a flow network model. The system comprises three components. Flow (gas/oil) is transmitted from $a$ to $b$. The components 1 and 2 are binary, whereas component 3 can be in one out of three states: $0, 1$, or $2$. The states of the components are interpreted as flow capacity rates for the components. The state/level of the system is defined as the maximum flow that can be transmitted from $a$ to $b$, i.e.,

$$\Phi = \Phi(\mathbf{x}) = \min\{x_1 + x_2, x_3\}.$$

If, for example, the component states are $x_1 = 0$, $x_2 = 1$, and $x_3 = 2$, then the flow throughput equals 1, i.e., $\Phi = \Phi(0, 1, 2) = 1$. The possible system levels are $0, 1$, and $2$. We see that $\Phi$ is a multistate monotone system. The

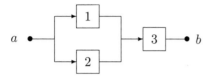

FIGURE 2.6. A Simple Example of a Flow Network

minimal cut vectors and path vectors are as follows:

System level 2
Minimal cut vectors: $(0,1,2)$, $(1,0,2)$, and $(1,1,1)$
Minimal path vectors : $(1,1,2)$

System level 1
Minimal cut vectors: $(0,0,2)$ and $(1,1,0)$
Minimal path vectors : $(0,1,1)$ and $(1,0,1)$.

## Computing System Reliability

Assume that the state $X_i$ of the $i$th component is a random variable, $i = 1, 2, \ldots, n$. Let

$$
\begin{aligned}
p_{ij} &= P(X_i = x_{ij}), \\
h_j &= P(\Phi(\mathbf{X}) \geq \Phi_j), \\
a &= E\Phi(\mathbf{X})/\Phi_M = \sum_j \Phi_j P(\Phi(\mathbf{X}) = \Phi_j)/\Phi_M.
\end{aligned}
$$

We call $h_j$ the reliability of the system at system level $j$. For the flow network example above, $a$ represents the expected throughput (flow) relatively to the maximum throughput (flow) level.

The problem is to compute $h_j$ for one or more values of $j$, and $a$, based on the probabilities $p_{ij}$. We assume that the random variables $X_i$ are independent.

**Example 17** (Continuation of Example 16). Assume that

$$
\begin{aligned}
p_{i1} &= 1 - p_{i0} = 0.96, i = 1, 2, \\
p_{32} &= 0.97,\ p_{31} = 0.02,\ p_{30} = 0.01.
\end{aligned}
$$

Then by simple probability calculus we find that

$$
\begin{aligned}
h_2 &= P(X_1 = 1, X_2 = 1, X_3 = 2) \\
&= 0.96 \cdot 0.96 \cdot 0.97 = 0.894; \\
h_1 &= P(X_1 = 1 \cup X_2 = 1, X_3 \geq 1) \\
&= P(X_1 = 1 \cup X_2 = 1) \, P(X_3 \geq 1) \\
&= \{1 - P(X_1 = 0) \, P(X_2 = 0)\} \, P(X_3 \geq 1) \\
&= 0.9984 \cdot 0.99 = 0.988; \\
a &= (0.094 \cdot 1 + 0.894 \cdot 2)/2 = 0.941.
\end{aligned}
$$

For the above example it is easy to calculate the system reliability directly by using elementary probability rules. For larger systems it will be very time-consuming (in some cases impossible) to perform these calculations if special techniques or algorithms are not used. If the minimal cut vectors or path vectors for a specific level are known, the system reliability for this level can be computed exactly, using, for example, the algorithm described in [20]. For highly reliable systems, which are most common in practice, simple approximations can be used as described in the following.

Analogous to the binary case, approximations can be established based on the inclusion–exclusion method. For example, we have

$$
1 - h_j = \sum_r \prod_{i=1}^{n} P(X_i \leq z_i^r) - \epsilon, \tag{2.13}
$$

where $(z_1^r, z_2^r, \ldots, z_n^r)$ represents the $r$th cut vector for level $j$ and $\epsilon$ is a positive error term satisfying

$$
\epsilon \leq \sum_{r<l} \prod_{i=1}^{n} P(X_i \leq \min\{z_i^r, z_i^l\}).
$$

**Example 18** (Continuation of Example 17). If we use (2.13) to calculate $h_j$, we obtain

$$
\begin{aligned}
h_2 &\approx 1 - (0.04 \cdot 1 \cdot 1 + 1 \cdot 0.04 \cdot 1 + 1 \cdot 1 \cdot 0.03) &= 0.890, \\
h_1 &\approx 1 - (0.04 \cdot 0.04 \cdot 1 + 1 \cdot 1 \cdot 0.01) &= 0.988, \\
a &\approx (1 \cdot 0.098 + 2 \cdot 0.890)/2 &= 0.939.
\end{aligned}
$$

We can conclude that the approximations are quite good for this example.

The problem of determining the probabilities $p_{ij}$ will, as in the binary case, depend on the particular situation considered. Often it will be appropriate to define $p_{ij}$ by the limiting availabilities of the component, cf. Chapter 4.

## Discussion

The traditional reliability theory based on a binary approach has recently been generalized by allowing components and system to have an arbitrary finite number of states. For most reliability applications, binary modeling should be sufficiently accurate, but for certain types of applications, such as gas and oil production and transportation systems and telecommunication, a multistate approach is usually required for the system and components. In a gas transportation system, for example, the state of the system is defined as the rate of delivered gas, and in most cases a binary model (100%, 0%) would be a poor representation of the system. A component in such a system may represent a compressor station comprising a certain number $(M)$ of compressor units in parallel. The states of the component equal the capacity levels corresponding to $M$ compressor units running, $M - 1$ compressor units running, and so on.

There also exists a number of reliability importance measures for multistate systems (see Bibliographic Notes, p. 45). Many of these measures represent natural generalizations of importance measures of binary systems. We see, for example, that the measure $I^A$ can easily be extended to multistate models. For the Birnbaum measure, it is not so straightforward to generalize the measure. Several measures have been proposed as, for example, the $r, s$-reliability importance $I_i^{r,s}$ of component $i$, which is given by

$$I_i^{r,s} = P(\Phi(r_i, \mathbf{X}) \geq \Phi_k) - P(\Phi(s_i, \mathbf{X}) \geq \Phi_k), \qquad (2.14)$$

where $\Phi(j_i, \mathbf{X})$ equals the state of the system given that $X_i = x_{ij}$.

## 2.2 Basic Notions of Aging

In this section we introduce and recapitulate some properties of lifetime distributions. Let $T$ be a positive random variable with distribution function $F : T \sim F$, i.e., $P(T \leq t) = F(t)$. If $F$ has a density $f$, then $\lambda(t) = f(t)/\bar{F}(t)$ is the failure or hazard rate, where as usual $\bar{F}(t) = 1 - F(t)$ denotes the survival probability. Here and in the following we sometimes simplify the notation and define a mapping by its values to avoid constructions like $\lambda : D \to \mathbb{R}_+, D \subset \mathbb{R} \setminus \{t \in \mathbb{R}_+ : \bar{F}(t) = 0\}, t \mapsto \lambda(t) = f(t)/\bar{F}(t)$, if there is no fear of ambiguity. Interpreting $T$ as the lifetime of some component or system, the failure rate measures the proneness to failure at time $t$ : $\lambda(t) \triangle t \approx P(T \leq t + \triangle t | T > t)$. The well-known relation

$$\bar{F}(t) = \exp\{-\int_0^t \lambda(s)ds\}$$

shows that $F$ is uniquely determined by the failure rate. One notion of aging could be an increasing failure rate (IFR). However, this IFR property is in some cases too strong and other intuitive notions of aging have been suggested. Among them are the increasing failure rate average (IFRA) property and the notions of new better than used (NBU) and new better than used in expectation (NBUE). In the following subsection these concepts are introduced formally and the relationships among them are investigated.

Furthermore, these notions should be applied to complex systems. If we consider the time dynamics of such systems, we want to investigate how the reliability of the whole system changes in time if the components have one of the mentioned aging properties.

Another question is how different lifetime (random) variables and their corresponding distributions can be compared. This leads to notions of stochastic ordering. The comparison of the lifetime distribution with the exponential distribution leads to useful estimates of the system reliability.

### 2.2.1 *Nonparametric Classes of Lifetime Distributions*

We first define the IFR and DFR properties of a lifetime distribution $F$ by means of the conditional survival probability

$$P(T > t + x | T > t) = \bar{F}(t + x)/\bar{F}(t).$$

**Definition 9** *Let $T$ be a positive random variable with $T \sim F$.*

*(i) $F$ is an increasing failure rate (IFR) distribution if $\bar{F}(t + x)/\bar{F}(t)$ is nonincreasing in $t$ on the domain of the distribution for each $x \geq 0$.*

*(ii) $F$ is a decreasing failure rate (DFR) distribution if $\bar{F}(t + x)/\bar{F}(t)$ is nondecreasing in $t$ on the domain of the distribution for each $x \geq 0$.*

In the following we will restrict attention to the "increasing" part in the definition of the aging notion. The "decreasing part" can be treated analogously. The IFR property says that with increasing age the probability of surviving $x$ further time units decreases. This definition does not make use of the existence of a density $f$ (failure rate $\lambda$). But if a density exists, then the IFR property is equivalent to a nondecreasing failure rate, which can immediately be seen as follows. From

$$\lambda(t) = \lim_{x \to 0+} \frac{1}{x} \left\{ 1 - \frac{\bar{F}(t + x)}{\bar{F}(t)} \right\}$$

we obtain that the IFR property implies that $\lambda$ is nondecreasing. Conversely, if $\lambda$ is nondecreasing, then we can conclude that

$$P(T > t + x | T > t) = \exp\{- \int_t^{t+x} \lambda(s)ds\}$$

is nonincreasing, i.e., $F$ is IFR. If $F$ has the IFR property, then it is continuous for all $t < t^* = \sup\{t \in \mathbb{R}_+ : \bar{F}(t) > 0\}$ (possibly $t^* = \infty$) and a jump can only occur at $t^*$ if $t^* < \infty$. This can be directly deduced from the IFR definition.

It seems reasonable that the aging properties of the components of a monotone structure are inherited by the system. However, the example of a parallel structure with two independent components, the lifetimes of which are distributed $\text{Exp}(\lambda_1)$ and $\text{Exp}(\lambda_2)$, respectively, shows that in this respect the IFR property is too strong. As was pointed out in Example 2, p. 6, for $\lambda_1 \neq \lambda_2$, the failure rate of the system lifetime is increasing in $(0, t^*)$ and decreasing in $(t^*, \infty)$ for some $t^* > 0$, i.e., constant component failure rates lead in this case to a nonmonotone system failure rate. To characterize the class of lifetime distributions of systems with IFR components we are led to the IFRA property. We use the notation

$$\Lambda(t) = \int_0^t \frac{dF(s)}{1 - F(s-)} \ ,$$

which is the accumulated failure rate. The distribution function $F$ is uniquely determined by $\Lambda$ and the relation is given by

$$\bar{F}(t) = \exp\{-\Lambda^c(t)\} \prod_{s \leq t}(1 - \Delta\Lambda(s))$$

for all $t$ such that $\Lambda(t) < \infty$, where $\Delta\Lambda(s) = \Lambda(s) - \Lambda(s-)$ is the jump height at time $s$ and $\Lambda^c(t) = \Lambda(t) - \sum_{s \leq t} \Delta\Lambda(s)$ is the continuous part of $\Lambda$ (cf. [2], p. 91 or [126], p. 436). In the case that $F$ is continuous, we obtain the simple exponential formula $\bar{F}(t) = \exp\{-\Lambda(t)\}$ or $\Lambda(t) = -\ln \bar{F}(t)$.

**Definition 10** *A distribution $F$ is increasing failure rate average (IFRA) if $-(1/t)\ln \bar{F}(t)$ is nondecreasing in $t > 0$ on $\{t \in \mathbb{R}_+: \bar{F}(t) > 0\}$.*

**Remark 1** (i) The "decreasing" analog is denoted DFRA.
    (ii) If $F$ is IFRA, then $(\bar{F}(t))^{1/t}$ is nonincreasing, which is equivalent to

$$\bar{F}(\alpha t) \geq (\bar{F}(t))^\alpha$$

for $0 \leq \alpha \leq 1$ and $t \geq 0$.

Next we will introduce two aging notions that are related to the residual lifetime of a component of age $t$. Let $T \sim F$ be a positive random variable with finite expectation. Then the distribution of the remaining lifetime after $t \geq 0$ is given by

$$P(T - t > x | T > t) = \frac{\bar{F}(x + t)}{\bar{F}(t)}$$

with expectation

$$\mu(t) = E[T - t | T > t] = \frac{1}{\bar{F}(t)} \int_0^\infty \bar{F}(x + t)dx = \frac{1}{\bar{F}(t)} \int_t^\infty \bar{F}(x)dx \quad (2.15)$$

for $0 \leq t < t^* = \sup\{t \in \mathbb{R}_+ : \bar{F}(t) > 0\}$. The conditional expectation $\mu(t)$ is called mean residual life at time $t$.

**Definition 11** *Let $T \sim F$ be a positive random variable.*
*(i) $F$ is new better than used (NBU), if*

$$\bar{F}(x+t) \leq \bar{F}(x)\bar{F}(t) \text{ for } x, t \geq 0.$$

*(ii) $F$ is new better than used in expectation (NBUE), if $\mu = ET < \infty$*
*and*

$$\mu(t) \leq \mu \text{ for } 0 \leq t < t^*.$$

**Remark 2** (i) The corresponding notions for "better" replaced by "worse," NWU and NWUE, are obtained by reversing the inequality signs.

(ii) These properties are intuitive notions of aging. $F$ is NBU means that the probability of surviving $x$ further time units for a component of age $t$ decreases in $t$. For NBUE distributions the expected remaining lifetime for a component of age $t$ is less than the expected lifetime of a new component.

Now we want to establish the relations between these four notions of aging.

**Theorem 2** *Let $T \sim F$ be a positive random variable with finite expectation. Then we have*

$$F \text{ IFR} \Rightarrow F \text{ IFRA} \Rightarrow F \text{ NBU} \Rightarrow F \text{ NBUE}.$$

**Proof.** $F$ **IFR**$\Rightarrow F$ **IFRA**: Since an IFR distribution $F$ is continuous for all $t < t^* = \sup\{t \in \mathbb{R}_+ : \bar{F}(t) > 0\}$, the simple exponential formula $\bar{F}(t) = \exp\{-\Lambda(t)\}$ holds true and we see that the IFR property implies that $\exp\{\Lambda(t+x) - \Lambda(t)\}$ is increasing in $t$ for all positive $x$. Therefore $\Lambda$ is convex, i.e., $\Lambda(\alpha t + (1-\alpha)u) \leq \alpha\Lambda(t) + (1-\alpha)\Lambda(u), 0 \leq \alpha \leq 1$. Taking the limit $u \to 0-$ we have $\Lambda(0-) = 0$ and $\Lambda(\alpha t) \leq \alpha\Lambda(t)$, which amounts to $\bar{F}(\alpha t) \geq (\bar{F}(t))^\alpha$. But this is equivalent to the IFRA property (see Remark 1 above).

$F$ **IFRA**$\Rightarrow F$ **NBU**: With the abbreviations $a = -(1/x)\ln\bar{F}(x)$ and $b = -(1/y)\ln\bar{F}(y)$ we obtain from the IFRA property for positive $x, y$ that $-(1/(x+y))\ln\bar{F}(x+y) \geq a \vee b = \max\{a, b\}$ and

$$-\ln\bar{F}(x+y) \geq (a \vee b)(x+y) \geq ax + by = -\ln\bar{F}(x) - \ln\bar{F}(y).$$

But this is the NBU property $\bar{F}(x+y) \leq \bar{F}(x)\bar{F}(y)$.

$F$ **NBU**$\Rightarrow F$ **NBUE**: This inequality follows by integrating the NBU inequality

$$\bar{F}(t)\mu(t) = \int_0^\infty \bar{F}(x+t)dx \leq \bar{F}(t)\int_0^\infty \bar{F}(x)dx = \bar{F}(t)\mu,$$

which completes the proof. ∎

Examples can be constructed which show that none of the above implications can be reversed.

### 2.2.2   Closure Theorems

In the previous subsection it was mentioned that the lifetime of a monotone system with IFR components need not be of IFR type. This gave rise to the definition of the IFRA class of lifetime distributions, and we will show that this class is *closed* under forming monotone structures. There are also other reliability operations, among them mixtures of distributions or forming the sum of random variables, and the question arises whether certain distribution classes are closed under these operations. For example, convolutions arise in connection with the addition of lifetimes and cold reserves.

Before we come to the IFRA Closure Theorem we need a preparatory lemma to prove a property of the reliability function $h(\mathbf{p}) = P(\Phi(\mathbf{X}) = 1)$ of a monotone structure.

**Lemma 3** *Let $h$ be the reliability function of a monotone structure. Then $h$ satisfies the inequality*

$$h(\mathbf{p}^\alpha) \geq h^\alpha(\mathbf{p}) \text{ for } 0 < \alpha \leq 1,$$

*where $\mathbf{p}^\alpha = (p_1^\alpha, ..., p_n^\alpha)$.*

**Proof.** We prove the result for binary structures, which are nondecreasing in each argument (nondecreasing structures) but not necessarily satisfy $\Phi(\mathbf{0}) = 0$ and $\Phi(\mathbf{1}) = 1$. We use induction by $n$, the number of components in the system. For $n = 1$ the assertion is obviously true. The induction step is carried out by means of the pivotal decomposition formula:

$$h(\mathbf{p}^\alpha) = p_n^\alpha h(1_n, \mathbf{p}^\alpha) + (1 - p_n^\alpha)h(0_n, \mathbf{p}^\alpha).$$

Now $h(1_n, \mathbf{p}^\alpha)$ and $h(0_n, \mathbf{p}^\alpha)$ define reliability functions of nondecreasing structures with $n - 1$ components. Therefore we have $h(\cdot_n, \mathbf{p}^\alpha) \geq h^\alpha(\cdot_n, \mathbf{p})$ and also

$$h(\mathbf{p}^\alpha) \geq p_n^\alpha h^\alpha(1_n, \mathbf{p}) + (1 - p_n^\alpha)h^\alpha(0_n, \mathbf{p}).$$

The last step is to show that

$$p_n^\alpha h^\alpha(1_n, \mathbf{p}) + (1 - p_n^\alpha)h^\alpha(0_n, \mathbf{p}) \geq (p_n h(1_n, \mathbf{p}) + (1 - p_n)h(0_n, \mathbf{p}))^\alpha.$$

But since $v(x) = x^\alpha$ is a concave function for $x \geq 0$, we have

$$v(x + a) - v(x) \geq v(y + a) - v(y) \text{ for } 0 \leq x \leq y, 0 \leq a.$$

Setting $a = p_n(h(1_n, \mathbf{p}) - h(0_n, \mathbf{p}))$, $x = p_n h(0_n, \mathbf{p})$ and $y = h(0_n, \mathbf{p})$ yields the desired inequality. ∎

Now we can establish the IFRA Closure Theorem.

**Theorem 4** *If each of the independent components of a monotone struc-ture has an IFRA lifetime distribution, then the system itself has an IFRA lifetime distribution.*

**Proof.** Let $F, F_i, i = 1, ..., n$, be the distributions of the lifetimes of the sys-tem and the components, respectively. The IFRA property is characterized by

$$\bar{F}_i(\alpha t) \geq (\bar{F}_i(t))^\alpha$$

for $0 \leq \alpha \leq 1$ and $t \geq 0$. The distribution $F$ is related to the $F_i$ by the reliability function $h$ :

$$\bar{F}(t) = h(\bar{F}_1(t), ..., \bar{F}_n(t)).$$

By Lemma 3 above using the monotonicity of $h$ we can conclude that

$$\begin{aligned} \bar{F}(\alpha t) &= h(\bar{F}_1(\alpha t), ..., \bar{F}_n(\alpha t)) \geq h(\bar{F}_1^\alpha(t), ..., \bar{F}_n^\alpha(t)) \\ &\geq h^\alpha(\bar{F}_1(t), ..., \bar{F}_n(t)) = \bar{F}^\alpha(t) \end{aligned}$$

for $0 < \alpha \leq 1$. For $\alpha = 0$ this inequality holds true since $F(0) = 0$. This proves the IFRA property of $F$.                                                     ∎

We know that independent IFR components form an IFRA monotone system and hence, if the components have exponentially distributed life-times, the system lifetime is of IFRA type. Since constant failure rates are also included in the DFR class, one cannot hope for a corresponding closure theorem for DFRA distributions. However, considering other reli-ability operations things may change. For example, let $\{F_k : k \in \mathbb{N}\}$ be a family of distributions and $F = \sum_{k=1}^\infty p_k F_k$ be its mixture with respect to some probability distribution $(p_k)$. Then it is known that the DFR and the DFRA property are preserved, i.e., if all $F_k$ are DFR(A), then the mixture $F$ is also DFR(A) (for a proof of a slightly more general result see [34] p. 103). Of course, by the same argument as above a closure theorem for mixtures cannot hold true for IFRA distributions.

Finally, we state a closure theorem for convolutions. Since a complete proof is lengthy (and technical), we do not present it here; we refer to [34], p. 100, and [149], p. 23.

**Theorem 5** *Let $X$ and $Y$ be two independent random variables with IFR distributions. Then $X + Y$ has an IFR distribution.*

By induction this property extends to an arbitrary finite number of ran-dom variables. This shows, for example, that the Erlang distribution is of IFR type because it is the distribution of the sum of exponentially dis-tributed random variables.

## 2.2.3  Stochastic Comparison

There are many possibilities to compare random variables or their distributions, respectively, with each other, and a rich literature treats various ways of defining stochastic orders. One of the most important in reliability is the stochastic order. Let $X$ and $Y$ be two random variables. Then $X$ is said to be *smaller in the stochastic order*, denoted $X \leq_{st} Y$, if $P(X > t) \leq P(Y > t)$ for all $t \in \mathbb{R}_+$. In reliability terms we say that $X$ is stochastically smaller than $Y$, if the probability of surviving a given time $t$ is smaller for $X$ than for $Y$ for all $t$. Note that the stochastic order compares two distributions, the random variables could even be defined on different probability spaces. One main point is now to compare a given lifetime distribution with the exponential one. The reason why we choose the exponential distribution is its simplicity and the special role it plays on the border between the IFR(A) and the DFR(A) classes. However, it turns out that in general a random variable with an IFR(A) distribution is not stochastically smaller than an exponentially distributed one, but their distributions cross at most once.

**Lemma 6** *Let $T$ be a positive random variable with IFRA distribution $F$ and $x_p$ be fixed such that $F(x_p) = p$ (p-quantile). Then for $0 < p < 1$*

$$\bar{F}(t) \geq e^{-\alpha t} \text{ for } 0 \leq t < x_p \text{ and}$$
$$\bar{F}(t) \leq e^{-\alpha t} \text{ for } x_p \leq t$$

*holds true, where $\alpha = -\frac{1}{x_p} \ln(1-p)$.*

**Proof.** For an IFRA distribution $v(t) = (-\ln \bar{F}(t))/t$ is nondecreasing. Therefore the result follows by noting that $v(t) \leq v(x_p) = \alpha$ for $t < x_p$ and $v(t) \geq \alpha$ for $t \geq x_p$. ∎

The last lemma compares an IFRA distribution with an exponential distribution with the same $p$-quantile. It is also of interest to compare $F$ having expectation $\mu$ with a corresponding $\text{Exp}(1/\mu)$ distribution. The easiest way seems to be to set $\alpha = 1/\mu$ in the above lemma. But an IFRA distribution function may have jumps so that there might be no $t$ with $v(t) = 1/\mu$. If, on the other hand, $F$ has the stronger IFR property, then it is continuous for $t < t^* = \sup\{t \in \mathbb{R}_+ : \bar{F}(t) > 0\}$ (possibly $t^* = \infty$) and a jump can only occur at $t^*$ if $t^* < \infty$. So we find a value $t_\mu$ with $v(t_\mu) = 1/\mu$ excluding the degenerate case $\bar{F}(\mu) = 0$, i.e., $t^* = \mu$. This leads to the following result.

**Lemma 7** *Let $T$ be a positive random variable with IFR distribution $F$, mean $\mu$ and let $t_\mu = \inf\{t \in \mathbb{R}_+ : -\frac{1}{t} \ln \bar{F}(t) \geq \frac{1}{\mu}\}$. Then*

$$\bar{F}(t) \geq e^{-\frac{t}{\mu}} \text{ for } 0 \leq t < t_\mu,$$
$$\bar{F}(t) \leq e^{-\frac{t}{\mu}} \text{ for } t_\mu \leq t$$

*and $t_\mu \geq \mu$ hold true.*

**Proof.** The inequality for the survival probability follows from Lemma 6 with $\alpha = 1/\mu$, where in the degenerate case $t^* = \mu$ we have $t_\mu = t^* = \mu$. It remains to show $t_\mu \geq \mu$. To this end we first confine ourselves to the continuous case and assume that $F$ has no jump at $t^*$. Then $F(T)$ has a uniform distribution on $[0,1]$ and we obtain $E[\ln \bar F(T)] = -1$. Now

$$\frac{\bar F(t+x)}{\bar F(t)} = \exp\{-(\Lambda(t+x) - \Lambda(t))\}$$

is nonincreasing in $t$ for all $x \geq 0$, which implies that $\Lambda(t) = -\ln \bar F(t)$ is convex, and we can apply J. L. Jensen's inequality to yield

$$1 = E[-\ln \bar F(T)] \geq -\ln \bar F(\mu).$$

This is tantamount to $-\frac{1}{\mu}\ln \bar F(\mu) \leq \frac{1}{\mu}$ and hence $t_\mu \geq \mu$, which proves the assertion for continuous $F$.

In case $F$ has a jump at $t^*$ we can approximate $F$ by continuous distributions. Then $t^*$ is finite and all considerations can be carried over to the limit. We omit the details. ∎

**Example 19** Let $T$ follow a Weibull distribution $\bar F(t) = \exp\{-t^\beta\}$ with mean $\mu = \Gamma(1 + 1/\beta)$, where $\Gamma$ is the Gamma function. Then clearly $F$ is IFR, if $\beta > 1$. Lemma 7 yields $\bar F(t) \geq \exp\{-t/\mu\}$ for $0 \leq t < t_\mu = (1/\mu)^{1/(\beta-1)}$ and $t_\mu \geq \mu$. Note that in this case $t_\mu > \mu$, which extends slightly the well-known result $\bar F(t) \geq \exp\{-t/\mu\}$ for $0 \leq t < \mu$ (see [34] Theorem 6.2, p. 111).

A lot of other bounds for the survival probability can be set up under various conditions (see the references listed in the Bibliographic Notes). Next we want to give one example of how such bounds can be carried over to monotone systems. As an immediate consequence of the last lemma we obtain the following corollary.

**Corollary 8** *Let $h$ be the reliability function of a monotone system with lifetime distribution $F$. If the components are independent with IFR distributions $F_i$ and mean $\mu_i, i = 1, ..., n$, then we have*

$$\bar F(t) \geq h(e^{-t/\mu_1}, ..., e^{-t/\mu_n}) \text{ for } t < \min\{\mu_1, ..., \mu_n\}.$$

Actually the inequality holds true for $t < \min\{t_{\mu_1}, ..., t_{\mu_n}\}$. The idea of this inequality is to give a bound on the reliability of the system at time $t$ only based on $h$ and $\mu_i$ and the knowledge that the $F_i$ are of IFR type. If the reliability function $h$ is unknown, then it could be replaced by that of a series system to yield

$$\bar{F}(t) \geq h(e^{-t/\mu_1}, ..., e^{-t/\mu_n}) \geq \prod_{i=1}^{n} e^{-t/\mu_i} = \exp\{-t \sum_{i=1}^{n} \frac{1}{\mu_i}\}$$

for $t < \min\{\mu_1, ..., \mu_n\}$.

These few examples given here indicate how aging properties lead to bounds on the reliability or survival probability of a single component and how these affect the lifetime of a system comprising independent components.

**Bibliographic Notes.** The basic reliability theory of complex systems was developed in the 1960s and 1970s, and is to large extent covered by the two books of Barlow and Proschan [33] and [34]. Some more recent books in this field are Aven [16] and Høyland and Rausand [99]. Our presentation is based on Aven [16], which also includes theory of multistate monotone systems. This theory was developed in the 1980s. Refer to Natvig [137] and Aven [20] for further details and references.

For specific references to methods (algorithms) for reliability computations, see [142] and the many papers on this topic appearing in reliability journals every year.

Birnbaum's reliability importance measure presented in Section 2.1.1 was introduced by Birnbaum [45]. The improvement potential measure has been used in different contexts, see, e.g., [16, 31]. The measure (2.14) was proposed by Butler [54]. For other references on reliability importance measures, see [16, 31, 41, 87, 95, 99, 136].

Section 2.2, which presents some well-known properties of lifetime distributions, is based on Barlow and Proschan [33], [34], Gertsbakh [82], and Shaked and Shanthikumar [149]. We have not dealt with stochastic comparisons and orders in detail. An overview of this topic with applications in reliability can be found in the book of Shaked and Shanthikumar [149].

# 3
# Stochastic Failure Models

A general set-up should include all basic failure time models, should take into account the time-dynamic development, and should allow for different information and observation levels. Thus, one is led in a natural way to the theory of stochastic processes in continuous time, including (semi-) martingale theory, in the spirit of Arjas [3], [4] and Koch [119]. As was pointed out in Chapter 1, this theory is a powerful tool in reliability analysis. It should be stressed, however, that the purpose of this chapter is to present and introduce ideas rather than to give a far reaching excursion into the theory of stochastic processes. So the mathematical technicalities are kept to the minimum level necessary to develop the tools to be used. Also, a number of remarks and examples are included to illustrate the theory. Yet, to benefit from reading this chapter a solid basis in stochastics is required. Section 3.1 summarizes the mathematics needed. For a more comprehensive and in-depth presentation of the mathematical basis, we refer to Appendix A and to monographs such as by Brémaud [52], Dellacherie and Meyer [67], [68], Kallenberg [112], or Rogers and Williams [143].

## 3.1 Notation and Fundamentals

Let $(\Omega, \mathcal{F}, P)$ be the basic probability space. The information up to time $t$ is represented by the pre-$t$-history $\mathcal{F}_t$, which contains all events of $\mathcal{F}$ that can be distinguished up to and including time $t$. The filtration $\mathbb{F} = (\mathcal{F}_t), t \in \mathbb{R}_+$, which is the family of increasing pre-$t$-histories, is assumed to follow the usual conditions of completeness and right continuity, i.e., $\mathcal{F}_t \subset \mathcal{F}$ contains all $P$-negligible sets of $\mathcal{F}$ and $\mathcal{F}_t = \mathcal{F}_{t+} = \bigcap_{s>t} \mathcal{F}_s$. We define $\mathcal{F}_\infty = \bigvee_{t \geq 0} \mathcal{F}_t$ as the smallest $\sigma$-algebra containing all events of $\mathcal{F}_t$ for all $t \in \mathbb{R}_+$.

If $\{X_j, j \in J\}$ is a family of random variables and $\{A_j, j \in J\}$ is a system of subsets in $\mathcal{F}$, then $\sigma(X_j, j \in J)$ and $\sigma(A_j, j \in J)$, respectively, denote the completion of the generated $\sigma$-field, i.e., the generated $\sigma$-field including all $P$-negligible sets of $\mathcal{F}$. In many cases the information is determined by a stochastic process $Z = (Z_t)$, $t \in \mathbb{R}_+$, and the corresponding filtration is the so-called natural or internal one, which is generated by this stochastic process and denoted $\mathbb{F}^Z = (\mathcal{F}_t^Z)$, $t \in \mathbb{R}_+$, $\mathcal{F}_t^Z = \sigma(Z_s, 0 \leq s \leq t)$. But since it is sometimes desirable to observe one stochastic process on different information levels, it seems more convenient to use filtrations as measures of information. On the basic filtered probability space we now consider a stochastic process $Z = (Z_t)$, which is adapted to a general filtration $\mathbb{F}$, i.e., on the $\mathbb{F}$-information level the process can be observed, or in mathematical terms: $\mathcal{F}_t^Z \subset \mathcal{F}_t$, which assures that $Z_t$ is $\mathcal{F}_t$-measurable for all $t \in \mathbb{R}_+$. All stochastic processes are, if not stated otherwise, assumed to be right-continuous and to have left limits.

A random variable $X$ is integrable if $E|X| < \infty$. If the $p$th power of a random variable $X$ is integrable, $E|X|^p < \infty$, $1 \leq p < \infty$, then it is sometimes said that $X$ is an element of $L^p$, the vector space of real-valued random variables with finite $p$th moment. A stochastic process $(X_t)$, $t \in \mathbb{R}_+$, is called integrable if all $X_t$ are integrable, i.e., $X_t \in L^1$ for all $t \in \mathbb{R}_+$. A family of random variables $(X_t)$, $t \in \mathbb{R}_+$, is called uniformly integrable, if

$$\lim_{c \to \infty} \sup_{t \in \mathbb{R}_+} E[|X_t| I(|X_t| \geq c)] = 0.$$

To simplify the notation, we assume that relations such as $\subset, =$ or $\leq, <, =$ between measurable sets and random variables, respectively, always hold with probability one, which means that the term $P$-a.s. is suppressed. For conditional expectations no difference is made between a version and the equivalence class of $P$-a.s. equal versions.

If we consider a stochastic process $X = (X_t)$ and do not demand that it is right-continuous, then expressions like $Y_t = \int_0^t X_s ds$ have no meaning unless $(X_t)$ fulfills some measurability condition in the argument $t$. One condition is the following.

**Definition 12** *A stochastic process $X$ is $\mathbb{F}$-progressive or progressively measurable, if for every $t$ the mapping $(s, \omega) \to X_s(\omega)$ on $[0, t] \times \Omega$ is measurable with respect to the product $\sigma$-algebra $\mathcal{B}([0, t]) \otimes \mathcal{F}_t$, where $\mathcal{B}([0, t])$ is the Borel $\sigma$-algebra on $[0, t]$.*

Every left- or right-continuous adapted process is progressively measurable. If $X$ is progressive, then so is $Y = (Y_t)$, $Y_t = \int_0^t X_s ds$. A further measurability restriction is needed in connection with stochastic processes in continuous time. This is the fundamental concept of predictability.

**Definition 13** *Let* $\mathbb{F}$ *be a filtration on the basic probability space and let* $\mathcal{P}(\mathbb{F})$ *be the* $\sigma$-*algebra on* $(0, \infty) \times \Omega$ *generated by the system of sets*

$$(s, t] \times A, \ 0 \leq s < t, \ A \in \mathcal{F}_s, \ t > 0.$$

$\mathcal{P}(\mathbb{F})$ *is called the* $\mathbb{F}$-*predictable* $\sigma$-*algebra on* $(0, \infty) \times \Omega$. *A stochastic process* $X = (X_t)$ *is called* $\mathbb{F}$-*predictable, if* $X_0$ *is* $\mathcal{F}_0$-*measurable and the mapping* $(t, \omega) \rightarrow X_t(\omega)$ *on* $(0, \infty) \times \Omega$ *into* $\mathbb{R}$ *is measurable with respect to* $\mathcal{P}(\mathbb{F})$.

Every left-continuous process adapted to $\mathbb{F}$ is $\mathbb{F}$-predictable. In most applications we will be concerned with predictable processes that are left-continuous. Note that $\mathbb{F}$-predictable processes are also $\mathbb{F}$-progressive.

To get an impression of the meaning of the term predictable, we remark that for an $\mathbb{F}$-predictable process $X$ the value $X_t$ can be predicted from the information available "just" before time $t$, i.e., $X_t$ is measurable with respect to $\mathcal{F}_{t-} = \bigvee_{s<t} \mathcal{F}_s = \sigma(A_s, A_s \in \mathcal{F}_s, 0 \leq s < t)$. Processes of this kind are important elements in the framework of point processes. Additional information on these measurability concepts can be found in Appendix A.3, p. 223.

Some further important terms are introduced in the following definitions.

**Definition 14** *A random variable* $\tau$ *with values in* $\mathbb{R}_+ \cup \{\infty\}$ *is called an* $\mathbb{F}$-*stopping time if* $\{\tau \leq t\} \in \mathcal{F}_t$ *for all* $t \in \mathbb{R}_+$.

Thus a stopping time is related to the given information in that at any time $t$ it is possible to decide whether $\tau$ has happened up to time $t$ or not, only using information of the past and present but not anticipating the future.

If $\mathbb{F} = (\mathcal{F}_t)$ is a filtration and $\tau$ an $\mathbb{F}$-stopping time, then the information up to the random time $\tau$ is given by $\mathcal{F}_\tau = \{A \in \mathcal{F}_\infty : A \cap \{\tau \leq t\} \in \mathcal{F}_t$ for all $t \in \mathbb{R}_+\}$. To understand the meaning of this definition, we specialize to a deterministic stopping time $\tau = t^* \in \mathbb{R}_+$. Then $A \in \mathcal{F}_{t^*}$ is equivalent to $A \cap \{t^* \leq t\} \in \mathcal{F}_t$ for all $t \in \mathbb{R}_+$, where $\{t^* \leq t\}$ stands for $\Omega$ if $t^* \leq t$ and for $\emptyset$ otherwise, i.e., for $t = t^*$ the event must be in $\mathcal{F}_{t^*}$ and then it is in $\mathcal{F}_t$ for all $t > t^*$ because the filtration is monotone.

**Definition 15** *An integrable* $\mathbb{F}$-*adapted process* $(X_t), t \in \mathbb{R}_+$, *is called a martingale (submartingale, supermartingale), if for all* $s > t, s, t \in \mathbb{R}_+$,

$$E[X_s | \mathcal{F}_t] = (\geq, \leq) X_t.$$

In the following we denote by $\mathcal{M}$ the set of martingales with paths that are right-continuous and have left-hand limits and by $\mathcal{M}_0$ the set of martingales $M \in \mathcal{M}$ with $M_0 = 0$.

### 3.1.1   The Semimartingale Representation

Semimartingale representations of stochastic processes play a key role in our set-up. They allow the process to be decomposed into a drift or regression part and an additive random fluctuation described by a martingale.

**Definition 16** *A stochastic process $Z = (Z_t), t \in \mathbb{R}_+$, is called a smooth semimartingale (SSM) if it has a decomposition of the form*

$$Z_t = Z_0 + \int_0^t f_s ds + M_t, \tag{3.1}$$

*where $f = (f_t), t \in \mathbb{R}_+$, is a progressively measurable stochastic process with $E \int_0^t |f_s| ds < \infty$ for all $t \in \mathbb{R}_+$, $E|Z_0| < \infty$ and $M = (M_t) \in \mathcal{M}_0$. Short notation: $Z = (f, M)$.*

A martingale is the mathematical model of a fair game with constant expectation function $EM_0 = 0 = EM_t$ for all $t \in \mathbb{R}_+$. The drift term is an integral over a stochastic process. To give this integral meaning, $(f_t)$ should also be measurable in the argument $t$, which is ensured, for example, if $f$ has right-continuous paths or, more general, if $f$ is progressively measurable. Since the drift part in the above decomposition is continuous, a process $Z$, which admits such a representation, is called a smooth semimartingale or smooth F-semimartingale if we would like to emphasize that $Z$ is adapted to the filtration $\mathbb{F}$. For some additional details concerning smooth semimartingales, see the Appendix A.6, p. 235.

Below we formulate conditions under which a process $Z$ admits a semimartingale representation and show how this decomposition can be found. To this end we denote $D(t, h) = h^{-1} E[Z_{t+h} - Z_t | \mathcal{F}_t], t, h \in \mathbb{R}_+$.

**C1** For all $t, h \in \mathbb{R}_+$, versions of the conditional expectation $E[Z_{t+h}|\mathcal{F}_t]$ exist such that the limit

$$f_t = \lim_{h \to 0+} D(t, h)$$

exists $P$-a.s. for all $t \in \mathbb{R}_+$ and $(f_t), t \in \mathbb{R}_+$, is F-progressively measurable with $E \int_0^t |f_s| ds < \infty$ for all $t \in \mathbb{R}_+$.

**C2** For all $t \in \mathbb{R}_+, (hD(t, h)), h \in \mathbb{R}_+$, has $P$-a.s. paths, which are absolutely continuous.

**C3** For all $t \in \mathbb{R}_+$, a constant $c > 0$ exists such that $\{D(t, h) : 0 < h \le c\}$ is uniformly integrable.

The following theorem shows that these conditions are sufficient for a smooth semimartingale representation.

**Theorem 9** *Let $Z = (Z_t), t \in \mathbb{R}_+$, be a stochastic process on the probability space $(\Omega, \mathcal{F}, P)$, adapted to the filtration $\mathbb{F}$. If C1, C2, and C3 hold true, then $Z$ is an SSM with representation $Z = (f, M)$, where $f$ is the limit defined in C1 and $M$ is an $\mathbb{F}$-martingale given by*

$$M_t = Z_t - Z_0 - \int_0^t f_s ds.$$

**Proof.** We have to show that with $(f_t)$ from condition C1 the right-continuous process $M_t = Z_t - Z_0 - \int_0^t f_s ds$ is an $\mathbb{F}$-martingale, i.e., that for all $A \in \mathcal{F}_t$ and $s \geq t, s, t \in \mathbb{R}_+, E[I_A M_s] = E[I_A M_t]$, where $I_A$ denotes the indicator variable. This is equivalent to

$$E[I_A(M_s - M_t)] = \int_A \left( Z_s - Z_t - \int_t^s f_u du \right) dP = 0.$$

For all $r, t \leq r \leq s$, and $A \in \mathcal{F}_t$, $I_A$ is $\mathcal{F}_r$-measurable. This yields

$$\frac{1}{h} E[I_A(Z_{r+h} - Z_r)] = \frac{1}{h} E\left[ E[I_A(Z_{r+h} - Z_r)|\mathcal{F}_r] \right]$$

$$= E\left[ I_A \frac{1}{h} E[Z_{r+h} - Z_r | \mathcal{F}_r] \right] = E[I_A D(r, h)].$$

From C1 it follows that $D(r, h) \to f_r$ as $h \to 0+$ and therefore also $I_A D(r, h) \to I_A f_r$ as $h \to 0+$ $P$-a.s. Now $I_A D(r, h)$ is uniformly integrable by C3, which ensures that

$$\lim_{h \to 0+} E[I_A D(r, h)] = \lim_{h \to 0+} \frac{1}{h} E[I_A(Z_{r+h} - Z_r)] = E[I_A f_r]. \qquad (3.2)$$

Because of C2 there exists a process $(g_t)$ such that

$$E[I_A(Z_s - Z_t)] = E\left[ I_A \int_t^s g_u du \right] = \int_t^s E[I_A g_u] du, \qquad (3.3)$$

where the second equality follows from Fubini's theorem. Then (3.2) and (3.3) together yield

$$E[I_A(Z_s - Z_t)] = \int_t^s E[I_A f_u] du = E\left[ I_A \int_t^s f_u du \right],$$

which proves the assertion. ∎

**Remark 3** (i) In the terminology of Dellacherie and Meyer [68] an SSM $Z = (f, M)$ is a special semimartingale because the drift term $\int_0^t f_s ds$ is continuous and therefore predictable. Hence the decomposition of $Z$ is unique $P$-a.s., because a second decomposition $Z = (f', M')$ leads to the continuous and therefore predictable martingale $M - M'$ of integrable variation, which is identically 0 (cf. Appendix A.5, Lemma 102, p. 233).

(ii) It can be shown that if $Z = (f, M)$ is an SSM and for some constant $c > 0$ the family of random variables $\{|h^{-1} \int_t^{t+h} f_s ds| : 0 < h \leq c\}$ is bounded by some integrable random variable $Y$, then the conditions C1–C3 hold true, i.e., C1–C3 are under this boundedness condition not only sufficient but also necessary for a semimartingale representation. The proof of the main part (C2) is based on the Radon/Nikodym theorem. The details are of technical nature, and they are therefore omitted and left to the interested reader.

(iii) For applications it is often of interest to find an SSM representation for *point processes*, i.e., to determine the compensator of such a process (cf. Definition 3.4 on p. 53). For such and other more specialized processes, specifically adapted methods to find the compensator can be applied, see below and [19, 52, 64, 114, 126].

One of the simplest examples of a process with an SSM representation is the Poisson process $(N_t), t \in \mathbb{R}_+$, with constant rate $\lambda > 0$. It is well-known and easy to see from the definition of a martingale that $M_t = N_t - \lambda t$ defines a martingale with respect to the internal filtration $\mathcal{F}_t^N = \sigma(N_s, 0 \leq s \leq t)$. If we consider conditions C1–C3, we find that $D(t, h) = \lambda$ for all $t, h \in \mathbb{R}_+$ because the Poisson process has independent and stationary increments: $E[N_{t+h} - N_t | \mathcal{F}_t^N] = E[N_{t+h} - N_t] = E N_h = h\lambda$. Therefore, we see that C1–C3 are satisfied with $f_t = \lambda$ for all $\omega \in \Omega$ and all $t \in \mathbb{R}_+$, which results in the representation $N_t = \int_0^t \lambda ds + M_t = \lambda t + M_t$.

The Poisson process is a point process as well as an example of a Markov process, and the question arises under which conditions point and Markov processes admit an SSM representation.

## Point and Counting Processes

A point process over $\mathbb{R}_+$ can be described by an increasing sequence of random variables or by a purely atomic random measure or by means of its corresponding counting process. Since we want to use the semimartingale structure of point processes, we will mostly use the last description of a point process. A (univariate) point process is an increasing sequence $(T_n), n \in \mathbb{N}$, of positive random variables, which may also take the value $+\infty : 0 < T_1 \leq T_2 \leq \ldots$ . The inequality is strict unless $T_n = \infty$. We always assume that $T_\infty = \lim_{n \to \infty} T_n = \infty$, i.e., that the point process is *nonexplosive*.

This point process is also completely characterized by the random measure $\mu$ on $(0, \infty)$ defined by

$$\mu(\omega, A) = \sum_{k \geq 1} I(T_k(\omega) \in A)$$

for all Borel sets $A$ of $(0, \infty)$.

Another equivalent way to describe a point process is by a *counting process* $N = (N_t), t \in \mathbb{R}_+$, with

$$N_t(\omega) = \sum_{k \geq 1} I(T_k(\omega) \leq t),$$

which is, for each realization $\omega$, a right-continuous step function with jumps of magnitude 1 and $N_0(\omega) = 0$. $N_t$ counts the number of time points $T_n$, which occur up to time $t$. Since $(N_t), t \in \mathbb{R}_+$, and $(T_n), n \in \mathbb{N}$, obviously carry the same information, the associated counting process is sometimes also called a point process.

A slight generalization is the notion of a multivariate point process. Let $(T_n), n \in \mathbb{N}$, be a point process as before and $(V_n), n \in \mathbb{N}$, a sequence of random variables with values in a finite set $\{a_1, ..., a_m\}$. Then the sequence of pairs $(T_n, V_n), n \in \mathbb{N}$, is called a multivariate point process and the associated $m$-variate counting process $N_t = (N_t(1), ..., N_t(m))$ is defined by

$$N_t(i) = \sum_{k \geq 1} I(T_k \leq t) I(V_k = a_i), \ i \in \{1, ..., m\}.$$

Let us now consider a univariate point process $(T_n), n \in \mathbb{N}$, and its associated counting process $(N_t), t \in \mathbb{R}_+$, with $EN_t < \infty$ for all $t \in \mathbb{R}_+$ on a filtered probability space $(\Omega, \mathcal{F}, \mathbb{F}, P)$. The traditional definition of the *compensator* of a point process is the following.

**Definition 17** *Let $N$ be an integrable point process adapted to the filtration $\mathbb{F}$. The unique $\mathbb{F}$-predictable increasing process $A = (A_t)$, such that*

$$E \int_0^\infty C_s dN_s = E \int_0^\infty C_s dA_s \qquad (3.4)$$

*is fulfilled for all nonnegative $\mathbb{F}$-predictable processes $C$, is called the compensator of $N$ with respect to $\mathbb{F}$.*

The existence and the uniqueness of the compensator can be proved by the so-called dual predictable projection. We refer to the work of Jacod [102]. The following martingale characterization of the compensator links the dynamical view of point processes with the semimartingale set-up (for a proof, see [114], p. 60).

**Theorem 10** *Let $N$ be an integrable point process adapted to the filtration $\mathbb{F}$. Then $A$ is the $\mathbb{F}$-compensator of $N$ if and only if the difference process $N - A$ is an $\mathbb{F}$-martingale of $\mathcal{M}_0$.*

**Proof.** (Sketch). Let $A$ be the compensator and $C$ be the predictable process defined as the indicator of the set $(t, s] \times B$, where $s > t, B \in \mathcal{F}_s$.

Then the definition of the compensator yields

$$E[I_B(N_s - N_t)] = E[I_B(A_s - A_t)], \qquad (3.5)$$

which gives

$$E[I_B(N_s - A_s)] = E[I_B(N_t - A_t)].$$

Hence, $N - A$ is a martingale.

Conversely, if $N - A$ is a martingale, then $A$ is integrable and we obtain (3.5). In the general case, (3.4) can be established using the monotone class theorem.  ∎

If we view the compensator as a random measure $A(dt)$ on $(0, \infty)$, then we can interpret this measure in an infinitesimal form by the heuristic expression

$$A(dt) = E[dN_t | \mathcal{F}_{t-}].$$

So, by an increment $dt$ in time from $t$ on, the increment $A(dt)$ is what we can predict from the information gathered in $[0, t)$ about the increase of $N_t$, and $dM_t = dN_t - A(dt)$ is what remains unforeseen. Thus, sometimes $M$ is called an *innovation martingale* and $A(dt)$ the (dual) *predictable projection*.

In many cases (which are those we are mostly interested in) the $\mathbb{F}$-compensator $A$ of a counting process $N$ can be represented as an integral of the form

$$A_t = \int_0^t \lambda_s ds$$

with some nonnegative ($\mathbb{F}$-progressively measurable) stochastic process $(\lambda_t)$, $t \in \mathbb{R}_+$, i.e., $N$ has an SSM representation $N = (\lambda, M)$.

**Definition 18** *Let $N$ be an integrable counting process with an $\mathbb{F}$-SSM representation*

$$N_t = A_t + M_t = \int_0^t \lambda_s ds + M_t,$$

*where $(\lambda_t), t \in \mathbb{R}_+$, is a nonnegative process. Then $\lambda$ is called the $\mathbb{F}$-intensity of $N$.*

**Remark 4** (i) To speak of *the* intensity is a little bit misleading (but harmless) because it is not unique. It can be shown (see Brémaud [52], p. 31) that if one can find a predictable intensity, then it is unique except on a set of measure 0 with respect to the product measure of $P$ and Lebesgue measure. On the other hand, if there exists an intensity, then one can always find a predictable version.

(ii) The heuristic interpretation

$$\lambda_t dt = E[dN_t | \mathcal{F}_{t-}]$$

is very similar to the ordinary failure or hazard rate of a random variable.

Theorem 10 and Definition 18 link the point process to the semimartingale representation, and using the definition of the compensator, it is possible to verify formally that a process $\lambda$ is the $\mathbb{F}$-intensity of the point process $N$. We have to show that

$$E \int_0^\infty C_s dN_s = E \int_0^\infty C_s \lambda_s ds$$

for all nonnegative $\mathbb{F}$-predictable processes $C$. Another way to verify that a process $A$ is the compensator is to check the general conditions C1–C3 on page 50 or to use the conditions given by Aven [19].

To go one step further we now specialize to the internal filtration $\mathbb{F}^N = (\mathcal{F}_t^N)$, $\mathcal{F}_t^N = \sigma(N_s, 0 \le s \le t)$, and determine the $\mathbb{F}^N$-compensator of $N$ in an explicit form. The proof of the following theorem can be found in Jacod [102] and in Brémaud [52], p. 61. Regular conditional distributions are introduced in Appendix A.2, p. 221.

**Theorem 11** *Let $N$ be an integrable point process and $\mathbb{F}^N$ its internal filtration. For each $n$ let $G_n(\omega, B)$ be the regular conditional distribution of the interarrival time $U_{n+1} = T_{n+1} - T_n$, $n \in \mathbb{N}_0$, $T_0 = 0$, given the past $\mathcal{F}_{T_n}^N$ at the $\mathbb{F}^N$-stopping time $T_n : G_n(\omega, B) = P(U_{n+1} \in B | \mathcal{F}_{T_n}^N)(\omega)$.*
*(i) Then for $T_n < t \le T_{n+1}$ the compensator $A$ is given by*

$$A_t = A_{T_n} + \int_0^{t-T_n} \frac{G_n(dx)}{G_n([x, \infty))}.$$

*(ii) If the conditional distribution $G_n$ admits a density $g_n$ for all $n$, then the $\mathbb{F}^N$-intensity $\lambda$ is given by*

$$\lambda_t = \sum_{n \ge 0} \frac{g_n(t - T_n)}{1 - \int_0^{t-T_n} g_n(x) dx} I(T_n < t \le T_{n+1}).$$

Note that expressions of the form "$\frac{0}{0}$" are always set equal to 0.

**Example 20** (Renewal process). Let the interarrival times $U_{n+1} = T_{n+1} - T_n$, $n \in \mathbb{N}_0$, $T_0 = 0$, be i.i.d. random variables with common distribution function $F$, density $f$ and failure rate $r$: $r(t) = f(t)/(1 - F(t))$. Then it follows from Theorem 11 that with respect to the internal history $\mathcal{F}_t^N = \sigma(N_s, 0 \le s \le t)$ the intensity on $\{T_n < t \le T_{n+1}\}$ is given by $\lambda_t = r(t - T_n)$. This results in the SSM representation $N = (\lambda, M)$,

$$N_t = \int_0^t \lambda_s ds + M_t$$

with the intensity

$$\lambda_t = \sum_{n \ge 0} r(t - T_n) I(T_n < t \le T_{n+1}).$$

This corresponds to our supposition that the intensity at time $t$ is the failure rate of the last renewed item before $t$ at an age of $t - T_n$.

**Example 21** (Markov-modulated Poisson process). A Poisson process can be generalized by replacing the constant intensity with a randomly varying intensity, which takes one of the $m$ values $\lambda_i, 0 < \lambda_i < \infty, i \in S = \{1, ..., m\}, m \in \mathbb{N}$. The changes are driven by a homogeneous Markov chain $Y = (Y_t), t \in \mathbb{R}_+$, with values in $S$ and infinitesimal parameters $q_i$, the rate to leave state $i$, and $q_{ij}$, the rate to reach state $j$ from state $i$:

$$q_i = \lim_{h \to 0+} \frac{1}{h} P(Y_h \neq i | Y_0 = i),$$

$$q_{ij} = \lim_{h \to 0+} \frac{1}{h} P(Y_h = j | Y_0 = i), i, j \in S, i \neq j,$$

$$q_{ii} = -q_i = -\sum_{j \neq i} q_{ij}.$$

The point process $(T_n)$ corresponds to the counting process $N = (N_t), t \in \mathbb{R}_+$, with

$$N_t = \sum_{n=1}^{\infty} I(T_n \leq t).$$

It is assumed that $N$ has a stochastic intensity $\lambda_{Y_t}$ with respect to the filtration $\mathbb{F}$, generated by $N$ and $Y$:

$$\mathcal{F}_t = \sigma(N_s, Y_s, 0 \leq s \leq t).$$

Then $N$ is called a Markov-modulated Poisson process with SSM representation

$$N_t = \int_0^t \lambda_{Y_s} ds + M_t.$$

Roughly spoken, in state $i$ the point process is Poisson with rate $\lambda_i$. But note that the ordinary failure rate of $T_1$ is not constant. If we cannot observe the Markov chain $Y$, but only the point process $(T_n)$, then we look for an intensity with respect to the subfiltration $\mathbb{A} = (\mathcal{A}_t), t \in \mathbb{R}_+$, $\mathcal{A}_t = \sigma(N_s, 0 \leq s \leq t)$. For this we have to estimate the current state of the Markov chain, involving the infinitesimal parameters $q_i, q_{ij}$. For this we refer to Sections 3.2.4 and 5.4.2.

## Markov Processes

The question whether Markov processes admit semimartingale representations can generally be answered in the affirmative: (most) Markov processes and bounded functions of such processes have an SSM representation.

Let $(X_t), t \in \mathbb{R}_+$, be a right-continuous homogeneous Markov process on $(\Omega, \mathcal{F}, P^x)$ with respect to the (internal) filtration $\mathcal{F}_t = \sigma(X_s, 0 \leq s \leq t)$

with values in a measurable space $(S, \mathcal{B}(S))$. For applications we will often confine ourselves to $S = \mathbb{R}$ with its Borel $\sigma$-field $\mathcal{B}$. Here $P^x$, $x \in S$, denotes the probability measure on the set of paths, which start in $X_0 = x$: $P^x(X_0 = x) = 1$.

Let $\mathbb{B}$ denote the set of bounded, measurable functions on $S$ with values in $\mathbb{R}$ and let $E^x$ denote expectation with respect to $P^x$. Then the infinitesimal generator $\mathcal{A}$ is defined as follows: If for $f \in \mathbb{B}$ the limit

$$\lim_{h \to 0+} \frac{1}{h}(E^x f(X_h) - f(x)) = g(x)$$

exists for all $x \in S$ with $g \in \mathbb{B}$, then we set $\mathcal{A}f = g$ and say that $f$ belongs to the domain $\mathbb{D}(\mathcal{A})$ of the infinitesimal generator $\mathcal{A}$. It is known that if $f \in \mathbb{D}(\mathcal{A})$, then

$$M_t^f = f(X_t) - f(X_0) - \int_0^t \mathcal{A}f(X_s)ds$$

defines a martingale (cf., e.g., [112], p. 328). This shows that a function $Z_t = f(X_t)$ of a homogeneous Markov process has an SSM representation if $f \in \mathbb{D}(\mathcal{A})$.

**Example 22 (Markov pure jump process).** A homogeneous Markov process $X = (X_t)$ with right-continuous paths, which are constant between isolated jumps, is called a Markov pure jump process. As before, $P^x$ denotes the probability law conditioned on $X_0 = x$ and $\tau_x = \inf\{t \in \mathbb{R}_+ : X_t \neq x\}$ the exit time of state $x$. It is known that $\tau_x$ follows an $\text{Exp}(\lambda(x))$ distribution if $0 < \lambda(x) < \infty$ and that $P^x(\tau_x = \infty) = 1$ if $\lambda(x) = 0$, for some suitable mapping $\lambda$ on the set of possible outcomes of $X_0$ with values in $\mathbb{R}_+$. Let $v(x, \cdot)$ be the jump law or transition probability at $x$, defined by $v(x, B) = P^x(X_{\tau_x} \in B)$ for $\lambda(x) > 0$. If $f$ belongs to the domain of $\mathbb{D}(\mathcal{A})$ of the infinitesimal generator, then we obtain (cf. Métivier [134])

$$\mathcal{A}f(x) = \lambda(x) \int (f(y) - f(x))v(x, dy). \tag{3.6}$$

Let us now consider some particular cases.

(i) Poisson process $N = (N_t)$ with parameter $\lambda > 0$. In this case we have jumps of height 1, i.e., $v(x, \{x + 1\}) = 1$. For $f(x) = x$ we get $\mathcal{A}f(x) \equiv \lambda$. This again shows that $N_t - \lambda t$ is a martingale. If we take $f(x) = x^2$, then we obtain $\mathcal{A}f(x) = \lambda(2x + 1)$ and for $N^2$ we have the SSM representation

$$N_t^2 = f(N_t) = \int_0^t \lambda(2N_s + 1)ds + M_t^f.$$

(ii) Compound Poisson process $X = (X_t)$. Let $N$ be a Poisson process with an intensity $\lambda : \mathbb{R} \to \mathbb{R}_+$, $0 < \lambda(x) < \infty$, and $(Y_n), n \in \mathbb{N}$, a sequence of i.i.d. random variables with finite mean $\mu$. Then

$$X_t = \sum_{n=1}^{N_t} Y_n$$

defines a Markov pure jump process with $\nu(x, B) = P^x(X_{\tau_x} \in B) = P(Y_1 \in B - x)$. By formula (3.6) for the infinitesimal generator we get the SSM representation

$$X_t = \int_0^t \lambda(X_s)\mu ds + M_t.$$

We now return to the general theory of Markov processes. The so-called Dynkin formula states that for a stopping time $\tau$ we have

$$E^x g(X_\tau) = g(x) + E^x \int_0^\tau Ag(X_s)ds$$

if $E^x\tau < \infty$ and $g \in \mathbb{D}(\mathcal{A})$ (see Dynkin [74], p. 133). This formula can now be extended to the more general case of smooth semimartingales. If $Z = (f, M)$ is an $\mathbb{F}$-SSM with ($P$-a.s.) bounded $Z$ and $f$, then for all $\mathbb{F}$-stopping times $\tau$ with $E\tau < \infty$ we obtain

$$EZ_\tau = EZ_0 + E \int_0^\tau f_s ds.$$

Here $EM_\tau = 0$ is a consequence of the *Optional Sampling Theorem* (see Appendix A.5, Theorem 98, p. 231). The following example shows how the Dynkin formula can be applied to determine the expectation of a stopping time.

**Example 23** Let $B = (B_t)$ be a $k$-dimensional *Brownian motion* with initial point $B_0 = x$ and $g$ a bounded twice continuously differentiable function on $\mathbb{R}^k$ with bounded derivatives. Then we obtain (cf. Métivier [134], p. 201) the SSM representation for $g(B_t)$ :

$$g(B_t) = g(x) + \frac{1}{2}\int_0^t \sum_{i,j=1}^k \frac{\partial^2 g}{\partial x_i \partial x_j}(B_s)ds + M_t^g.$$

For some $R > 0$ and $|x| < R$ we consider the stopping time $\sigma = \inf\{t \in \mathbb{R}_+ : |B_t| \geq R\}$ with respect to the internal filtration, which is the first exit time of the ball $K_R = \{y \in \mathbb{R}^k : |y| < R\}$. By means of the Dynkin formula we can determine the expectation $E^x\sigma$ in the following way. Let us assume $E^x\sigma < \infty$ and choose $g(x) = |x|^2$. Dynkin's formula then yields

$$\begin{aligned} E^x g(B_\sigma) &= R^2 = |x|^2 + \frac{1}{2}E^x \int_0^\sigma 2k\, ds \\ &= |x|^2 + kE^x\sigma, \end{aligned}$$

which is tantamount to $E^x \sigma = k^{-1}(R^2 - |x|^2)$. To show $E^x \sigma < \infty$ we may replace $\sigma$ by $\tau_n = n \wedge \sigma$ in the above formula: $E^x \tau_n \leq k^{-1}(R^2 - |x|^2)$ and together with the monotone convergence theorem the result is established.

### 3.1.2 Transformations of Smooth Semimartingales

Next we want to investigate under which conditions certain transformations of SSMs again lead to SSMs and leave the SSM property unchanged.

### Random Stopping

One example is the stopping of a process $Z$, i.e., the transformation from $Z = (Z_t)$ to the process $Z^\zeta = (Z_{t \wedge \zeta})$, where $\zeta$ is some stopping time. If $Z = (f, M)$ is an $\mathbb{F}$-SSM and $\zeta$ is an $\mathbb{F}$-stopping time, then $Z^\zeta$ is again an $\mathbb{F}$-SSM with representation

$$Z_t^\zeta = Z_0 + \int_0^t I(\zeta > s) f_s ds + M_{t \wedge \zeta}, \ t \in \mathbb{R}_+.$$

This result is an immediate consequence of the fact that a stopped martingale is a martingale.

### A Product Rule

A second example of a transformation is the product of two SSMs. To see under which conditions such a product of two SSMs again forms an SSM, some further notations and definitions are required, which are presented in Appendix A. Here we only give the general result. For the conditions and a detailed proof we refer to Appendix A.6, Theorem 108, p. 239.

Let $Z = (f, M)$ and $Y = (g, N)$ be $\mathbb{F}$-SSMs with $M, N \in \mathcal{M}_0^2$ and $MN \in \mathcal{M}_0$. Then, under suitable integrability conditions, $ZY$ is an $\mathbb{F}$-SSM with representation

$$Z_t Y_t = Z_0 Y_0 + \int_0^t (Y_s f_s + Z_s g_s) ds + R_t,$$

where $R = (R_t)$ is a martingale in $\mathcal{M}_0$.

**Remark 5** (i) If $Z = (f, M)$ and $Y = (g, N)$ are two SSMs and $f$ and $g$ are considered as "derivatives," then $Yf + Zg$ is the "derivative" of the product $ZY$ in accordance with the ordinary product rule.

(ii) Martingales $M, N$, for which $MN$ is a martingale are called orthogonal. This property can be interpreted in the sense that the increments of the martingales are "conditionally uncorrelated," i.e.,

$$E[(M_t - M_s)(N_t - N_s)|\mathcal{F}_s] = 0$$

for all $0 \leq s \leq t$.

## A Change of Filtration

Another transformation is a certain change of the filtration, which allows the observation of a stochastic process on different information levels.

**Definition 19** *Let* $\mathbb{A} = (\mathcal{A}_t), t \in \mathbb{R}_+,$ *and* $\mathbb{F} = (\mathcal{F}_t), t \in \mathbb{R}_+,$ *be two filtrations on the same probability space* $(\Omega, \mathcal{F}, P)$. *Then* $\mathbb{A}$ *is called a subfiltration of* $\mathbb{F}$ *if* $\mathcal{A}_t \subset \mathcal{F}_t$ *for all* $t \in \mathbb{R}_+$.

In this case $\mathbb{F}$ can be viewed as the complete information filtration and $\mathbb{A}$ as the actual observation filtration on a lower level. If $Z = (f, M)$ is an SSM with respect to the filtration $\mathbb{F}$, then the projection to the observation filtration $\mathbb{A}$ is given by the conditional expectation $\hat{Z}$ with $\hat{Z}_t = E[Z_t|\mathcal{A}_t]$. The following projection theorem states that $\hat{Z}$ is an $\mathbb{A}$-semimartingale. Different versions of this theorem are proved in the literature. The version presented here for SSMs is based on, [52], pp. 87, 108, [111], p. 202 and [173].

**Theorem 12 (Projection Theorem)** *Let* $Z = (f, M)$ *be an* $\mathbb{F}$-*SSM and* $\mathbb{A}$ *a subfiltration of* $\mathbb{F}$. *Then* $\hat{Z}$ *with*

$$\hat{Z}_t = \hat{Z}_0 + \int_0^t \hat{f}_s ds + \bar{M}_t \qquad (3.7)$$

*is an* $\mathbb{A}$-*SSM, where*
  *(i)* $\hat{Z}$ *is* $\mathbb{A}$-*adapted with a.s. right-continuous paths with left-hand limits and* $\hat{Z}_t = E[Z_t|\mathcal{A}_t]$ *for all* $t \in \mathbb{R}_+$;
  *(ii)* $\hat{f}$ *is* $\mathbb{A}$-*progressively measurable with* $\hat{f}_t = E[f_t|\mathcal{A}_t]$ *for almost all* $t \in \mathbb{R}_+$ *(Lebesgue measure)*;
  *(iii)* $\bar{M}$ *is an* $\mathbb{A}$-*martingale.*
  *If in addition* $Z_0, \int_0^\infty |f_s|ds \in L^2$ *and* $M \in \mathcal{M}_0^2$, *then* $\hat{Z}_0, \int_0^\infty |\hat{f}_s|ds \in L^2$ *and* $\bar{M} \in \mathcal{M}_0^2$.

Unfortunately, monotonicity properties of $Z$ and $f$ do not in general extend to $\hat{Z}$ and $\hat{f}$, respectively. So if, for example, $f$ has monotone paths, this need not be true for the corresponding process $\hat{f}$. Whether $\hat{f}$ has monotone paths depends on the path properties of $f$ as well as on the subfiltration $\mathbb{A}$. If $f$ is already adapted to the subfiltration $\mathbb{A}$, then it is obvious that $\hat{f} = f$. In this case projecting onto the subfiltration only filters information out, which does not affect the drift term.

The Projection Theorem will mainly be applied to solve optimal stopping problems on different information levels in the following manner. Let $Z = (f, M)$ be an $\mathbb{F}$-SSM and let $\hat{Z} = (\hat{f}, \bar{M})$ be the corresponding $\mathbb{A}$-SSM with respect to a subfiltration $\mathbb{A}$ of $\mathbb{F}$. To determine the maximum of $EZ_\tau$ in the set $C^{\mathbb{A}}$ of $\mathbb{A}$-stopping times $\tau$, i.e., to solve the optimal stopping problem on the lower $\mathbb{A}$-information level, we can use the rule of successive conditioning for conditional expectations (cf. Appendix A.2, p. 219) to obtain

$$\sup\{EZ_\tau : \tau \in C^{\mathbb{A}}\} = \sup\{E\hat{Z}_\tau : \tau \in C^{\mathbb{A}}\}.$$

In Section 5.2.1, Theorem 71, p. 175, conditions are given under which the stopping problem for an SSM $Z$ can be solved. If these conditions apply to $\hat{Z}$, then we can solve this optimal stopping problem on the $\mathbb{A}$-level according to Theorem 71. Could the stopping problem be solved on the $\mathbb{F}$-level, then we get a bound for the stopping value on the $\mathbb{A}$-level in view of the inequality

$$\sup\{E\hat{Z}_\tau : \tau \in C^{\mathbb{A}}\} \leq \sup\{EZ_\tau : \tau \in C^{\mathbb{F}}\}.$$

## 3.2   A General Lifetime Model

First let us consider the simple indicator process $Z_t = I(T \leq t)$, where $T$ is the lifetime random variable defined on the basic probability space. Obviously $Z$ is the counting process corresponding to the simple point process $(T_n)$ with $T = T_1$ and $T_n = \infty$ for $n \geq 2$. The paths of this indicator process $Z$ are constant, except for one jump from 0 to 1 at $T$. Let us assume that this indicator process has a smooth $\mathbb{F}$-semimartingale representation with an $\mathbb{F}$-martingale $M \in \mathcal{M}_0$ and a nonnegative stochastic process $\lambda = (\lambda_t)$:

$$I(T \leq t) = \int_0^t I(T > s)\lambda_s ds + M_t, \ t \in \mathbb{R}_+. \tag{3.8}$$

The general lifetime model is then defined by the filtration $\mathbb{F}$ and the corresponding $\mathbb{F}$-SSM representation of the indicator process.

**Definition 20** *The process $\lambda = (\lambda_t), t \in \mathbb{R}_+$, in the SSM-representation (3.8) is called the $\mathbb{F}$-failure rate or the $\mathbb{F}$-hazard rate process and the compensator $\Lambda_t = \int_0^t I(T > s)\lambda_s ds$ is called the $\mathbb{F}$-hazard process.*

We drop $\mathbb{F}$, when it is clear from the context. As was mentioned before (cf. Remark 4 on p. 54), the intensity of the indicator (point) process is not unique. If one $\mathbb{F}$-failure rate $\lambda$ is known, we may pass to a left-continuous version $(\lambda_{t-})$ to obtain a predictable, unique intensity:

$$I(T \leq t) = \int_0^t I(T \geq s)\lambda_{s-} ds + M_t.$$

Before investigating under which conditions such a representation exists, some examples are given.

**Example 24** If the failure rate process $\lambda$ is deterministic, forming expectations leads to the integral equation

$$F(t) = P(T \leq t) = EI(T \leq t) = \int_0^t P(T > s)\lambda_s ds = \int_0^t (1 - F(s))\lambda_s ds.$$

The unique solution

$$\bar{F}(t) = 1 - F(t) = \exp\{-\int_0^t \lambda_s ds\} \qquad (3.9)$$

is just the well-known relation between the standard failure rate and the distribution function. This shows that if the hazard rate process $\lambda$ is deterministic, it coincides with the ordinary failure rate.

**Example 25** In continuation of Example 1, p. 2, we consider a three-component system with one component in series with a two-component parallel system. It is assumed that the component lifetimes $T_1, T_2, T_3$ are i.i.d. exponentially distributed with parameter $\alpha > 0$. What is the failure rate process corresponding to the system lifetime $T = T_1 \wedge (T_2 \vee T_3)$? This depends on the information level, i.e., on the filtration $\mathbb{F}$.

- $\mathcal{F}_t = \sigma(\mathbf{X}_s, 0 \leq s \leq t)$, where $\mathbf{X}_s = (X_s(1), X_s(2), X_s(3))$ and $X_s(i) = I(T_i > s), i = 1, 2, 3$. Observing on the component level means that $\mathcal{F}_t$ is generated by the indicator processes of the component lifetimes up to time $t$. It can be shown (by means of the results of the next section) that the failure rate process of the system lifetime is given by $\lambda_t = \alpha\{1 + (1 - X_t(2)) + (1 - X_t(3))\}$ on $\{T > t\}$. As long as all components work, the rate is $\alpha$ due to component 1. When one of the two parallel components 2 or 3 fails first, then the rate switches to $2\alpha$.

- $\mathcal{F}_t = \sigma(I(T \leq s), 0 \leq s \leq t)$. If only the system lifetime can be observed, the failure rate process diminishes to the ordinary deterministic failure rate

$$\lambda_t = \alpha \left(1 + 2\frac{1 - e^{-\alpha t}}{2 - e^{-\alpha t}}\right).$$

**Example 26** Consider the damage threshold model in which the deterioration is described by the Wiener process $X_t = \sigma B_t + \mu t$, where $B$ is standard Brownian motion and $\sigma, \mu > 0$ are constants. In this case, whether and in what way the lifetime $T = \inf\{t \in \mathbb{R}_+ : X_t \geq K\}, K \in \mathbb{R}_+$, can be characterized by a failure rate process, also depends on the available information.

- $\mathcal{F}_t = \sigma(B_s, 0 \leq s \leq t)$. Observing the actual state of the system proves to be too informative to be described by a failure rate process. The martingale part is identically 0, the drift part or the predictable compensator is the indicator process $I(T \leq t)$ itself. No semimartingale representation (3.8) exists because the lifetime is predictable, as we will see in the following section.

- $\mathcal{F}_t = \sigma(I(T \leq s), 0 \leq s \leq t)$. If only the system lifetime can be observed, conditions change completely. A representation (3.8) exists. The first hitting time $T$ of the barrier $K$ is known to follow a so-called inverse Gaussian distribution (cf. [143], p. 26). The failure rate process is then the ordinary failure rate corresponding to the density

$$f(t) = \frac{K}{\sqrt{2\pi\sigma^2 t^3}} \exp\left\{-\frac{(K - \mu t)^2}{2\sigma^2 t}\right\}, t > 0.$$

### 3.2.1  Existence of Failure Rate Processes

It is possible to formulate rather general conditions on $Z$ to ensure a semimartingale representation (3.8) as shown by Theorem 9, p. 51. But in reliability models we often have more specific processes $V_t = I(T \leq t)$ for which a representation (3.8) has to be found. Whether such a representation exists should depend on the random variable $T$ (or on the probability measure $P$) and on the filtration $\mathbb{F}$. If $T$ is a stopping time with respect to the filtration $\mathbb{F}$, then a representation (3.8) only exists for stopping times which are *totally inaccessible* in the following sense:

**Definition 21** *An $\mathbb{F}$-stopping time $\tau$ is called*

- *predictable if an increasing sequence $(\tau_n), n \in \mathbb{N}$, of $\mathbb{F}$-stopping times $\tau_n < \tau$ exists such that $\lim_{n\to\infty} \tau_n = \tau$;*

- *totally inaccessible if $P(\tau = \sigma < \infty) = 0$ for all predictable $\mathbb{F}$-stopping times $\sigma$.*

Roughly speaking, a stopping time $\tau$ is predictable, if it is announced by a sequence of (observable) stopping times, $\tau$ is totally inaccessible if it occurs "suddenly" without announcement. For example, a random variable $T$ with an absolutely continuous distribution has the representation

$$V_t = I(T \leq t) = \int_0^t I(T > s)\lambda(s)ds + M_t, \ t \in \mathbb{R}_+$$

with respect to the filtration $\mathbb{F}^T = (\mathcal{F}_t)$ generated by $T$: $\mathcal{F}_t = \sigma(T \wedge t)$, where $\lambda$ is the ordinary failure rate. One might guess that $T$ is totally inaccessible.

In general it can be shown that, if $V$ has a smooth semimartingale representation (3.8), then $T$ is a totally inaccessible stopping time. On the other hand, if $T$ is totally inaccessible, then there is a (unique) decomposition $V = \Lambda + M$ in which the process $\Lambda$ is ($P$-a.s.) continuous. We state this result without proof (cf. [68], p. 137 and [134], p. 113).

**Lemma 13** *Let $(\Omega, \mathcal{F}, \mathbb{F}, P)$ be a filtered probability space and $T$ an $\mathbb{F}$-stopping time.*

*(i) If the process $V = (V_t)$, $V_t = I(T \leq t)$, has an SSM representation*

$$V_t = \int_0^t I(T > s)\lambda_s ds + M_t, t \in \mathbb{R}_+,$$

*then $T$ is a totally inaccessible stopping time and the martingale $M$ is square integrable, $M \in \mathcal{M}_0^2$.*

*(ii) If $T$ is a totally inaccessible stopping time, then the process $V = (V_t)$, $V_t = I(T \leq t)$, has a unique ($P$-a.s.) decomposition $V = \Lambda + M$, where $M$ is a uniformly integrable martingale and $\Lambda$ is continuous ($P$-a.s., the predictable compensator).*

"Most" continuous functions are absolutely continuous (except some pathological special cases). Therefore, we can conclude from Lemma 13 that the class of lifetime models with a compensator $\Lambda$ of the form $\Lambda_t = \int_0^t I(T > s)\lambda_s ds$ is rich enough to include models for most real-life systems in continuous time. In view of Example 26 the condition that $V$ admits an SSM representation seems a natural restriction, because if the lifetime could be predicted by an announcing sequence of stopping times, maintenance actions would make no sense, they could be carried out "just" before a failure. In Example 26 $\tau_n = \inf\{t \in \mathbb{R}_+ : X_t = K - \frac{1}{n}\}$ is such an announcing sequence with respect to $\mathcal{F}_t = \sigma(B_s, 0 \leq s \leq t)$ (compare also Figure 1.1, p. 6). In addition, Example 26 shows that one and the same random variable $T$ can be predictable or totally inaccessible depending on the corresponding information filtration.

How can the failure rate process $\lambda$ be ascertained or identified for a given information level $\mathbb{F}$? In general, we can determine $\lambda$ under the conditions of Theorem 9 as the limit

$$I(T > t)\lambda_t = \lim_{h \to 0+} \frac{1}{h} P(t < T \leq t + h | \mathcal{F}_t)$$

in the sense of almost sure convergence. Another way to verify whether a given process $\lambda$ is the failure rate is to show that the corresponding hazard process defines the compensator of $I(T \leq t)$. In some special cases $\lambda$ can be represented in a more explicit form, as for example for complex systems. This will be carried out in some detail in the next section.

### 3.2.2  Failure Rate Processes in Complex Systems

In the following we want to derive the hazard rate process for the lifetime $T$ of a complex system under fairly general conditions. We make no independence assumption concerning the component lifetimes, and we allow two or more components to fail at the same time with positive probability.

Let $T_i, i = 1, ..., n$, be $n$ positive random variables that describe the component lifetimes of a monotone complex system with structure function $\Phi$. Our aim is to derive the failure rate process for the lifetime

$$T = \inf\{t \in \mathbb{R}_+ : \Phi(\mathbf{X}_t) = 0\}$$

with respect to the filtration $\mathbb{F}$ given by $\mathcal{F}_t = \sigma(\mathbf{X}_s, 0 \leq s \leq t)$, where as before $\mathbf{X}_s = (X_s(1), ..., X_s(n))$ and $X_s(i) = I(T_i > s), i = 1, ..., n$. We call this filtration the complete information filtration or filtration on the component level.

For a specific outcome $\omega$ let $m(\omega)$ be the number of different failure time points $0 < T_{(1)} < T_{(2)} < ... < T_{(m)}$ and $J_{(k)} = \{i : T_i(\omega) = T_{(k)}(\omega)\}$ the set of components that fail at $T_{(k)}$. For completeness we define

$$T_{(r)} = \infty, J_{(r)} = \emptyset \text{ for } r \geq m+1.$$

Thus, the sequence $(T_{(k)}, J_{(k)}), k \in \mathbb{N}$, forms a multivariate point process. Now we fix a certain failure pattern $J \subset \{1, ..., n\}$ and consider the time $T_J$ of occurrence of this pattern, i.e.,

$$T_J = \begin{cases} T_{(k)} & \text{if} \quad J_{(k)} = J \text{ for some } k \\ \infty & \text{if} \quad J_{(k)} \neq J \text{ for all } k. \end{cases}$$

The corresponding counting process $V_t(J) = I(T_J \leq t)$ has a compensator $A_t(J)$ with respect to $\mathbb{F}$, which is assumed to be absolutely continuous such that $\lambda_t(J)$ is the $\mathbb{F}$-failure rate process:

$$V_t(J) = \int_0^t I(T > s)\lambda_s(J)ds + M_t(J).$$

In the case $P(T_J = \infty) = 1$, we set $\lambda_t(J) = 0$ for $t \in \mathbb{R}_+$.

**Example 27** If we assume that the component lifetimes are independent random variables, the only interesting (nontrivial) failure patterns are those consisting of only one single component $J = \{j\}, j \in \{1, ..., n\}$. In this case the $\mathbb{F}$-failure rate processes $\lambda_t(\{j\})$ are merely the ordinary failure rates $\lambda_t(j)$ corresponding to $T_j$.

**Example 28** We now consider the special case $n = 2$ in which $(T_1, T_2)$ follows the bivariate exponential distribution of Marshall and Olkin (cf. [132]) with parameters $\beta_1, \beta_2 > 0$ and $\beta_{12} \geq 0$. A plausible interpretation of this distribution is as follows. Three independent exponential random variables $Z_1, Z_2, Z_{12}$ with corresponding parameters $\beta_1, \beta_2, \beta_{12}$ describe the time points when a shock causes failure of component 1 or 2 or all intact components at the same time, respectively. Then the component lifetimes are given by $T_1 = Z_1 \wedge Z_{12}$ and $T_2 = Z_2 \wedge Z_{12}$, and the joint survival probability is seen to be

$$P(T_1 > t, T_2 > s) = \exp\{-\beta_1 t - \beta_2 s - \beta_{12}(t \vee s)\}, s, t \in \mathbb{R}_+.$$

The three different patterns to distinguish are $\{1\}, \{2\}, \{1, 2\}$. Note that $T_{\{1\}} \neq T_1$ as we have for example $T_{\{1\}} = \infty$ on $\{T_1 = T_2\}$, i.e., on $\{Z_{12} < Z_1 \wedge Z_2\}$. Calculations then yield

$$\lambda_t(\{1\}) = \begin{cases} \beta_1 & \text{on} & \{T_1 > t, T_2 > t\} \\ \beta_1 + \beta_{12} & \text{on} & \{T_1 > t, T_2 \leq t\} \\ 0 & \text{elsewhere}, \end{cases}$$

$\lambda_t(\{2\})$ is given by obvious index interchanges, and

$$\lambda_t(\{1, 2\}) = \begin{cases} \beta_{12} & \text{on} & \{T_1 > t, T_2 > t\} \\ 0 & \text{elsewhere}. \end{cases}$$

Now we have the $\mathbb{F}$-failure rate processes $\lambda(J)$ at hand for each pattern $J$. We are interested in deriving the $\mathbb{F}$-failure rate process $\lambda$ of $T$. The next theorem shows how this process $\lambda$ is composed of the single processes $\lambda(J)$ on the component observation level $\mathbb{F}$. Here we remind the reader of some notation introduced in Chapter 2. For $\mathbf{x} \in \mathbb{R}^n$ and $J = \{j_1, ..., j_r\} \subset \{1, ..., n\}$, the vectors $(1_J, \mathbf{x})$ and $(0_J, \mathbf{x})$ denote those $n$-dimensional state vectors in which the components $x_{j_1}, ..., x_{j_r}$ of $\mathbf{x}$ are replaced by 1's and 0's, respectively. Let $D(t)$ be the set of components that have failed up to time $t$, formally

$$D(t) = \begin{cases} J_{(1)} \cup ... \cup J_{(k)} & \text{if} & T_{(k)} \leq t < T_{(k+1)} \\ \emptyset & \text{if} & t < T_{(1)}. \end{cases}$$

Then we define a pattern $J$ to be critical at time $t \geq 0$ if

$$I(J \cap D(t) = \emptyset) \, (\Phi(1_J, \mathbf{X}_t) - \Phi(0_J, \mathbf{X}_t)) = 1$$

and denote by

$$\Gamma_\Phi(t) = \{J \subset \{1, ..., n\} : I(J \cap D(t) = \emptyset) \, (\Phi(1_J, \mathbf{X}_t) - \Phi(0_J, \mathbf{X}_t)) = 1\}$$

the collection of all such patterns critical at $t$.

**Theorem 14** Let $(\lambda_t(J))$ be the $\mathbb{F}$-failure rate process corresponding to $T_J$, $J \subset \{1, ..., n\}$. Then for all $t \in \mathbb{R}_+$ on $\{T > t\}$ :

$$\lambda_t = \sum_{J \subset \{1, ..., n\}} I(J \cap D(t) = \emptyset)(\Phi(1_J, \mathbf{X}_t) - \Phi(0_J, \mathbf{X}_t))\lambda_t(J) = \sum_{J \in \Gamma_\Phi(t)} \lambda_t(J).$$

**Proof.** By Definition 17, p. 53, a predictable increasing process $(A_t)$ is the compensator of the counting process $(V_t), V_t = I(T \leq t)$, if

$$E \int_0^\infty C_s dV_s = E \int_0^\infty C_s dA_s$$

holds true for every nonnegative $\mathbb{F}$-predictable process $C$. Thus, we have to show that

$$E \int_0^\infty C_s dV_s = E \int_0^\infty C_s I(T > s) \sum_{J \in \Gamma_\Phi(s)} \lambda_s(J) ds \qquad (3.10)$$

for all nonnegative predictable processes $C$. Since $(\lambda_t(J))$ are the $\mathbb{F}$-failure rate processes corresponding to $T_J$, we have for all $J \subset \{1, ..., n\}$

$$E \int_0^\infty C_s(J) dV_s(J) = E \int_0^\infty C_s(J) I(T_J > s) \lambda_s(J) ds$$

and therefore

$$E \int_0^\infty \sum_{J \subset \{1,...,n\}} C_s(J) dV_s(J) = E \int_0^\infty \sum_{J \subset \{1,...,n\}} C_s(J) I(T_J > s) \lambda_s(J) ds$$
$$(3.11)$$

holds true for all nonnegative predictable processes $(C_t(J))$. If we especially choose for some nonnegative predictable process $C$

$$C_t(J) = C_t f_{t-},$$

where $f_{t-}$ is the left-continuous version of $f_t = I(J \in \Gamma_\Phi(t))$, we see that (3.11) reduces to (3.10), noting that under the integral sign we can replace $f_{t-}$ by $f_t$, and the proof is complete. ∎

**Remark 6** (i) The proof follows the lines of Arjas (Theorem 4.1 in [6]) except the definition of the set $\Gamma_\Phi(t)$ of the critical failure patterns at time $t$. In [6] this set includes on $\{T > t\}$ all cut sets, whereas in our definition those cut sets $J$ are excluded for which at time $t$ "it is known" that $T_J = \infty$. However, this deviation is harmless because in [6] only extra zeros are added.

(ii) We now have a tool that allows us to determine the failure rate process corresponding to the lifetime $T$ of a complex system in an easy way: Add at time $t$ the failure rates of those patterns that are critical at $t$.

As an immediate consequence we obtain the following corollary.

**Corollary 15** *Let $T_i, i = 1, ..., n$, be independent random variables that have absolutely continuous distributions with ordinary failure rates $\lambda_t(i)$.*

*Then the* F-*failure rate processes* $\lambda(\{i\})$ *are deterministic,* $\lambda_t(\{i\}) = \lambda_t(i)$ *and on* $\{T > t\}$

$$\lambda_t = \sum_{i=1}^{n}(\Phi(1_i, \mathbf{X}_t) - \Phi(0_i, \mathbf{X}_t))\lambda_t(i) = \sum_{\{i\} \in \Gamma_\Phi(t)} \lambda_t(i), \; t \in \mathbb{R}_+. \quad (3.12)$$

In the case of independent component lifetimes we only have to add the ordinary failure rates of those components critical at $t$ to obtain the F-failure rate of the system at time $t$. It is *not* enough to require that $P(T_i = T_j) = 0$ for $i \neq j$ if we drop the independence assumption as the following example shows.

**Example 29** Let $U_1, U_2$ be i.i.d. random variables from an $\mathrm{Exp}(\beta)$ distribution and $T_1 = U_1, T_2 = U_1 + U_2$ be the component lifetimes of a two-component series system. Then we obviously have $P(T_1 = T_2) = 0$, but the F-failure rate of $T_{\{2\}} = T_2$ on $\{T_2 > t\}$

$$\lambda_t(\{2\}) = \beta I(T_1 \leq t)$$

is not deterministic. The system F-failure rate is seen to be

$$I(T > t)\lambda_t = I(T_1 > t)\beta.$$

To see how formula (3.12) can be used we resume Example 25, p. 62.

**Example 30** Again we consider the three-component system with one component in series with a two-component parallel system such that the lifetime of the system is given by $T = T_1 \wedge (T_2 \vee T_3)$. It is assumed that the component lifetimes $T_1, T_2, T_3$ are i.i.d. exponentially distributed with parameter $\alpha > 0$. If at time $t$ all three components work, then only component 1 belongs to $\Gamma_\Phi(t)$ and $I(T > t)\lambda_t = \alpha I(T_1 > t)$ on $\{T_2 > t, T_3 > t\}$. If one of the components 2 or 3 has failed first before time $t$, say component 2, then $\Gamma_\Phi(t) = \{\{1\}, \{3\}\}$ and $I(T > t)\lambda_t = \alpha(I(T_1 > t) + I(T_3 > t))$ on $\{T_2 \leq t\}$. Combining these two formulas yields the failure rate process on $\{T > t\}$

$$\lambda_t = \alpha(1 + I(T_2 \leq t) + I(T_3 \leq t))$$

given in Example 25.

**Example 31** We now go back to the pair $(T_1, T_2)$ of random variables, which follows the bivariate exponential distribution of Marshall and Olkin with parameters $\beta_1, \beta_2 > 0$ and $\beta_{12} \geq 0$ and consider a parallel system with lifetime $T = T_1 \vee T_2$. Then on $\{T > t\}$ the critical patterns are

$$\Gamma_\Phi(t) = \begin{cases} \{1,2\} & \text{on} \quad \{T_1 > t, T_2 > t\} \\ \{1\} & \text{on} \quad \{T_1 > t, T_2 \leq t\} \\ \{2\} & \text{on} \quad \{T_1 \leq t, T_2 > t\}. \end{cases}$$

Using the results of Example 28, p. 65, the $\mathbb{F}$-failure rate process of the system lifetime is seen to be

$$I(T > t)\lambda_t = \beta_{12}I(T_1 > t, T_2 > t) + (\beta_1 + \beta_{12})I(T_1 > t, T_2 \le t)$$
$$+ (\beta_2 + \beta_{12})I(T_1 \le t, T_2 > t),$$

which can be reduced to

$$I(T > t)\lambda_t = \beta_{12}I(T > t) + \beta_1 I(T_1 > t, T_2 \le t) + \beta_2 I(T_1 \le t, T_2 > t).$$

### 3.2.3 Monotone Failure Rate Processes

We have investigated under which conditions failure rate processes exist and how they can be determined explicitly for complex systems. In reliability it plays an important role whether failure rates are monotone increasing or decreasing. So it is quite natural to extend such properties to $\mathbb{F}$-failure rates in the following way.

**Definition 22** *Let an $\mathbb{F}$-SSM representation (3.8) hold true for the positive random variable $T$ with failure rate process $\lambda$. Then $\lambda$ is called $\mathbb{F}$-increasing ($\mathbb{F}$-IFR, increasing failure rate) or $\mathbb{F}$-decreasing ($\mathbb{F}$-DFR, decreasing failure rate), if $\lambda$ has $P$-a.s. nondecreasing or nonincreasing paths, respectively, for $t \in [0, T)$.*

**Remark 7** (i) Clearly, monotonicity properties of $\lambda$ are only of importance on the random interval $[0, T)$. On $[T, \infty)$ we can specify $\lambda$ arbitrarily.

(ii) In the case of complex systems the above definition reflects both, the information level $\mathbb{F}$ and the structure function $\Phi$. An alternative definition, which is derived from notions of multivariate aging terms, is given by Arjas [5]; see also Shaked and Shanthikumar [150].

In the case of a complex system with independent component lifetimes, the following closure lemma can be established.

**Proposition 16** *Assume that in a monotone system the component lifetimes $T_i, i = 1, ..., n$, are independent random variables with absolutely continuous distributions and ordinary nondecreasing failure rates $\lambda_t(i)$ and let $\mathbb{F}$ be the filtration on the component level. Then the $\mathbb{F}$-failure rate process $\lambda$ corresponding to the system lifetime $T$ is $\mathbb{F}$-IFR.*

**Proof.** Under the assumptions of the lemma no patterns with two or more components are critical. Since the system is monotone, the number of elements in $\Gamma_\Phi(t)$ is nondecreasing in $t$. So from (3.12), p. 68, it can be seen that if all component failure rates are nondecreasing, the $\mathbb{F}$-failure rate process $\lambda$ is also nondecreasing for $t \in [0, T)$. ∎

Such a closure theorem does not hold true for the ordinary failure rate of the lifetime $T$ as can be seen from simple counterexamples (see Section 2.2.1 or [34], p. 83). From the proof of Proposition 16 it is evident that we cannot draw an analogous conclusion for decreasing failure rates.

## 3.2.4  Change of Information Level

One of the advantages of the semimartingale technique is the possibility of studying the random evolution of a stochastic process on different information levels. This was described in general in Section 3.1.2 by the projection theorem, which says in which way an SSM representation changes when changing the filtration from $\mathbb{F}$ to a subfiltration $\mathbb{A}$. This projection theorem can be applied to the lifetime indicator process

$$V_t = I(T \leq t) = \int_0^t I(T > s)\lambda_s ds + M_t. \tag{3.13}$$

If the lifetime can be observed, i.e., $\{T \leq s\} \in \mathcal{A}_s$ for all $0 \leq s \leq t$, then the change of the information level from $\mathbb{F}$ to $\mathbb{A}$ leads from (3.13) to the representation

$$\hat{V}_t = E[I(T \leq t)|\mathcal{A}_t] = I(T \leq t) = \int_0^t I(T > s)\hat{\lambda}_s ds + \bar{M}_t, \tag{3.14}$$

where $\hat{\lambda}_t = E[\lambda_t|\mathcal{A}_t]$. Note that, in general, this formula only holds for almost all $t \in \mathbb{R}_+$. In all our examples we can find $\mathbb{A}$-progressive versions of the conditional expectations. The projection theorem shows that it is possible to obtain the failure rate on a lower information level merely by forming conditional expectations under some mild technical conditions.

**Remark 8** Unfortunately, monotonicity properties are in general not preserved when changing the observation level. As was noted above (see Proposition 16), if all components of a monotone system have independent lifetimes with increasing failure rates, then $T$ is $\mathbb{F}$-IFR on the component observation level. But switching to a subfiltration $\mathbb{A}$ may lead to a non-monotone failure rate process $\hat{\lambda}$.

The following example illustrates the role of partial information.

**Example 32** Consider a two-component parallel system with i.i.d. random variables $T_i$, $i = 1, 2$, describing the component lifetimes, which follow an exponential distribution with parameter $\alpha > 0$. Then the system lifetime is $T = T_1 \vee T_2$ and the complete information filtration is given by

$$\mathcal{F}_t = \sigma(I(T_1 > s), I(T_2 > s), 0 \leq s \leq t).$$

In this case the $\mathbb{F}$-semimartingale representation (3.13) is given by

$$
I(T \leq t) = \int_0^t I(T > s)\alpha\{I(T_1 \leq s) + I(T_2 \leq s)\}ds + M_t
$$

$$
= \int_0^t I(T > s)\lambda_s ds + M_t.
$$

Now several subfiltrations can describe different lower information levels where it is assumed that the system lifetime $T$ can be observed on all observation levels. Examples of partial information and the formal description via subfiltrations $\mathbb{A}$ and $\mathbb{A}$-failure rates are as follows:

a) Information about component lifetimes with time lag $h > 0$. Up to time $h$ only system failure can be observed, after $h$ additional information about failure of a component is available with a time lag of $h$, i.e., it is always known whether the system works or not but failures of single components which do not cause a system failure are only reported to the observer with a delay of $h$ time units.

$$
\mathcal{A}_t^a = \begin{cases} \sigma(I(T > s), 0 \leq s \leq t) & \text{for } 0 \leq t < h \\ \sigma(I(T > s), I(T_i > u), i = 1, 2, s \leq t, u \leq t - h) & \text{for } t \geq h, \end{cases}
$$

$$
\hat{\lambda}_t^a = \begin{cases} 2\alpha(1 - (2 - e^{-\alpha t})^{-1}) & \text{for } 0 \leq t < h \\ \alpha(2 - I(T_1 > t - h)e^{-\alpha h} - I(T_2 > t - h)e^{-\alpha h}) & \text{for } t \geq h. \end{cases}
$$

b) Information about $T$ until $h$, after $h$ complete information.

$$
\mathcal{A}_t^b = \begin{cases} \sigma(I(T \leq s), 0 \leq s \leq t) & \text{for } 0 \leq t < h \\ \mathcal{F}_t & \text{for } t \geq h, \end{cases}
$$

$$
\hat{\lambda}_t^b = \begin{cases} 2\alpha(1 - (2 - e^{-\alpha t})^{-1}) & \text{for } 0 \leq t < h \\ \lambda_t & \text{for } t \geq h. \end{cases}
$$

c) Information about component lifetime $T_1$ and $T$:

$$
\mathcal{A}_t^c = \sigma(I(T \leq s), I(T_1 \leq s), 0 \leq s \leq t),
$$
$$
\hat{\lambda}_t^c = \alpha(I(T_1 \leq t) + I(T_1 > t)P(T_2 \leq t)).
$$

d) Information about $T$ only:

$$
\mathcal{A}_t^d = \sigma(I(T \leq s), 0 \leq s \leq t),
$$
$$
\hat{\lambda}_t^d = 2\alpha(1 - (2 - e^{-\alpha t})^{-1}).
$$

The failure rate corresponding to $\mathbb{A}^d$ of this example is the standard deterministic failure rate, because $\{T > t\}$ is an atom of $\mathcal{A}_t^d$ (there is no subset of $\{T > t\}$ in $\mathcal{A}_t^d$ of positive probability) so that $\hat{\lambda}^d$ can always be chosen to be deterministic on $\{T > t\}$. This corresponds to our intuition because on this information level we cannot observe any other random event before $T$. Example 24 shows that such deterministic failure rates satisfy the well-known exponential formula (3.9), p. 62. An interesting question to ask is then: Under what conditions will such an exponential formula also extend to random failure rate processes? This question is referred to briefly in [4] and answered in [177] to some extent. The following treatment differs slightly in that the starting point is the basic lifetime model of this section. The failure rate process $\lambda$ is assumed to be observable on some level $\mathbb{A}$, i.e., $\lambda$ is adapted to that filtration. This observation level can be somewhere between the trivial filtration $\mathbb{G} = (\mathcal{G}_t), t \in \mathbb{R}_+$, $\mathcal{G}_t = \{\emptyset, \Omega\}$, which does not allow for any random information, and the basic complete information filtration $\mathbb{F}$. So $T$ itself need not be observable at level $\mathbb{A}$ (and should not, if we want to arrive at an exponential formula). Using the projection theorem we obtain

$$E[I(T \le t)|\mathcal{A}_t] = 1 - \bar{F}_t = \int_0^t \bar{F}_s \lambda_s ds + \bar{M}_t, \qquad (3.15)$$

where $\bar{F}$ denotes the conditional survival probability,

$$\bar{F}_t = E[I(T > t)|\mathcal{A}_t] = P(T > t|\mathcal{A}_t),$$

and $\bar{M}$ is an $\mathbb{A}$-martingale. In general, $\bar{F}$ need not be monotone and can be rather irregular. But if $\bar{F}$ has continuous paths of bounded variation, then the martingale $\bar{M}$ is identically 0 and the solution of the resulting integral equation is

$$\bar{F}_t = \exp\left\{ -\int_0^t \lambda_s ds \right\}, \qquad (3.16)$$

which is a generalization of formula (3.9). If $\mathbb{A}$ is the trivial filtration $\mathbb{G}$, then (3.16) coincides with (3.9). For (3.16) to hold, it is necessary that the observation of $\lambda$ and other events on level $\mathbb{A}$ only have "smooth" influence on the conditional survival probability.

**Remark 9** This is a more technical remark to show how one can proceed if $\bar{F}$ is not continuous. Let $(\bar{F}_{t-}), t \in \mathbb{R}_+$, be the left-continuous version of $\bar{F}$. Equation (3.15) can be rewritten as

$$\bar{F}_t = 1 - \int_0^t \bar{F}_{s-} \lambda_s ds - \bar{M}_t.$$

Under mild conditions an $\mathbb{A}$-martingale $L$ can be found such that $\bar{M}$ can be represented as the (stochastic) integral $\bar{M}_t = \int_0^t \bar{F}_{s-} dL_s$, take

$$L_t = \int_0^t \frac{I(\bar{F}_{s-} > 0)}{\bar{F}_{s-}} d\bar{M}_s.$$

With the semimartingale $Z$, $Z_t = -\int_0^t \lambda_s ds - L_t$, equation (3.15) becomes

$$\bar{F}_t = 1 + \int_0^t \bar{F}_{s-} dZ_s.$$

The unique solution of this integral equation is given by the so-called Doléans exponential

$$\begin{aligned}
\bar{F}_t &= \mathcal{E}(Z_t) = \exp\{Z_t^c\} \prod_{0 < s \leq t} (1 + \Delta Z_s) \\
&= \exp\left\{-\int_0^t \lambda_s ds\right\} \exp\{-L_t^c\} \prod_{0 < s \leq t} (1 - \Delta L_s),
\end{aligned}$$

where $Z^c(L^c)$ denotes the continuous part of $Z(L)$ and $\Delta Z_s = Z_s - Z_{s-}(\Delta L_s = L_s - L_{s-})$ denotes the jump height at $s$. This extended exponential formula shows that possible jumps of the conditional survival probability are not caused by jumps of the failure rate process but by (unpredictable) jumps of the martingale part.

## 3.3 Point Processes in Reliability – Failure Time and Repair Models

A number of models in reliability are described by point processes and their corresponding counting processes. As examples we can think of shock models, in which shocks affecting a technical system arrive at random time points $T_n$ according to a point process causing some damage of random amount $V_n$, or we can think of repair models, in which failures occur at random time points $T_n$ causing random repair costs $V_n$. In both cases the sequence $(T_n, V_n)$ is a multivariate or marked point process to be introduced as follows.

**Definition 23** *Let $(T_n), n \in \mathbb{N}$, be a point process and $(V_n), n \in \mathbb{N}$, a sequence of random variables taking values in a measurable space $(S, \mathcal{S})$. Then a marked point process $(T_n, V_n), n \in \mathbb{N}$, is the ordered sequence of time points $T_n$ and marks $V_n$ associated with the time points, and $(S, \mathcal{S})$ is called the mark space.*

The mark $V_n$ describes the event occurring at time $T_n$, for example the magnitude of the shock arriving at a system at time $T_n$ (see Figure 3.1). For each $A \in \mathcal{S}$ we associate the counting process $(N_t(A)), t \in \mathbb{R}_+$,

$$N_t(A) = \sum_{n=1}^{\infty} I(V_n \in A) I(T_n \leq t),$$

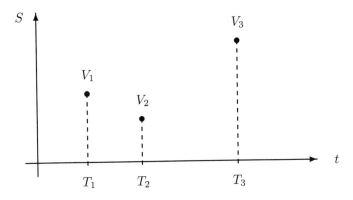

FIGURE 3.1. Marked Point Process

which counts the number of marked points up to time $t$ with marks in $A$. This family of counting processes $N$ carries the same information as the sequence $(T_n, V_n)$ and is therefore an equivalent description of the marked point process.

**Example 33** A point process $(T_n)$ can be viewed as a marked point process for which $S$ consists of a single point. Another link between point and marked point processes is given by the counting process $N = (N_t)$, $N_t = N_t(S)$, which corresponds to the sequence $(T_n)$.

**Example 34** (Alternating Renewal Process). Consider a system, which is repaired or replaced after failure (models of this kind are treated in detail in Section 4.2). Let $U_k$ represent the length of the $k$th operation period and $R_k$ the length of the $k$th repair/replacement time. Assume that $(U_k)$ and $(R_k)$, $k \in \mathbb{N}$, are independent i.i.d. sequences of positive random variables. Let the mark space be $S = \{0, 1\}$, where 0 and 1 stand for "failure" and "repair/replacement completed," respectively. Then the random time points $T_n$ are

$$T_n = \sum_{k=1}^{[\frac{n+1}{2}]} U_k + \sum_{k=1}^{[\frac{n}{2}]} R_k, n = 1, 2, ...,$$

where $[a]$ denotes the integer part of $a$. The mark sequence is deterministic and alternating between 0 and 1:

$$V_n = \frac{1}{2}(1 + (-1)^n).$$

We see that $N_t(\{0\})$ counts the number of failures and $N_t(\{1\})$ the number of completed repairs up to time $t$.

We now want to extend the concept of stochastic intensities from point processes to marked point processes. The internal filtration $\mathbb{F}^N$ of $(T_n, V_n)$

is defined by

$$\mathcal{F}_t^N = \sigma(N_s(A), 0 \le s \le t, A \in \mathcal{S}).$$

This filtration is equivalently generated by the history $\{(T_n, V_n), T_n \le t\}$ of the marked point process.

**Definition 24** *Let* $\mathbb{F}$ *be some filtration including* $\mathbb{F}^N \colon \mathcal{F}_t^N \subset \mathcal{F}_t, t \in \mathbb{R}_+$. *A stochastic process* $(\lambda_t(A), t \in \mathbb{R}_+, A \in \mathcal{S})$ *is called the stochastic intensity of the marked point process* $N$, *if*
(i) *for each* $t$, $A \to \lambda_t(A)$ *is a random measure on* $\mathcal{S}$;
(ii) *for each* $A \in \mathcal{S}$, $N_t(A)$ *admits the* $\mathbb{F}$-*intensity* $\lambda_t(A)$.

We can now formulate the extension of Theorem 11, p. 55, to marked point processes (cf. [52], p. 238, [102], [126], p. 22).

**Theorem 17** *Let* $N$ *be an integrable marked point process and* $\mathbb{F}^N$ *its internal filtration. Suppose that for each* $n$ *there exists a regular conditional distribution of* $(U_{n+1}, V_{n+1}), U_{n+1} = T_{n+1} - T_n$, *given the past* $\mathcal{F}_{T_n}^N$ *of the form*

$$
\begin{aligned}
G_n(\omega, A, B) &= P(U_{n+1} \in A, V_{n+1} \in B | \mathcal{F}_{T_n}^N)(\omega) \\
&= \int_A g_n(\omega, s, B) ds,
\end{aligned}
$$

*where* $g_n(\omega, s, B)$ *is, for fixed* $B$, *a measurable function and, for fixed* $(\omega, s)$, *a finite measure on* $(S, \mathcal{S})$. *Then the process given by*

$$\lambda_t(C) = \frac{g_n(t - T_n, C)}{G_n([t - T_n, \infty), S)} = \frac{g_n(t - T_n, C)}{1 - \int_0^{t - T_n} g_n(s, S) ds}$$

*on* $(T_n, T_{n+1}]$ *is a stochastic intensity of* $N$ *and for each* $C \in \mathcal{S}$,

$$N_t(C) - \int_0^t \lambda_s(C) ds$$

*is an* $\mathbb{F}^N$-*martingale.*

To find the SSM representation of a stochastic process, which is derived from a marked point process, we can make use of the intensity of the latter. The following theorem is proved in Brémaud [52]. For the formulation of this result it is more convenient to use a slightly different notation for the process $N_t(C)$, namely,

$$N_t(C) = N(t, C) = \sum_{n=1}^{\infty} I(V_n \in C) I(T_n \le t).$$

**Theorem 18** *Let $(N(t, C)), t \in \mathbb{R}_+, C \in \mathcal{S}$, be an integrable marked point process admitting the intensity $\lambda_t(C)$ with respect to some filtration $\mathbb{F}$. Let $H(t, z)$ be an $S$-marked $\mathbb{F}$-predictable process, such that, for all $t \in \mathbb{R}_+$, we have*

$$E \int_0^t \int_S |H(s, z)| \lambda_s(dz) ds < \infty.$$

*Then, defining $M(ds, dz) = N(ds, dz) - \lambda_s(dz) ds$,*

$$\int_0^t \int_S H(s, z) M(ds, dz)$$

*is an $\mathbb{F}$-martingale.*

In the following subsections we consider some examples and particular cases. As was mentioned in Example 33 a point process $(T_n)$ and its associated counting process $(N_t)$ are special cases of marked point processes. Point process models in our SSM set-up require the assumption that the counting process $(N_t), t \in \mathbb{R}_+$, on a filtered probability space $(\Omega, \mathcal{F}, \mathbb{F}, P)$ has an absolutely continuous compensator or, what amounts to the same, admits an $\mathbb{F}$-SSM representation

$$N_t = \int_0^t \lambda_s ds + M_t. \tag{3.17}$$

This point process model is consistent with the general lifetime model considered in Section 3.2. If the process $N$ is stopped at $T_1$, then (3.17) reduces to (3.13):

$$\begin{aligned}
N_{t \wedge T_1} &= I(T_1 \le t) = \int_0^{t \wedge T_1} \lambda_s ds + M_{t \wedge T_1} \\
&= \int_0^t I(T_1 > s) \lambda_s ds + M_t',
\end{aligned}$$

where $M'$ is the stopped martingale $M, M_t' = M_{t \wedge T_1}$. The time to first failure or shock corresponds to the lifetime $T = T_1$.

In general, $N$ is determined by its compensator or by its intensity $\lambda$, and it is possible to construct a point process $N$ (and a corresponding probability measure) from a given intensity $\lambda$ (these problems are considered in some detail in [102], see also [126] Chapter 8). This allows us to define point process models in reliability by considering a given intensity.

### 3.3.1  Alternating Renewal Processes – One-Component Systems with Repair

We resume Example 34, p. 74, and assume that the operating times $U_k$ follow a distribution $F$ with density $f$ and failure rate $\rho(t) = f(t)/\bar{F}(t)$, whereas the repair times follow a distribution $G$ with density $g$ and hazard rate $\eta(t) = g(t)/\bar{G}(t)$. Note that the failure/hazard rate is always set to 0 outside the support of the distribution. Then $N_t(\{0\})$ counts the number of failures up to time $t$ with an intensity $\lambda_t(\{0\}) = \rho(t - T_n)X(t)$ on $(T_n, T_{n+1}]$, where $X(t) = V_n$ on $(T_n, T_{n+1}]$ indicates whether the system is up or down at $t$. The corresponding internal intensity for $N_t(\{1\})$ is $\lambda_t(\{1\}) = \eta(t - T_n)(1 - X(t))$. If the operating times are exponentially distributed with rate $\rho > 0$, the expected number of failures up to time $t$ is given by

$$EN_t(\{0\}) = \rho \int_0^t EX(s)ds.$$

## 3.3.2  Number of System Failures for Monotone Systems

We now consider a monotone system comprising $m$ independent components. For each component we define an alternating renewal process, indexed by "$i$." The operating and repair times $U_{ik}$ and $R_{ik}$, respectively, are independent i.i.d. sequences with distributions $F_i$ and $G_i$. We make the assumption that the up-time distributions $F_i$ are absolutely continuous with failure rates $\lambda_t(i)$. The point process $(T_n)$ is the superposition of the $m$ independent alternating renewal processes $(T_{in}), i = 1, ..., m$, and the associated counting process is merely the sum of the single counting processes. Since we are only interested in the occurrence of failures now, we denote by $N_t(i)$ the number of failures of component $i$ (omitting the argument $\{0\}$) and the total number of component failures by $N_t = \sum_{i=1}^m N_t(i)$. The time $T_n$ records the occurrence of a component failure or completion of a repair. As in Chapter 2, $\Phi : A \to \{0, 1\}$ is the structure function, where $A = \{0, 1\}^m$, and the process $\mathbf{X}_t = (X_t(1), ..., X_t(m))$ denotes the vector of component states at time $t$ with values in $A$. The mark space is $S = A \times A$ and the value of $V_n = (\mathbf{X}_{T_n-}, \mathbf{X}_{T_n})$ describes the change of the component states occurring at time $T_n$, where we set $V_0 = \{(1, ..., 1), (1, ..., 1)\}$, i.e., we start with intact components at $T_0 = 0$. Note that $V_n = (\mathbf{x}, \mathbf{y})$ means that $\mathbf{y} = (0_i, \mathbf{x})$ or $\mathbf{y} = (1_i, \mathbf{x})$ for some $i \in \{1, ..., m\}$, because we have absolutely continuous up-time distributions so that at time $T_n$ only one component changes its status. Combining Corollary 15, p. 67, and Theorem 17, p. 75, we get the following result.

**Corollary 19** *Let* $\Gamma = \{(\mathbf{x}, \mathbf{y}) \in S : \Phi(\mathbf{x}) = 1, \Phi(\mathbf{y}) = 0, \mathbf{y} = (0_j, \mathbf{x})$ *for some* $j \in \{1, ..., m\}\}$ *be the set of marks indicating a system failure. Then the process*

$$N_t(\Gamma) = \sum_{i=1}^m \int_0^t \{\Phi(1_i, \mathbf{X}_s) - \Phi(0_i, \mathbf{X}_s)\}dN_s(i)$$

*counting the number of system failures up to time t admits the intensity*

$$\lambda_t(\Gamma) = \sum_{i=1}^{m} \{\Phi(1_i, \mathbf{X}_t) - \Phi(0_i, \mathbf{X}_t)\} \rho_t(i) X_t(i)$$

*with respect to the internal filtration, where*

$$\rho_t(i) = \sum_{k=0}^{\infty} \lambda_{t-T_{ik}}(i) I(T_{ik} < t \leq T_{i,k+1}).$$

**Proof.** We know that $\rho_t(i) X_t(i)$ are intensities of $N_t(i)$ and thus

$$M_t(i) = N_t(i) - \int_0^t \rho_s(i) X_s(i) ds$$

defines a martingale (also with respect to the internal filtration of the superposition because of the independence of the component processes). Define

$$\Delta\Phi_t(i) = \Phi(1_i, \mathbf{X}_t) - \Phi(0_i, \mathbf{X}_t)$$

and let $\Delta\Phi_{t-}(i)$ be the left-continuous and therefore predictable version of this process. Since at a jump of $N_t(i)$ no other components change their status ($P$-a.s.), we have

$$\int_0^t \Delta\Phi_s(i) dN_s(i) = \int_0^t \Delta\Phi_{s-}(i) dN_s(i).$$

It follows that

$$\begin{aligned} N_t(\Gamma) - \int_0^t \lambda_s(\Gamma) ds &= \int_0^t \sum_{i=1}^{m} \Delta\Phi_s(i) dM_s(i) \\ &= \int_0^t \sum_{i=1}^{m} \Delta\Phi_{s-}(i) dM_s(i). \end{aligned}$$

But the last integral of a bounded, predictable process is a martingale, which proves the assertion. ∎

To determine the expected number of system failures up to time $t$, we observe that $EM_t(i) = 0$, i.e., $EN_t(i) = \int_0^t m_s(i) ds$ with $m_s(i) = E\rho_s(i) X_s(i)$, and that $\Delta\Phi_t(i)$ and $\rho_t(i) X_t(i)$ are stochastically independent. This results in

$$EN_t(\Gamma) = \int_0^t \sum_{i=1}^{m} E[\Delta\Phi_s(i)] m_s(i) ds. \tag{3.18}$$

### 3.3.3 *Compound Point Process – Shock Models*

Let us now assume that a system is exposed to shocks at random times $(T_n)$. A shock occurring at $T_n$ causes a random amount of damage $V_n$ and these damages accumulate. The marked point process $(T_n, V_n)$ with mark space $(\mathbb{R}, \mathcal{B}(\mathbb{R}))$ describes this shock process. To avoid notational difficulties we write in this subsection $N(t, C)$ for the associated counting processes, describing the number of shocks up to time $t$ with amounts in $C$. We are interested in the so-called compound point process

$$X_t = \sum_{n=1}^{N(t)} V_n$$

with $N(t) = N(t, \mathbb{R})$, which gives the total damage up to $t$, and we want to derive the infinitesimal characteristics or the "intensity" of this process, i.e., to establish an SSM representation. We might also think of repair models, in which failures occur at random time points $T_n$. Upon failure, repair is performed. If the cost for the $n$th repair is $V_n$, then $X_t$ describes the accumulated costs up to time $t$.

To derive an SSM representation of $X$, we first assume that we are given a general intensity $\lambda_t(C)$ of the marked point process with respect to some filtration $\mathbb{F}$. The main point now is to observe that

$$X_t = \int_0^t \int_S z N(ds, dz).$$

Then we can use Theorem 18, p. 76, with the predictable process $H(s, z) = z$ to see that

$$M_t^{\mathbb{F}} = \int_0^t \int_S z(N(ds, dz) - \lambda_s(dz)ds)$$

is a martingale if $E \int_0^t \int_S |z| \lambda_s(dz)ds < \infty$. Equivalently, we see that $X$ has the $\mathbb{F}$-SSM representation $X = (f, M^{\mathbb{F}})$, with

$$f_s = \int_S z \lambda_s(dz).$$

To come to a more explicit representation we make the following assumptions (**A**):

- The filtration is the internal one $\mathbb{F}^N$;

- $U_{n+1} = T_{n+1} - T_n$ is independent of $\mathcal{F}_{T_n}^N \vee \sigma(V_{n+1})$;

- $U_{n+1}$ has absolutely continuous distribution with density $g_n(t)$ and (ordinary) failure or hazard rate $r_n(t)$;

- $V_{n+1}$ is a positive random variable, independent of $\mathcal{F}_{T_n}^N$, with finite mean $EV_{n+1}$.

Under these assumptions we get by Theorem 17, p. 75,

$$\lambda_t(C) = \sum_{n=0}^{\infty} r_n(t - T_n)P(V_{n+1} \in C)I(T_n < t \le T_{n+1})$$

and therefore the SSM representation

$$X_t = \int_0^t \sum_{n=0}^{\infty} E[V_{n+1}]r_n(s - T_n)I(T_n < s \le T_{n+1})ds + M_t^{\mathbb{F}^N}.$$

In the case of constant expectations $EV_n = EV_1$ we have

$$f_s = E[V_1]\lambda_s(\mathbb{R}).$$

### 3.3.4   Shock Models with State-Dependent Failure Probability

Now we introduce a failure mechanism in which the marks $V_n = (Y_n, W_n)$ are pairs of random variables, where $Y_n, Y_n > 0$, represents the amount of damage caused by the $n$th shock and $W_n$ equals 1 or 0 according to whether the system fails or not at the $n$th shock. So the marks $V_n$ take values in $S = \mathbb{R}_+ \times \{0, 1\}$. The associated counting process is $N(t, \mathbb{R}_+ \times \{0, 1\})$, and $\tilde{N}(t) = N(t, \mathbb{R}_+ \times \{1\})$ counts the number of failures up to time $t$. The accumulated damage is described by

$$X_t = \sum_{n=1}^{N(t,S)} Y_n.$$

In addition to (**A**), p. 79, we now assume

- $Y_{n+1}$ is independent of $\mathcal{F}_{T_n}^N$ with distribution

$$F_{n+1}(y) = P(Y_{n+1} \le y);$$

- For each $k \in \mathbb{N}_0$ there exists a measurable function $p_k(x)$ such that $0 \le p_k(x) \le 1$ and

$$P(W_{n+1} = 1|\mathcal{F}_{T_n}^N \vee \sigma(Y_{n+1})) = p_{\tilde{N}(T_n)}(X_{T_n} + Y_{n+1}). \qquad (3.19)$$

Note that $\mathcal{F}_{T_n}^N = \sigma((T_i, Y_i, W_i), i = 1, ..., n)$ and that

$$\tilde{N}(T_n) = \sum_{i=1}^{n} W_i, \quad X_{T_n} = \sum_{i=1}^{n} Y_i.$$

The assumption (3.19) can be interpreted as follows: If the accumulated damage is $x$ and $k$ failures have already occurred, then an additional shock of magnitude $y$ causes the system to fail with probability $p_k(x + y)$.

To derive the compensator of $N(t, \mathbb{R}_+ \times \{1\})$, the number of failures up to time $t$, we observe that

$$P(U_{n+1} \in A, Y_{n+1} \in \mathbb{R}_+, W_{n+1} = 1 | \mathcal{F}_{T_n}^N)$$
$$= P(U_{n+1} \in A) P(W_{n+1} = 1 | \mathcal{F}_{T_n}^N)$$
$$= P(U_{n+1} \in A) E \left[ p_{\tilde{N}(T_n)}(X_{T_n} + Y_{n+1}) | \mathcal{F}_{T_n}^N \right].$$

Then Theorem 17 yields the intensity on $\{T_n < t \le T_{n+1}\}$:

$$\lambda_t(\mathbb{R}_+ \times \{1\}) = r_n(t - T_n) E \left[ p_{\tilde{N}(T_n)}(X_{T_n} + Y_{n+1}) | \mathcal{F}_{T_n}^N \right].$$

**Example 35** As a shock arrival process we now consider a Poisson process with rate $\nu, 0 < \nu < \infty$, and an i.i.d. sequence of shock amounts with common distribution $F$. Then we get

$$\lambda_t(\mathbb{R}_+ \times \{1\}) = \nu \int_0^\infty p_{\tilde{N}(t)}(X_t + y) dF(y).$$

If the failure probability does not depend on the number of failures $\tilde{N}$ and the shock magnitudes are deterministic, $Y_n = 1$, then we have

$$\lambda_t(\mathbb{R}_+ \times \{1\}) = \nu p(N_t + 1).$$

To derive a semimartingale description of the first time to failure

$$T = \inf\{T_n : W_n = 1\},$$

we simply stop the counting process $\tilde{N}$ at the $\mathbb{F}^N$-stopping time $T$ and get

$$
\begin{aligned}
I(T \le t) &= \tilde{N}(t \wedge T) = \int_0^{t \wedge T} \lambda_s(\mathbb{R}_+ \times \{1\}) ds + M_{t \wedge T} \\
&= \int_0^t I(T > s) \lambda_s(\mathbb{R}_+ \times \{1\}) ds + M_{t \wedge T},
\end{aligned}
$$

where $M$ is a martingale. The time to first failure admits a failure rate process, which is just the intensity of the counting process $\tilde{N}$.

## 3.3.5  Shock Models with Failures of Threshold Type

The situation is as above; we only change the failure mechanism in that the first time to failure $T$ is defined as the first time the accumulated damage reaches or exceeds a given threshold $K \in \mathbb{R}_+$:

$$T = \inf\{t \in \mathbb{R}_+ : \sum_{i=1}^{N(t,S)} Y_i \ge K\} = \inf\{T_n : \sum_{i=1}^n Y_i \ge K\}.$$

This is the hitting time of the set $[K, \infty)$.

This failure model seems to be quite different from the previous one. However, we see that it is just a special case setting the failure probability function $p_k(x)$ of (3.19) for all $k$ equal to the indicator of the interval $[K, \infty)$ :

$$p_k(x) = p(x) = I_{[K,\infty)}(x).$$

Then we get

$$
\begin{aligned}
P(W_{n+1} = 1|\mathcal{F}_{T_n}^N) &= E[p(X_{T_n} + Y_{n+1})|\mathcal{F}_{T_n}^N] \\
&= P(Y_{n+1} + X_{T_n} \geq K|\mathcal{F}_{T_n}^N) \\
&= 1 - F_{n+1}((K - X_{T_n})-).
\end{aligned}
$$

This can be interpreted as follows: If the accumulated damage after $n$ shocks is $x$, then the system fails with probability $P(Y_{n+1} \geq K - x)$ when the next shock occurs, which is the probability that the total damage hits the threshold $K$. Obviously, all shocks after $T$ are counted by $\tilde{N}(t) = N(t, \mathbb{R}_+ \times \{1\})$. The failure counting process $\tilde{N}$ has on $\{T_n < t \leq T_{n+1}\}$ the intensity

$$\lambda_t(\mathbb{R}_+ \times \{1\}) = r_n(t - T_n)\{1 - F_{n+1}((K - X_{T_n})-)\}. \tag{3.20}$$

The first time to failure is described by

$$I(T \leq t) = \int_0^t I(T > s)\lambda_s(\mathbb{R}_+ \times \{1\})ds + M_t,$$

with a suitable martingale $M$.

**Example 36** Let us again consider the compound Poisson case with shock arrival rate $\nu$ and $F_n = F$ for all $n \in \mathbb{N}_0$. Since $r_n(s - T_n) = \nu$ and $(K - X_{T_n}) = (K - X_t)$ on $\{T_n < t < T_{n+1}\}$, we get

$$I(T \leq t) = \int_0^t I(T > s)\nu \bar{F}((K - X_s)-)ds + M_t.$$

## 3.3.6   Minimal Repair Models

In the literature covering repair models special attention has been given to so-called minimal repair models. Instead of replacing a failed system by a new one, a repair restores the system to a certain degree. These minimal repairs are often verbally described (and defined) as in the following:

- "The ... assumption is made that the system failure rate is not disturbed after performing minimal repair. For instance, after replacing a single tube in a television set, the set as a whole will be about as prone to failure after the replacement as before the tube failure" (Barlow and Hunter [32]).

- "A minimal repair is one which leaves the unit in precisely the condition it was in immediately before the failure" (Phelps [139]).

The definition of the state of the system immediately before failure depends to a considerable degree on the information one has about the system. So it makes a difference whether all components of a complex system are observed or only failure of the whole system is recognized. In the first case the lifetime of the repaired component (tube of TV set) is associated with the residual system lifetime. In the second case the only information about the condition of the system immediately before failure is the age. So a minimal repair in this case would mean replacing the system (the whole TV set) by another one of the same age that as yet has not failed. Minimal repairs of this kind are also called *black box* or *statistical* minimal repairs, whereas the component-wise minimal repairs are also called *physical* minimal repairs.

**Example 37** We consider a simple two-component parallel system with independent $\text{Exp}(1)$ distributed component lifetimes $X_1, X_2$ and allow for exactly one minimal repair.

- Physical minimal repair. After failure at $T = T_1 = X_1 \vee X_2$ the component that caused the system to fail is repaired minimally. Since the component lifetimes are exponentially distributed, the additional lifetime is given by an $\text{Exp}(1)$ random variable $X_3$ independent of $X_1$ and $X_2$. The total lifetime $T_1 + X_3$ has distribution

$$P(T_1 + X_3 > t) = e^{-t}(2t + e^{-t}).$$

- Black box minimal repair. The lifetime $T = T_1 = X_1 \vee X_2$ until the first failure of the system has distribution $P(T_1 \leq t) = (1 - e^{-t})^2$ and failure rate $\lambda(t) = 2\frac{1 - \exp(-t)}{2 - \exp(-t)}$. The additional lifetime $T_2 - T_1$ until the second failure is assumed to have conditional distribution

$$P(T_2 - T_1 \leq x | T_1 = t) = P(T_1 \leq t + x | T_1 > t) = 1 - e^{-x}\frac{2 - e^{-(t+x)}}{2 - e^{-t}}.$$

Integrating leads to the distribution of the total lifetime $T_2$ :

$$P(T_2 > t) = e^{-t}(2 - e^{-t})(1 + t - \ln(2 - e^{-t})).$$

It is (perhaps) no surprise that the total lifetime after a black box minimal repair is stochastically greater than after a physical minimal repair:

$$P(T_2 > t) \geq P(T_1 + X_3 > t), \text{ for all } t \geq 0.$$

Below we summarize some typical categories of minimal repair models, and give some further examples. Let $(T_n)$ be a point process describing the failure times at which instantaneous repairs are carried out and let $N = (N_t), t \in \mathbb{R}_+$, be the corresponding counting process

$$N_t = \sum_{n=1}^{\infty} I(T_n \le t).$$

We assume that $N$ is adapted to some filtration $\mathbb{F}$ and has $\mathbb{F}$-intensity $(\lambda_t)$. Different types of repair processes are characterized by different intensities $\lambda$. The repairs are minimal if the intensity $\lambda$ is not affected by the occurrence of failures or, in other words, if one cannot determine the failure time points from the observation of $\lambda$. More formally, minimal repairs can be characterized as follows.

**Definition 25** *Let $(T_n), n \in \mathbb{N}$, be a point process with an integrable counting process $N$ and corresponding $\mathbb{F}$-intensity $\lambda$. Suppose that $\mathbb{F}^\lambda = (\mathcal{F}_t^\lambda), t \in \mathbb{R}_+$, is the filtration generated by $\lambda$: $\mathcal{F}_t^\lambda = \sigma(\lambda_s, 0 \le s \le t)$. Then the point process $(T_n)$ is called a minimal repair process (MRP) if none of the variables $T_n, n \in \mathbb{N}$, for which $P(T_n < \infty) > 0$ is an $\mathbb{F}^\lambda$-stopping time, i.e., for all $n \in \mathbb{N}$ with $P(T_n < \infty) > 0$ there exists $t \in \mathbb{R}_+$ such that $\{T_n \le t\} \notin \mathcal{F}_t^\lambda$.*

This is a rather general definition that comprises the well-known special case of a nonhomogeneous Poisson process as is seen below. A renewal process with a strictly increasing or decreasing hazard rate $r$ of the interarrival times has intensity (compare Example 20, p. 55)

$$\lambda_t = \sum_{n \ge 0} r(t - T_n) I(T_n < t \le T_{n+1}), T_0 = 0, \lambda_0 = r(0),$$

and is therefore not an MRP, because $N_t = |\{s \in \mathbb{R}_+ : 0 < s \le t, \lambda_s = \lambda_0\}|$. In the following we give some examples of (minimal) repair processes.

**a)** In the basic statistical minimal repair model the intensity is a time-dependent deterministic function $\lambda_t = \lambda(t)$, so that the process is a nonhomogeneous Poisson process. This means that the age (the failure intensity) is not changed as a result of a failure (minimal repair). Here $\mathcal{F}_t^\lambda = \{\Omega, \emptyset\}$ for all $t \in \mathbb{R}_+$, so clearly the failure times $T_n$ are no $\mathbb{F}^\lambda$-stopping times. The following special cases have been given much attention in the literature:

$$\begin{aligned} \lambda_p(t) &= \lambda\beta(\lambda t)^{\beta-1} \text{ (Power law)}, \\ \lambda_L(t) &= \lambda e^{\beta t} \text{ (Log linear model)}. \end{aligned}$$

For the parallel system in Example 37, one has $\lambda(t) = 2\frac{1-\exp(-t)}{2-\exp(-t)}$. If the intensity is a constant, $\lambda_t \equiv \lambda$, the times between successive repairs are independent $Exp(\lambda)$ distributed random variables. This is the case in which repairs have the same effect as replacements.

**b)** If in a) the intensity is not deterministic but a random variable $\lambda(\omega)$, which is known at the time origin ($\lambda$ is $\mathcal{F}_0$-measurable), or, more general, $\lambda = (\lambda_t)$ is a stochastic process such that $\lambda_t$ is $\mathcal{F}_0$-measurable for all $t \in \mathbb{R}_+$, i.e., $\mathcal{F}_0 = \sigma(\lambda_s, s \in \mathbb{R}_+)$ and $\mathcal{F}_t = \mathcal{F}_0 \vee \sigma(N_s, 0 \le s \le t)$, then the process is called a doubly stochastic Poisson process or a Cox process. The process generalizes the basic model a); the failure (minimal repair) times are no $\mathbb{F}^\lambda$-stopping times, since $\mathcal{F}_t^\lambda = \sigma(\lambda) \subset \mathcal{F}_0$ and $T_n$ is not $\mathcal{F}_0$-measurable.

Also the Markov-modulated Poisson process of Example 21, p. 56, where the intensity $\lambda_t = \lambda_{Y_t}$ is determined by a Markov chain $(Y_t)$, is an MRP. Indeed, it is a slight modification of a doubly stochastic Poisson process in that the filtration $\mathcal{F}_t = \sigma(N_s, Y_s, 0 \le s \le t)$ does not include the information about the paths of $\lambda$ in $\mathcal{F}_0$.

**c)** For the physical minimal repair in Example 37, $\lambda_t = I(X_1 \wedge X_2 \le t)$. In this case $\mathbb{F}^\lambda$ is generated by the minimum of $X_1$ and $X_2$. The first failure time of the system, $T_1$, equals $X_1 \vee X_2$, which is not an $\mathbb{F}^\lambda$-stopping time.

In the following we give another characterization of an MRP.

**Theorem 20** *Assume that $P(T_n < \infty) = 1$ for all $n \in \mathbb{N}$ and that there exist versions of conditional probabilities $F_t(n) = E[I(T_n \le t)|\mathcal{F}_t^\lambda]$ such that for each $n \in \mathbb{N}$ $(F_t(n)), t \in \mathbb{R}_+$, is an $(\mathbb{F}^\lambda$-progressive) stochastic process.*

*(i) Then the point process $(T_n)$ is an MRP if and only if for each $n \in \mathbb{N}$ there exists some $t \in \mathbb{R}_+$ such that*

$$P(0 < F_t(n) < 1) > 0.$$

*(ii) If furthermore $(F_t) = (F_t(1))$ has $P$-a.s. continuous paths of bounded variation on finite intervals, then*

$$1 - F_t = \exp\{-\int_0^t \lambda_s ds\}.$$

**Proof.** *(i)* To prove *(i)* we show that $P(F_t(n) \in \{0,1\}) = 1$ for all $t \in \mathbb{R}_+$ is equivalent to $T_n$ being an $\mathbb{F}^\lambda$-stopping time. Since we have $F_0(n) = 0$ and by the dominated convergence theorem for conditional expectations

$$\lim_{t \to \infty} F_t(n) = 1,$$

the assumption that $P(F_t(n) \in \{0,1\}) = 1$ for all $t \in \mathbb{R}_+$ is equivalent to $F_t(n) = I(T_n \le t)$ ($P$-a.s.). But as $(F_t(n))$ is adapted to $\mathbb{F}^\lambda$ this means that $T_n$ is an $\mathbb{F}^\lambda$-stopping time. This shows that under the given assumptions $P(0 < F_t(n) < 1) > 0$ is equivalent to $T_n$ being no $\mathbb{F}^\lambda$-stopping time.

*(ii)* For the second assertion we apply the exponential formula (3.16) as described on p. 72. ∎

**Example 38** In continuation of Example 37 of the two-component parallel system we allow for repeated physical minimal repairs. Let $(X_k), k \in \mathbb{N}$, be a sequence of i.i.d. random variables following an exponential distribution with parameter $1 : X_k \sim \text{Exp}(1)$. Then we define

$$T_1 = X_1 \vee X_2, T_{n+1} = T_n + X_{n+2}, n \in \mathbb{N}.$$

The intensity of the corresponding counting process $N$ is then $\lambda_t = I(X_1 \wedge X_2 \leq t)$. By elementary calculations it can be seen that

$$E[I(T_1 > t)|\mathcal{F}_t^\lambda] = P(T_1 > t|X_1 \wedge X_2 \wedge t)$$

is continuous and nonincreasing. According to Theorem 20 it follows that $(T_n)$ is an MRP and that the time to the first failure has conditional distribution

$$1 - F_t = \exp\{-\int_0^t I(X_1 \wedge X_2 \leq s)ds\} = \exp\{-(t - X_1 \wedge X_2)^+\}.$$

Now we want to illustrate the above definition of a minimal repair in a more complex situation. We consider the shock damage repair model described in Section 3.3.4. We now assume that the shock arrival process $(T_k^*)$ is a nonhomogeneous Poisson process with intensity function $\nu(t)$ and that $(V_k)$ with $V_k = (Y_k, W_k)$ is an i.i.d. sequence of pairs of random variables, independent of $(T_k^*)$. The common distribution of the positive variables $Y_k$ is denoted $F$. The failure mechanism is as before, but the probability of failure at the occurrence of a shock $p(x)$ if the accumulated damage is $x$, is independent of the number of previous failures. Then we obtain for the failure counting process the intensity

$$\lambda_t = \nu(t) \int_0^\infty p(X_{t-} + y)dF(y), \tag{3.21}$$

where

$$X_t = \sum_{k=1}^\infty Y_k I(T_k^* \leq t)$$

denotes the accumulated damage up to time $t$. The following theorem shows under which condition the failure point process is an MRP.

**Theorem 21** If $0 < p(x) < 1$ for all $x$ holds true, then the point process $(T_n)$ driven by the intensity (3.21) is an MRP.

**Proof.** The random variables $W_k$ equal 1 or 0 according to whether the system fails or not at the $k$th shock. The first failure time $T_1$ can then be represented by

$$T_1 = \inf\{T_k^* : W_k = 1\}.$$

At each occurrence of a shock a Bernoulli experiment is carried out with outcome $W_k$. The random variable $W_k$ is not measurable with respect to $\sigma(X_{T_k^*})$ because by the condition $0 < p(x) < 1$ it follows that

$$E[I(W_k = 1)|X_{T_k^*}] = P(W_k = 1|X_{T_k^*}) = p(X_{T_k^*}) \notin \{0, 1\}.$$

This shows that $T_1$ cannot be an $\mathbb{F}^X$-stopping time, where $\mathbb{F}^X$ is generated by the process $X = (X_t)$. Since we have $\mathcal{F}_t^\lambda \subset \mathcal{F}_t^X$, $T_1$ is no $\mathbb{F}^\lambda$-stopping time either. By induction via

$$T_{n+1} = \inf\{T_k^* > T_n : W_k = 1\}$$

we infer that none of the variables $T_n$ is an $\mathbb{F}^\lambda$-stopping time, which shows that $(T_n)$ is an MRP.                                                         ∎

**Remark 10** 1. In the case $p(x) = c$ for some $c, 0 < c \leq 1$, the process is a nonhomogeneous Poisson process with intensity $\lambda_t = \nu(t)c$ and therefore an MRP.

2. The condition $0 < p(x) < 1$ excludes the case of threshold models for which $p(x) = 1$ for $x \geq K$ and $p(x) = 0$ else for some constant $K > 0$. For such a threshold model we have

$$T_1 = \inf\{t \in \mathbb{R}_+ : \lambda_t \geq \nu(t)\},$$

if $P(Y_k \leq x) > 0$ for all $x > 0$. In this case $T_1$ is an $\mathbb{F}^\lambda$-stopping time and consequently $(T_n)$ is no MRP.

### 3.3.7 Comparison of Repair Processes for Different Information Levels

Consider a monotone system comprising $m$ independent components with lifetimes $Z_i, i = 1, ..., m$ and corresponding ordinary failure rates $\lambda_t(i)$. Its structure function $\Phi : \{0, 1\}^m \to \{0, 1\}$ represents the state of the system (1:intact, 0:failure), and the process $\mathbf{X}_t = (X_t(1), ..., X_t(m))$ denotes the vector of component states at time $t$ with values in $\{0, 1\}^m$. Example 37 suggests comparing the effects of minimal repairs on different information levels. However, it seems difficult to define such point processes for arbitrary information levels. One possible way is sketched in the following where considerations are restricted to the complete information $\mathbb{F}$-level (component-level) and the "black-box-level" $\mathbb{A}^T$ generated by $T = T_1, \mathcal{A}_t = \sigma(I(T_1 \leq s), 0 \leq s \leq t)$. Note that $T_1$ describes the time to first failure, i.e.,

$$T_1 = \inf\{t \in \mathbb{R}_+ : \Phi(\mathbf{X}_t) = 0\}.$$

This time to first system failure is governed by the hazard rate process $\lambda$ for $t \in [0, T)$ (cf. Corollary 15 on p. 67):

$$\lambda_t = \sum_{i=1}^{m} (\Phi(1_i, \mathbf{X}_t) - \Phi(0_i, \mathbf{X}_t))\lambda_t(i). \tag{3.22}$$

Our aim is to extend the definition of $\lambda_t$ also on $\{T_1 \leq t\}$. To this end we extend the definition of $X_t(i)$ on $\{Z_i \leq t\}$ following the idea that upon system failure the component which caused the failure is repaired minimally in the sense that it is restored and operates at the same failure rate as it had not failed before. So we define $X_t(i) = 0$ on $\{Z_i \leq t\}$ if the first failure of component $i$ caused no system failure, otherwise we set $X_t(i) = 1$ on $\{Z_i \leq t\}$ (Note that in the latter case the value of $X_t(i)$ is redefined for $t = Z_i$). In this way we define $\mathbf{X}_t$ and by (3.22) the process $\lambda_t$ for all $t \in \mathbb{R}_+$. This completed intensity $\lambda_t$ induces a point process $(N_t)$ which counts the number of minimal repairs on the component level. The corresponding complete information filtration $\mathbb{F} = (\mathcal{F}_t), t \in \mathbb{R}_+$, is given by

$$\mathcal{F}_t = \sigma(N_s, I(Z_i \leq s), 0 \leq s \leq t, i = 1, ..., m).$$

To investigate whether the process $(N_t)$ is an MRP we define the random variables

$$Y_i = \inf\{t \in \mathbb{R}_+ : \Phi(1_i, \mathbf{X}_t) - \Phi(0_i, \mathbf{X}_t) = 1\}, i = 1, ..., m, \ \inf \emptyset = \infty,$$

which describe the time when component $i$ becomes critical, i.e., the time from which on a failure of component $i$ would lead to system failure. It follows that

$$\lambda_t = \sum_{i=1}^m I(Y_i \leq t)\lambda_t(i),$$
$$\mathcal{F}_t^\lambda = \sigma(I(Y_i \leq s), 0 \leq s \leq t, i = 1, ..., m).$$

Obviously on $\{Y_i < \infty\}$ we have $Z_i > Y_i$ and it can be shown that $Z_i$ is not measurable with respect to $\sigma(Y_1, ..., Y_m)$. For a two component parallel system this means that $Z_1 \vee Z_2$ is not measurable with respect to $\sigma(Z_1 \wedge Z_2)$, which holds true observing that $E[I(Z_1 \vee Z_2 > z)|Z_1 \wedge Z_2] \notin \{0, 1\}$ for some $z$ (Note that the random variables $Z_i$ are assumed to be independent). The extension to the general case is intuitive but the details of a formal, lengthy proof are omitted. We state that the time to the first failure

$$T_1 = \min_{i=1,...,m} Z_i I(Y_i < \infty)$$

is no $\mathbb{F}^\lambda$-stopping time. By induction it can be seen that also $T_n$ is no $\mathbb{F}^\lambda$-stopping time and $(T_n)$ is an MRP.

Now we want to consider the same system on the "black-box-level". The change to the $\mathbb{A}^T$-level by conditioning leads to the failure rate $\hat{\lambda}, \hat{\lambda}_t = E[\lambda_t|\mathcal{A}_t]$. This failure rate $\hat{\lambda}$ can be chosen to be deterministic,

$$\hat{\lambda}_t = E[\lambda_t|T_1 > t],$$

it is the ordinary failure rate of $T_1$. For the time to the first system failure we have the two representations

$$I(T_1 \leq t) = \int_0^t I(T_1 > s)\lambda_s ds + M_t \quad \mathbb{F} - \text{level}$$

$$= \int_0^t I(T_1 > s)\hat{\lambda}_s ds + \bar{M}_t \quad \mathbb{A}^T - \text{level}.$$

From the deterministic failure rate $\hat{\lambda}$ a nonhomogeneous Poisson process $(T'_n)_{n \in \mathbb{N}}, 0 < T'_1 < T'_2 < \dots$ can be constructed where $T_1$ and $T'_1$ have the same distribution. This nonhomogeneous Poisson process with

$$N'_t = \sum_{n=1}^\infty I(T'_n \leq t) = \int_0^t \hat{\lambda}_s ds + M'_t$$

describes the MRP on the $\mathbb{A}^T$-level. Comparing these two information levels, Example 37 suggests $EN_t \geq EN'_t$ for all positive $t$. A general comparison, also for arbitrary subfiltrations, seems to be an open problem (cf. [4] and [135]).

**Example 39** In the two-component parallel system of Example 37 we have the failure rate process $\lambda_t = I(X_1 \wedge X_2 \leq t)$ on the component level and $\hat{\lambda}_t = 2\frac{1-\exp(-t)}{2-\exp(-t)}$ on the black-box level. So one has two descriptions of the same random lifetime $T = T_1$

$$I(T_1 \leq t) = \int_0^t I(T_1 > s)I(X_1 \wedge X_2 \leq s)ds + M_t$$

$$= \int_0^t I(T_1 > s)2\frac{1-e^{-s}}{2-e^{-s}}ds + \bar{M}_t.$$

The process $N$ counts the number of minimal repairs on the component level:

$$N_t = \int_0^t I(X_1 \wedge X_2 \leq s)ds + M_t.$$

This is a delayed Poisson process, the (repair) intensity of which is equal to 1 after the first component failure. The process $N'$ counts the number of minimal repairs on the black-box level:

$$N'_t = \int_0^t 2\frac{1-e^{-s}}{2-e^{-s}}ds + M'_t.$$

This is a nonhomogeneous Poisson process with an intensity which corresponds to the ordinary failure rate of $T_1$. Elementary calculations yield indeed

$$EN_t = t - \frac{1}{2}(1 - e^{-2t}) \geq EN'_t = t - \ln(2 - e^{-t}).$$

To interpret this result one should note that on the component level only the critical component which caused the system to fail is repaired. A black box repair, which is a replacement by a system of the same age that has not yet failed, could be a replacement by a system with both components working.

### 3.3.8  Repair Processes with Varying Degrees of Repair

As in the minimal repair section, let $(T_n)$ be a point process describing failure times at which instantaneous repairs are carried out and let $N = (N_t), t \in \mathbb{R}_+$, be the corresponding counting process. We assume that $N$ is adapted to some filtration $\mathbb{F}$ and has $\mathbb{F}$-intensity $(\lambda_t)$.

One way to model varying levels or degrees of repairs is the following. Consider a new item or system having lifetime distribution $F$ with failure rate $r(t)$. Assume that the $n$th repair has the effect that the distribution to the next failure is that of an unfailed item of age $A_n \geq 0$. Then $A_n = 0$ means complete repair (as good as new) or replacement and $A_n > 0$ can be interpreted as a partial repair which sets the item back to the functioning state. Theorem 11, p. 55, immediately yields the intensity of such a repair process with respect to the internal filtration $\mathbb{F}^N$ : Let $(A_n), n \in \mathbb{N}$, be a sequence of nonnegative random variables such that $A_n$ is $\mathbb{F}^N_{T_n}$-measurable, then the $\mathbb{F}^N$-intensity of $N$ is given by

$$\lambda_t = \sum_{n=0}^{\infty} r(t - T_n + A_n)I(T_n < t \leq T_{n+1}), A_0 = T_0 = 0.$$

The two extreme cases are:

1.  $A_n = 0$, for all $n \in \mathbb{N}$. Then $N$ is a renewal process with interarrival time distribution $F$, all repairs are complete restorations to the as good as new state.

2.  $A_n = T_n$ for all $n \in \mathbb{N}$. Then $N$ is a nonhomogeneous Poisson process with intensity $r(t)$, all repairs are (black box) minimal repairs.

In addition we can introduce random degrees $Z_n \leq 1$ of the $n$th repair. Starting with a new item the first failure occurs at $T_1$. A repair with degree $Z_1$ is instantaneously carried out and results in a virtual age of $A_1 = (1 - Z_1)T_1$. Continuing we can define the sequence of virtual ages recursively by

$$A_{n+1} = (1 - Z_{n+1})(A_n + T_{n+1} - T_n), \quad A_0 = 0.$$

Negative values of $Z_n$ may be interpreted as additional aging due to the $n$th failure or a clumsy repair. In the literature there exist many models describing different ways of generating or prescribing the random sequence of repair degrees, cf. Bibliographic Notes.

### 3.3.9 Minimal Repairs and Probability of Ruin

In this section we investigate a model that combines a certain reward and cost structure with minimal repairs. Consider a one-unit system that fails from time to time according to a point process. After failure a minimal repair is carried out that leaves the state of the system unchanged. The system can work in one of $m$ *unobservable* states. State "1" stands for new or in good condition and "$m$" is defective or in bad condition. Aging of the system is described by a link between the failure point process and the unobservable state of the system. The failure or minimal repair intensity may depend on the state of the system.

Starting with an initial capital of $u \geq 0$, there is some constant flow of income, on the one hand, and, on the other hand, each minimal repair incurs a random cost. The risk process $R = (R_t), t \in \mathbb{R}_+$, describes the difference between the income including the initial capital $u$ and the accumulated costs for minimal repairs up to time $t$. The time of ruin is defined as $\tau = \tau(u) = \inf\{t \in \mathbb{R}_+ : R_t \leq 0\}$. Since explicit formulas are rarely available, we are interested in bounds for $P(\tau < \infty)$ and $P(\tau \leq t)$, the infinite and the finite horizon ruin probabilities.

A related question is when to stop processing the system and carrying out an inspection or a renewal in order to maximize some reward functional. This problem is treated in Section 5.4.

For the mathematical formulation of the model, let the basic probability space $(\Omega, \mathcal{F}, P)$ be equipped with a filtration $\mathbb{F}$, the complete information level, to which all processes are adapted, and let $S = \{1, ..., m\}$ be the set of unobservable states. We assume that the time points of failures (minimal repairs) $0 < T_1 < T_2 < ...$ form a Markov-modulated Poisson process as described in Example 21, p. 56. Let us recapitulate the details:

- The changes of the states are driven by a homogeneous Markov process $Y = (Y_t), t \in \mathbb{R}_+$, with values in $S$ and infinitesimal parameters $q_i$, the rate to leave state $i$, and $q_{ij}$, the rate to reach state $j$ from state $i$:

$$
\begin{aligned}
q_i &= \lim_{h \to 0+} \frac{1}{h} P(Y_h \neq i | Y_0 = i), \\
q_{ij} &= \lim_{h \to 0+} \frac{1}{h} P(Y_h = j | Y_0 = i), i, j \in S, i \neq j, \\
q_{ii} &= -q_i = -\sum_{j \neq i} q_{ij}.
\end{aligned}
$$

- The time points $(T_n)$ form a point process and $N = (N_t), t \in \mathbb{R}_+$, is the corresponding counting process $N_t = \sum_{n \geq 1} I(T_n \leq t)$, which has a stochastic intensity $\lambda_{Y_t}$ depending on the unobservable state, i.e.,

$N$ admits the representation

$$N_t = \int_0^t \lambda_{Y_s} ds + M_t,$$

where $M$ is an $\mathbb{F}$-martingale and $0 < \lambda_i < \infty, i \in S$. Since the filtration $\mathbb{F}^\lambda (\mathbb{F}^\lambda = \mathbb{F}^Y$, if $\lambda_i \neq \lambda_j$ for $i \neq j)$ generated by the intensity does not include $\mathbb{F}^N$ as a subfiltration, it follows that $T_n, n \in \mathbb{N}$, is not an $\mathbb{F}^\lambda$-stopping time. Therefore, according to Definition 25, p. 84, $N$ is a minimal repair process.

- $(X_n), n \in \mathbb{N}$, is a sequence of positive i.i.d. random variables, independent of $N$ and $Y$, with common distribution $F$ and finite mean $\mu$. The cost caused by the $n$th minimal repair at time $T_n$ is described by $X_n$.

- There is an initial capital $u$ and an income of constant rate $c > 0$ per unit time.

Now the process $R$, given by

$$R_t = u + ct - \sum_{n=1}^{N_t} X_n$$

describes the available capital at time $t$ as the difference of the income and the total amount of costs for minimal repairs up to time $t$.

The process $R$ is commonly used in other branches of applied probability like queueing or collective risk theory. In risk theory one is mainly interested in the distribution of the time to ruin $\tau = \inf\{t \in \mathbb{R}_+ : R_t \leq 0\}$.

## The Failure Rate Process of the Ruin Time

We want to show that the indicator process $V_t = I(\tau(u) \leq t)$ has a semi-martingale representation

$$V_t = I(\tau \leq t) = \int_0^t I(\tau > s) h_s ds + M_t, t \in \mathbb{R}_+, \qquad (3.23)$$

where $M$ is a mean zero martingale with respect to the filtration $\mathbb{F} = (\mathcal{F}_t), t \in \mathbb{R}_+$, which is generated by all introduced random quantities:

$$\mathcal{F}_t = \sigma(N_s, Y_s, X_i, 0 \leq s \leq t, i = 1, ..., N_t).$$

The failure rate process $h = (h_t), t \in \mathbb{R}_+$, can be derived in the same way as was done for shock models with failures of threshold type (cf. p. 81).

Note that ruin can only occur at a failure time; therefore, the ruin time is a hitting time of a compound point process:

$$\tau = \inf\{t \in \mathbb{R}_+ : A_t = \sum_{n=1}^{N_t} B_n \geq u\} = \inf\{T_n : A_{T_n} \geq u\},$$

where $B_n = X_n - cU_n$ and $U_n = T_n - T_{n-1}, n = 1, 2, \ldots$ . Replacing $X_t$ by $A_t$, $r(t - T_n)$ by $\lambda_{Y_t}$, and the threshold $S$ by $u$ in formula (3.20) on p. 82, we get the following lemma.

**Lemma 22** *Let* $\tau = \tau(u)$ *be the ruin time and* $F$ *the distribution of the claim sizes,* $\bar{F}(x) = F((x, \infty)) = P(X_1 > x), x \in \mathbb{R}$. *Then the* $\mathbb{F}$-*failure rate process* $h$ *is given by*

$$h_t = \lambda_{Y_t} \bar{F}(R_t-) = \sum_{i=1}^{m} \lambda_i I(Y_t = i)\bar{F}(R_t-), t \in \mathbb{R}_+.$$

The failure rate processes $h$ is bounded above by $\max\{\lambda_i : i \in S\}$. If all claim arrival rates $\lambda_i$ coincide, $\lambda = \lambda_i, i \in S$, we have the classical Poisson case, and it is not surprising that the hazard rate decreases when the risk reserve increases and vice versa. Of course, the paths of $R$ are not monotone and so the failure rate processes do not have monotone paths either. But they have (stochastically) a tendency to increase or decrease in the following sense. As follows from the results of Section 3.3.3 the process $R$ has an $\mathbb{F}$-semimartingale representation

$$R_t = \int_0^t \sum_{i=1}^{m} I(Y_s = i)(c - \lambda_i \mu)ds + L_t$$

with a mean zero $\mathbb{F}$-martingale $L$. If we have positive drift in all environmental states, i.e., $c - \lambda_i \mu > 0, i = 1, \ldots, m$, then $R$ is a submartingale and it is seen that $h$ tends to 0 as $t \to \infty$ $(P - a.s.)$. On the other hand, if the claim rate $\lambda_{Y_t}$ is increasing $(P\text{–a.s.})$ and the drift is nonpositive for all states, i.e., $c - \lambda_i \mu \leq 0, i = 1, \ldots, m$, and $\bar{F}$ is convex on the support of the distribution, then $R$ is a supermartingale and it follows by Jensen's inequality for conditional expectations:

$$
\begin{aligned}
E[h_{t+s}|\mathcal{F}_t] &= E[\lambda_{Y_{t+s}}\bar{F}(R_{t+s}-)|\mathcal{F}_t] \geq E[\lambda_{Y_t}\bar{F}(R_{t+s}-)|\mathcal{F}_t] \\
&= \lambda_{Y_t}E[\bar{F}(R_{t+s}-)|\mathcal{F}_t] \geq \lambda_{Y_t}\bar{F}(E[R_{t+s} - |\mathcal{F}_t]) \\
&\geq \lambda_{Y_t}\bar{F}(R_t-) = h_t, \ t, s \in \mathbb{R}_+.
\end{aligned}
$$

This shows that $h$ is a submartingale, i.e., $h$ is stochastically increasing.

## Bounds for Finite Time Ruin Probabilities

Except in simple cases, such as Poisson arrivals of exponentially distributed claims (P/E case), the finite time ruin probabilities $\psi(u,t) = P(\tau(u) \leq t)$ cannot be expressed by the basic model parameters in an explicit form. So there is a variety of suggested bounds and approximations (see Asmussen [10] and Grandell [86] for overviews). In the following, bounds for the ruin probabilities in finite time will be derived that are based on the semimartingale representation given in Lemma 22. It turns out that especially for small values of $t$ known bounds can be improved.

From now on we assume that the claim arrival process is Poisson with rate $\lambda > 0$. Then Lemma 22 yields the representation

$$V_t = I(\tau(u) \leq t) = \int_0^t I(\tau(u) > s)\lambda \bar{F}(R_s)ds + M_t, t \in \mathbb{R}_+. \qquad (3.24)$$

Note that the paths of $R$ have only countable numbers of jumps such that under the integral sign $R_s-$ can be replaced by $R_s$. Taking expectations on both sides of (3.24) one gets by Fubini's theorem

$$
\begin{aligned}
\psi(u,t) &= \int_0^t E[I(\tau(u) > s)\lambda\bar{F}(R_s)]ds \qquad (3.25)\\
&= \int_0^t (1 - \psi(u,s))\lambda E[\bar{F}(R_s)|\tau(u) > s]ds.
\end{aligned}
$$

As a solution of this integral equation we have the following representation of the finite time ruin probability:

$$\psi(u,t) = 1 - \exp\{-\int_0^t \lambda E[\bar{F}(R_s)|\tau(u) > s]ds\}. \qquad (3.26)$$

This shows that the (possibly defective) distribution of $\tau(u)$ has the hazard rate

$$\lambda E[\bar{F}(R_t)|\tau(u) > t].$$

Now let $N^X$ be the renewal process generated by the sequence $(X_i), i \in \mathbb{N}, N_t^X = \sup\{k \in \mathbb{N}_0 : \sum_{i=1}^k X_i \leq t\}$, and $A(u,t) = \int_0^t a(u,s)ds$, where $a(u,s) = \lambda P(N_{u+cs}^X = N_s)$. Then bounds for $\psi(u,t)$ can be established.

**Theorem 23** *For all $u, t \geq 0$, the following inequality holds true:*

$$B(u,t) \leq \psi(u,t) \leq A(u,t),$$

*where $A$ is defined as above and $B(u,t) = 1 - \exp\{-\lambda \int_0^t \bar{F}(u + cs)ds\}$.*

**Proof.** For the lower bound we use the representation (3.26) and simply observe that $E[\bar{F}(R_s)|\tau(u) > s] \geq \bar{F}(u + cs)$.

For the upper bound we start with formula (3.24). Since $\{\tau(u) > t\} \subset \{R_t \geq 0\}$, we have

$$
\begin{aligned}
V_t &= \int_0^t I(\tau(u) > s)\lambda \bar{F}(R_s)ds + M_t \\
&\leq \int_0^t I(R_s \geq 0)\lambda \bar{F}(R_s)ds + M_t.
\end{aligned}
$$

Taking expectations on both sides of this inequality we get

$$
\psi(u,t) = EV_t \leq \int_0^t \lambda E[I(R_s \geq 0)\bar{F}(R_s)]ds.
$$

It remains to show that $a(u,t) = \lambda E[I(R_s \geq 0)\bar{F}(R_s)]$. Denoting the $k$-fold convolution of $F$ by $F^{*k}$ and $T_k = \sum_{i=1}^k X_i$ it follows by the independence of the claim arrival process and $(X_i), i \in \mathbb{N}$,

$$
E[I(R_t \geq 0)\bar{F}(u + ct - \sum_{i=1}^{N_t} X_i)]
$$

$$
\begin{aligned}
&= \sum_{k=0}^{\infty} E[I(u + ct - \sum_{i=1}^k X_i \geq 0)\bar{F}(u + ct - \sum_{i=1}^k X_i)]P(N_t = k) \\
&= \sum_{k=0}^{\infty} \int_0^{u+ct} \bar{F}(u + ct - x)dF^{*k}(x)P(N_t = k) \\
&= \sum_{k=0}^{\infty} \{F^{*k}(u + ct) - F^{*(k+1)}(u + ct)\}P(N_t = k) \\
&= \sum_{k=0}^{\infty} P(N_{u+ct}^X = k)P(N_t = k) \\
&= P(N_{u+ct}^X = N_t),
\end{aligned}
$$

which completes the proof.                                                      ∎

The bounds of the theorem seem to have several advantages: As numerical examples show, they perform well especially for small values of $t$ for which $\psi(u,t) \ll \psi(u,\infty)$ (see Aven and Jensen [27]). In addition no assumptions have been made about the tail of the claim size distribution $F$ and the drift of the risk reserve process, which are necessary for most of the asymptotic methods. This makes clear, on the other hand, that one cannot expect these bounds to perform well for $t \to \infty$.

**Bibliographic Notes.** The book of Brémaud [52] is one of the basic sources of the martingale dynamics of point process systems. The introduction (p. XV) also contains a sketch of the historical development.

The smooth semimartingale approach in connection with optimal stopping problems is considered by Jensen [109]. Comprehensive overviews over lifetime models in the martingale framework are those of Arjas [3], [4] and Koch [119]. An essential basis for the presentation of point processes in the martingale framework was laid by Jacod [102]. A number of books on point processes are available now. Among others, the martingale approach is exposed in Brémaud [52], Karr [114], and Daley and Vere-Jones [64], which also include the basic results about marked point processes. A full account on marked point processes can be found in the monograph of Last and Brandt [126].

Details on the theory of Markov processes, briefly mentioned in Section 3.1, can be found in the classic book of Dynkin [74] or in the more recent monographs on stochastic processes mentioned at the beginning of this chapter.

One of the first papers considering random hazard rates in lifetime models is that of Bergman [40]. Failure rate processes for multivariate reliability systems were introduced by Arjas in [6]. Shock processes have been investigated by a number of authors. Aven treated these processes in the framework of counting processes in some generality in [18]. Recent work on shock models of threshold type concentrates on deriving the distribution of the hitting (life-) time under general conditions. Wendt [175] considers a doubly stochastic Poisson shock arrival process, whereas Lehmann [130] investigates shock models with failure thresholds varying in time.

Models of minimal repairs have been considered by Barlow and Hunter [32], Aven [21], Bergman [41], Block et al. [50], Stadje and Zuckerman [163], Shaked and Shanthikumar [152], and Beichelt [37], among others. Our formulation of the minimal repair concept in a general counting process framework is taken from [26]. Varying degrees of repairs are investigated in a number of papers like Brown and Proschan [53], Kijima [118], and Last and Szekli [127], [128].

As was pointed out by Bergman [41], information plays an important role in minimal repair models. Further steps in investigating information-based minimal repair were carried out by Arjas and Norros [7] and Natvig [135].

General references to risk theory are among others the books of Grandell [85] and Rolski et al. [144]. Overviews over bounds and approximations of ruin probabilities can be found in Asmussen [10] and Grandell [86]. Most of the approximations are based on limit theorems for $\psi(u, t)$ as $u \to \infty, t \to \infty$. One of the exceptions is the inverse martingales technique used by Delbaen and Haezendonck [66].

# 4

# Availability Analysis of Complex Systems

In this chapter we establish methods and formulas for computing various performance measures of monotone systems of repairable components. Emphasis is placed on the point availability, the distribution of the number of failures in a time interval, and the distribution of downtime of the system. A number of asymptotic results are formulated and proved, mainly for systems having highly available components.

The performance measures are introduced in Section 4.1. In Sections 4.3−4.6 results for binary monotone systems are presented. Since many of these results are based on the one-component case, we first give in Section 4.2 a rather comprehensive treatment of this case. Section 4.7 presents generalizations and related models. Section 4.7.1 covers multistate monotone systems. In Sections 4.2−4.5 and 4.7.1 it is assumed that there are at least as many repair facilities (channels) as components. In Section 4.7.2 we consider a parallel system having $r$ repair facilities, where $r$ is less than the number of components. Attention is drawn to the case with $r = 1$. Finally, in Section 4.7.3 we present models for analysis of passive redundant systems.

In this chapter we focus on the situation that the components have exponential lifetime distributions. See Section 4.7.1, p. 156, and Bibliographic Notes, p. 166, for some comments concerning the more general case of nonexponential lifetimes.

## 4.1 Performance Measures

We consider a binary monotone system with state process $(\Phi_t) = (\Phi(\mathbf{X}_t))$, as described in Section 2.1. Here $\Phi_t$ equals 1 if the system is functioning at time $t$ and 0 if the system is not-functioning at time $t$, and $\mathbf{X}_t = (X_t(1), X_t(2), \ldots, X_t(n)) \in \{0,1\}^n$ describes the states of the components. The performance measures relate to one point in time $t$ or an interval $J$, which has the form $[0, u]$ or $(u, v]$, $0 < u < v$. To simplify notation, we simply write $u$ instead of $[0, u]$.

Emphasis will be placed on the following performance measures:

a) Point availability at time $t$, $A(t)$, given by

$$A(t) = E\Phi_t = P(\Phi_t = 1).$$

b) Let $N_J$ be equal to the number of system failures in the interval $J$. We consider the following performance measures

$$P(N_J \leq k), \; k \in \mathbb{N}_0,$$
$$M(J) = EN_J,$$
$$A[u, v] = P(\Phi_t = 1, \; \forall t \in [u, v])$$
$$= P(\Phi_u = 1, N_{(u,v]} = 0).$$

The performance measure $A[u, v]$ is referred to as the *interval reliability*.

c) Let $Y_J$ denote the downtime in the interval $J$, i.e.,

$$Y_J = \int_J (1 - \Phi_t) \, dt.$$

We consider the performance measures

$$P(Y_J \leq y), \; y \in \mathbb{R}_+,$$
$$A^D(J) = \frac{EY_J}{|J|},$$

where $|J|$ denotes the length of the interval $J$. The measure $A^D(J)$ is in the literature sometimes referred to as the *interval unavailability*, but we shall not use this term here.

The above performance measures relate to a fixed point in time or a finite time interval. Often it is more attractive, in particular from a computational point of view, to consider the asymptotic limit of the measure (as $t, u$ or $v \to \infty$), suitably normalized (in most cases such limits exist). In the following we shall consider both the above measures and suitably defined limits.

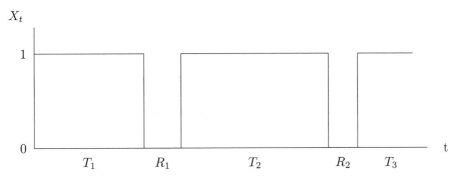

FIGURE 4.1. Time Evolution of a Failure and Repair Process for a One-Component System Starting at Time $t = 0$ in the Operating State

## 4.2   One-Component Systems

We consider in this section a one-component system. Hence $\Phi_t = X_t = X_t(1)$. If the system fails, it is repaired or replaced. Let $T_k$, $k \in \mathbb{N}$, represent the length of the $k$th operation period, and let $R_k$, $k \in \mathbb{N}$, represent the length of the $k$th repair/replacement time for the system; see Figure 4.1. We assume that $(T_k)$, $k \in \mathbb{N}$, and $(R_k)$, $k \in \mathbb{N}$, are independent i.i.d. sequences of positive random variables. We denote the probability distributions of $T_k$ and $R_k$ by $F$ and $G$, respectively, and assume that they have finite means, i.e.,

$$\mu_F < \infty, \quad \mu_G < \infty.$$

In reliability engineering $\mu_F$ and $\mu_G$ are referred to as the mean time to failure (MTTF) and the mean time to repair (MTTR), respectively.

To simplify the presentation, we also assume that $F$ is an absolutely continuous distribution, i.e., $F$ has a density function $f$ and failure rate function $\lambda$. We do not make the same assumption for the distribution function $G$, since that would exclude discrete repair time distributions, which are often used in practice.

In some cases we also need the variances of $F$ and $G$, denoted $\sigma_F^2$ and $\sigma_G^2$, respectively. In the following, when writing the variance of a random variable, or any other moment, it is tacitly assumed that these are finite.

The sequence

$$T_1, R_1, T_2, R_2, \cdots$$

forms an alternating renewal process.

We introduce the following variables

$$S_n = T_1 + \sum_{k=1}^{n-1}(R_k + T_{k+1}), \quad n \in \mathbb{N},$$

and

$$S_n^\circ = \sum_{k=1}^{n}(T_k + R_k), \quad n \in \mathbb{N}.$$

By convention, $S_0 = S_0^\circ = 0$, and sums over empty sets are zero. We see that $S_n$ represents the $n$th failure time, and $S_n^\circ$ represents the completion time of the $n$th repair.

The $S_n$ sequence generates a modified (delayed) renewal process $N$ with renewal function $M$. The first interarrival time has distribution $F$. All other interarrival times have distribution $F * G$ (convolution of $F$ and $G$), with mean $\mu_F + \mu_G$. Let $H^{(n)}$ denote the distribution function of $S_n$. Then

$$H^{(n)} = F * (F * G)^{*(n-1)},$$

where $B^{*n}$ denotes the $n$-fold convolution of a distribution $B$ and as usual $B^{*0}$ equals the distribution with mass of 1 at 0. Note that we have

$$M(t) = \sum_{n=1}^{\infty} H^{(n)}(t)$$

(cf. (B.2), p. 244, in Appendix B). The $S_n^\circ$ sequence generates an ordinary renewal process $N^\circ$ with renewal function $M^\circ$. The interarrival times, $T_k + R_k$, have distribution $F * G$, with mean $\mu_F + \mu_G$. Let $H^{\circ(n)}$ denote the distribution function of $S_n^\circ$. Then

$$H^{\circ(n)} = (F * G)^{*n}.$$

Let $\alpha_t$ denote the forward recurrence time at time $t$, i.e., the time from $t$ to the next event:

$$\alpha_t = S_{N_t+1} - t \quad \text{on} \quad \{X_t = 1\}$$

and

$$\alpha_t = S_{N_t^\circ+1}^\circ - t \quad \text{on} \quad \{X_t = 0\}.$$

Hence, given that the system is up at time $t$, the forward recurrence time $\alpha_t$ equals the time to the next failure time. If the system is down at time $t$, the forward recurrence time equals the time to complete the repair. Let $F_{\alpha_t}$ and $G_{\alpha_t}$ denote the conditional distribution functions of $\alpha_t$ given that $X_t = 1$ and $X_t = 0$, respectively. Then we have for $x \in \mathbb{R}$

$$F_{\alpha_t}(x) = P(\alpha_t \le x | X_t = 1) = P(S_{N_t+1} - t \le x | X_t = 1)$$

and

$$G_{\alpha_t}(x) = P(\alpha_t \le x | X_t = 0) = P(S_{N_t^\circ+1}^\circ - t \le x | X_t = 0).$$

Similarly for the backward recurrence time, we define $\beta_t$, $F_{\beta_t}$, and $G_{\beta_t}$. The backward recurrence time $\beta_t$ equals the age of the system if the system is up at time $t$ and the duration of the repair if the system is down at time $t$, i.e.,

$$\beta_t = t - S^\circ_{N^\circ_t} \quad \text{on} \quad \{X_t = 1\}$$

and

$$\beta_t = t - S_{N_t} \quad \text{on} \quad \{X_t = 0\}.$$

### 4.2.1 Point Availability

We will show that the point availability $A(t)$ is given by

$$A(t) = \bar{F}(t) + \int_0^t \bar{F}(t - x) dM^\circ(x) = \bar{F}(t) + \bar{F} * M^\circ(t). \qquad (4.1)$$

Using a standard renewal argument conditioning on the duration of $T_1 + R_1$, it is not difficult to see that $A(t)$ satisfies the following equation:

$$A(t) = \bar{F}(t) + \int_0^t A(t - x) d(F * G)(x)$$

(cf. the derivation of the renewal equation in Appendix B, p. 245). Hence, by using Theorem 111, p. 245, in Appendix B, formula (4.1) follows. Alternatively, we may use a more direct approach, writing

$$X_t = I(T_1 > t) + \sum_{n=1}^{\infty} I(S^\circ_n \le t, S^\circ_n + T_{n+1} > t),$$

which gives

$$
\begin{aligned}
A(t) = EX_t &= \bar{F}(t) + \sum_{n=1}^{\infty} \int_0^t \bar{F}(t - x) dH^{\circ(n)}(x) \\
&= \bar{F}(t) + \int_0^t \bar{F}(t - x) dM^\circ(x).
\end{aligned}
$$

The point unavailability $\bar{A}(t)$ is given by $\bar{A}(t) = 1 - A(t) = F(t) - \bar{F} * M^\circ(t)$.

In the case that $F$ is exponential with failure rate $\lambda$, it can be shown that

$$\bar{A}(t) \le \lambda \mu_G,$$

see Proposition 34, p. 107.

By the Key Renewal Theorem (Theorem 116, p. 247, in Appendix B), it follows that

$$\lim_{t \to \infty} A(t) = \frac{\mu_F}{\mu_F + \mu_G}, \qquad (4.2)$$

noting that the mean of $F * G$ equals $\mu_F + \mu_G$ and $\int_0^\infty \bar{F}(t) dt = \mu_F$. The right-hand side of (4.2) is called the *limiting availability* (or steady-state availability) and is for short denoted $A$. The *limiting unavailability* is defined as $\bar{A} = 1 - A$. Usually $\mu_G$ is small compared to $\mu_F$, so that

$$\bar{A} = \frac{\mu_G}{\mu_F} + o\left(\frac{\mu_G}{\mu_F}\right), \qquad \frac{\mu_G}{\mu_F} \to 0.$$

## 4.2.2   The Distribution of the Number of System Failures

Consider first the interval $[0, v]$. We see that

$$\{N_v \leq n\} = \{S_{n+1} > v\}, \quad n \in \mathbb{N}_0,$$

because if the number of failures in this interval is less than or equal to $n$, then the $(n+1)$th failure occurs after $v$, and vice versa. Thus, for $n \in \mathbb{N}_0$,

$$P(N_v \leq n) = 1 - (F * G)^{*n} * F(v). \tag{4.3}$$

Some closely related results are stated below in Propositions 24 and 25.

**Proposition 24** *The probability of $n$ failures occurring in $[0, v]$ and the system being up at time $v$ is given by*

$$P(N_v = n, X_v = 1) = \int_0^v \bar{F}(v - x)d(F * G)^{*n}(x), \ n \in \mathbb{N}_0.$$

**Proof.** The result clearly holds for $n = 0$. For $n \geq 1$, the result follows by observing that

$$\{N_v = n, X_v = 1\} = \{S_n^{\circ} + T_{n+1} > v, S_n^{\circ} \leq v\}. \qquad \blacksquare$$

**Proposition 25** *The probability of $n$ failures occurring in $[0, v]$ and the system being down at time $v$ is given by*

$$P(N_v = n, X_v = 0) = \begin{cases} \int_0^v \bar{G}(v - x)dH^{(n)}(x) & n \in \mathbb{N} \\ 0 & n = 0. \end{cases}$$

**Proof.** The proof is similar to the proof of Proposition 24. For $n \in \mathbb{N}$, it is seen that

$$\{N_v = n, X_v = 0\} = \{S_n + R_n > v, S_n \leq v\}. \qquad \blacksquare$$

From Propositions 24 and 25 we can deduce several results, for example, a formula for $P(N_u = n | X_u = 1)$ using that

$$P(N_u = n | X_u = 1) = \frac{P(N_u = n, X_u = 1)}{A(u)}.$$

In the theorem below we establish general formulas for $P(N_{(u,v]} \leq n)$ and $A[u, v]$.

**Theorem 26** *The probability that at most* $n$ $(n \in \mathbb{N}_0)$ *failures occur during the interval* $(u, v]$ *equals*

$$
\begin{aligned}
P(N_{(u,v]} \leq n) &= [1 - F_{\alpha_u} * (F * G)^{*n}(v - u)]A(u) \\
&\quad + [1 - G_{\alpha_u} * (F * G)^{*n} * F(v - u)]\bar{A}(u),
\end{aligned}
$$

*and*

$$
A[u, v] = \bar{F}_{\alpha_u}(v - u)A(u).
$$

**Proof.** To establish the formula for $P(N_{(u,v]} \leq n)$, we condition on the state of the system at time $u$:

$$
P(N_{(u,v]} \leq n) = \sum_{j=0}^{1} P(N_{(u,v]} \leq n | X_u = j)P(X_u = j).
$$

From this equality the formula follows trivially for $n = 0$. For $n \in \mathbb{N}$, we need to show that the following two equalities hold true:

$$
P(N_{(u,v]} > n | X_u = 1) = (F_{\alpha_u} * G) * (F * G)^{*(n-1)} * F(v - u), \quad (4.4)
$$
$$
P(N_{(u,v]} > n | X_u = 0) = G_{\alpha_u} * (F * G)^{*n} * F(v - u). \quad (4.5)
$$

But (4.4) follows directly from (4.3) with the forward recurrence time distribution given $\{X_u = 1\}$ as the first operating time distribution. Formula (4.5) is established analogously.

The formula for $A[u, v]$ is seen to hold observing that

$$
\begin{aligned}
A[u, v] &= P(X_u = 1, N_{(u,v]} = 0) \\
&= A(u)P(N_{(u,v]} = 0 | X_u = 1) \\
&= A(u)P(\alpha_u > v - u | X_u = 1).
\end{aligned}
$$

This completes the proof of the theorem. ∎

If the downtimes are much smaller then the uptimes in probability (which is the common situation in practice), then $N$ is close to a renewal process generated by all the uptimes. Hence, if the times to failure are exponentially distributed, the process $N$ is close to a homogeneous Poisson process. Formal asymptotic results will be established later, see Section 4.4.

In the following two propositions we relate the distribution of the forward and backward recurrence times and the renewal functions $M$ and $M^\circ$.

**Proposition 27** *The probability that the system is up (down) at time* $t$ *and the forward recurrence time at time* $t$ *is greater than* $w$ *is given by*

$$
\begin{aligned}
A[t, t + w] &= P(X_t = 1, \alpha_t > w) \\
&= \bar{F}(t + w) + \int_0^t \bar{F}(t - x + w)dM^\circ(x), \quad (4.6)
\end{aligned}
$$
$$
P(X_t = 0, \alpha_t > w) = \int_0^t \bar{G}(t - x + w)dM(x). \quad (4.7)
$$

**Proof.** Consider first formula (4.6). It is not difficult to see that

$$X_t I(\alpha_t > w) = \sum_{n=0}^{\infty} I(S_n^\circ \le t, S_n^\circ + T_{n+1} > t + w). \qquad (4.8)$$

By taking expectations we find that

$$
\begin{aligned}
P(X_t = 1, \alpha_t > w) &= \bar{F}(t + w) + \sum_{n=1}^{\infty} \int_0^t \bar{F}(t - x + w) dH^{\circ(n)}(x) \\
&= \bar{F}(t + w) + \int_0^t \bar{F}(t - x + w) dM^\circ(x).
\end{aligned}
$$

This proves (4.6). To prove (4.7) we use a similar argument writing

$$(1 - X_t) I(\alpha_t > w) = \sum_{n=1}^{\infty} I(S_n \le t, S_n + R_n > t + w). \qquad (4.9)$$

This completes the proof of the proposition. ∎

**Proposition 28** *The probability that the system is up (down) at time $t$ and the backward recurrence time at time $t$ is greater than $w$ is given by*

$$P(X_t = 1, \beta_t > w) = \begin{cases} \bar{F}(t) + \int_0^{t-w} \bar{F}(t - x) dM^\circ(x) & w \le t \\ 0 & w > t \end{cases} \qquad (4.10)$$

$$P(X_t = 0, \beta_t > w) = \begin{cases} \int_0^{t-w} \bar{G}(t - x) dM(x) & w \le t \\ 0 & w > t. \end{cases} \qquad (4.11)$$

**Proof.** The proof is similar to the proof of Proposition 27. Replace the indicator function in the sums in (4.8) and (4.9) by

$$I(S_n^\circ + T_{n+1} > t, S_n^\circ + w < t)$$

and

$$I(S_n + R_n > t, S_n + w < t),$$

respectively. ∎

**Theorem 29** *The asymptotic distributions of the state process $(X_t)$ and the forward (backward) recurrence times at time $t$ are given by*

$$
\begin{aligned}
\lim_{t \to \infty} P(X_t = 1, \alpha_t > w) &= \frac{\int_w^{\infty} \bar{F}(x)\, dx}{\mu_F + \mu_G} \\
\lim_{t \to \infty} P(X_t = 0, \alpha_t > w) &= \frac{\int_w^{\infty} \bar{G}(x)\, dx}{\mu_F + \mu_G} \\
\lim_{t \to \infty} P(X_t = 1, \beta_t > w) &= \frac{\int_w^{\infty} \bar{F}(x)\, dx}{\mu_F + \mu_G} \qquad (4.12) \\
\lim_{t \to \infty} P(X_t = 0, \beta_t > w) &= \frac{\int_w^{\infty} \bar{G}(x)\, dx}{\mu_F + \mu_G}.
\end{aligned}
$$

**Proof.** The results follow by applying the Key Renewal Theorem (see Appendix B, p. 247) to formulas (4.6), (4.7), (4.10), and (4.11). ∎

Let us introduce

$$F_\infty(w) = \frac{\int_0^w \bar{F}(x)\,dx}{\mu_F}, \tag{4.13}$$

$$G_\infty(w) = \frac{\int_0^w \bar{G}(x)\,dx}{\mu_G}. \tag{4.14}$$

The distribution $F_\infty$ ($G_\infty$) is the asymptotic limit distribution of the forward and backward recurrence times in a renewal process generated by the uptimes (downtimes) and is called the equilibrium distribution for $F$ ($G$), cf. Theorem 120, p. 249, in Appendix B. We would expect that $F_\infty$ and $G_\infty$ are equal to the asymptotic distributions of the forward and backward recurrence times in the alternating renewal process. As shown in the following proposition, this holds in fact true.

**Proposition 30** *The asymptotic distribution of the forward and backward recurrence times are given by*

$$\lim_{t\to\infty} \bar{F}_{\alpha_t}(w) = \lim_{t\to\infty} \bar{F}_{\beta_t}(w) = \bar{F}_\infty(w)$$

*and*

$$\lim_{t\to\infty} \bar{G}_{\alpha_t}(w) = \lim_{t\to\infty} \bar{G}_{\beta_t}(w) = \bar{G}_\infty(w). \tag{4.15}$$

**Proof.** To establish these formulas, we use (4.2) (see p. 101), Theorem 29, and identities like

$$P(\alpha_t > w | X_t = 1) = \frac{P(X_t = 1, \alpha_t > w)}{A(t)}. \qquad ∎$$

The following theorem expresses the asymptotic distribution of $N_{(t,t+w]}$ as a function of $F$, $G$, $F_\infty$, $G_\infty$ and $A$.

**Theorem 31** *For $n \in \mathbb{N}_0$,*

$$\lim_{t\to\infty} P(N_{(t,t+w]} \le n) = [1 - F_\infty * (F * G)^{*n}(w)]A +$$

$$+ [1 - G_\infty * (F * G)^{*n} * F(w)]\bar{A}.$$

**Proof.** The result follows from the expression for the distribution of the number of failures given in Theorem 26, p. 103, combined with the limiting availability formula (4.2), p. 101, and Proposition 30. ∎

If the lifetime distribution $F$ is exponential with failure rate $\lambda$, then we know that the forward recurrence time $\alpha_t$ has the same distribution for all $t$, and it is easily verified from the expression (4.13) for the equilibrium distribution for $F$ that $F_\infty(t) = F(t)$.

Next we consider an increasing interval $(t, t+w]$, $w \to \infty$. Then we can use the normal distribution to find an approximate value for the distribution of $N$. The asymptotic normality, as formulated in the following theorem, follows by applying the Central Limit Theorem for renewal processes, see Theorem 119, p. 248, in Appendix B. The notation $N(\mu, \sigma^2)$ is used for the normal distribution with mean $\mu$ and variance $\sigma^2$.

**Theorem 32** *The asymptotic distribution of $N_{(t,t+w]}$ as $w \to \infty$, is given by*

$$\frac{N_{(t,t+w]} - w/(\mu_F + \mu_G)}{[w(\sigma_F^2 + \sigma_G^2)/(\mu_F + \mu_G)^3]^{1/2}} \xrightarrow{D} N(0,1). \qquad (4.16)$$

The expected number of system failures can be found from the distribution function. Obviously, $M(v) \approx M^\circ(v)$ for large $v$. The exact relationship between $M(v)$ and $M^\circ(v)$ is given in the following proposition.

**Proposition 33** *The difference between the renewal functions $M(v)$ and $M^\circ(v)$ equals the unavailability at time $v$, i.e.,*

$$M(v) = M^\circ(v) + \bar{A}(v).$$

**Proof.**   Using that $P(N_v \le n) = 1 - (F * G)^{*n} * F(v)$ (by (4.3), p. 102) and the expression (4.1), p. 101, for the availability $A(t)$, we obtain

$$
\begin{aligned}
M(v) &= \sum_{n=1}^{\infty} P(N_v \ge n) \\
&= \sum_{n=0}^{\infty} (F * G)^{*n} * F(v) = F(v) + M^\circ * F(v) \\
&= M^\circ(v) + \bar{A}(v),
\end{aligned}
$$

which is the desired result.    ■

The number of system failures in $[0, v]$, $N_v$, generates a counting process with stochastic intensity process

$$\eta_v = \lambda(\beta_v) X_v, \qquad (4.17)$$

where $\lambda$ is the failure rate function and $\beta_v$ is the backward recurrence time at time $v$, i.e., the relative age of the system at time $v$, cf. Section 3.3.2, p. 77. We have $m(v) = E\eta_v$, where $m(v)$ is the renewal density of $M(v)$. Thus if the system has an exponential lifetime distribution with failure rate $\lambda$,

$$m(v) = \lambda A(v). \tag{4.18}$$

In general,

$$m(v) \le [\sup_{s \le v} \lambda(s)] A(v). \tag{4.19}$$

This bound can be used to establish an upper bound also for the unavailability $\bar{A}(t)$.

**Proposition 34** *The unavailability at time $t$, $\bar{A}(t)$, satisfies*

$$\bar{A}(t) \le \sup_{s \le t} \lambda(s) \int_0^t \bar{G}(u)du \le [\sup_{s \le t} \lambda(s)]\mu_G. \tag{4.20}$$

**Proof.** From (4.7), p. 103, we have

$$\bar{A}(t) = P(X_t = 0) = \int_0^t \bar{G}(t-x)dM(x) = \int_0^t \bar{G}(t-x)m(x)dx. \tag{4.21}$$

Using (4.19) this gives

$$\bar{A}(t) \le \int_0^t \bar{G}(t-x)[\sup_{s \le x} \lambda(s)]A(x)dx.$$

It follows that

$$
\begin{aligned}
\bar{A}(t) &\le \sup_{s \le t} \lambda(s) \int_0^t \bar{G}(t-x)dx \\
&= \sup_{s \le t} \lambda(s) \int_0^t \bar{G}(u)du \le [\sup_{s \le t} \lambda(s)]\mu_G,
\end{aligned}
$$

which proves (4.20). ∎

Hence, if the system has an exponential lifetime distribution with failure rate $\lambda$, then

$$\bar{A}(t) \le \lambda \int_0^t \bar{G}(s)ds \le \lambda\mu_G. \tag{4.22}$$

It is also possible to establish lower bounds on $\bar{A}(t)$. A simple bound is obtained by combining (4.21) and the fact that

$$t \le ES_{N_t+1} \le (\mu_F + \mu_G)(1 + M(t))$$

(cf. Appendix B, p. 249), giving

$$\bar{A}(t) \ge \bar{G}(t)M(t) \ge \bar{G}(t)(\frac{t}{\mu_F + \mu_G} - 1).$$

Now suppose at time $t$ that the system is functioning and the relative age is $u$. What can we then say about the intensity process at time $t + v$ ($v > 0$)? The probability distribution of $\eta_{t+v}$ is determined if we can find the distribution of the relative age at time $t + v$. But the relative age is given by (4.10), p. 104, slightly modified to take into account that the first uptime has distribution given by $F_u(x) = 1 - \bar{F}(u+x)/\bar{F}(u)$ for $0 \leq u \leq t$:

$$P(X_{t+v} = 1, \beta_{t+v} > w | X_t = 1, \beta_t = u)$$
$$= \begin{cases} \bar{F}_u(v) + \int_0^{v-w} \bar{F}(v - x) dM^\circ(x) & w \leq v \\ 0 & w > v. \end{cases}$$

The asymptotic distribution, as $v \to \infty$, is the same as in formula (4.12), p. 104.

The (modified) renewal process $(N_t)$ has cycle lengths $T_k + R_k$ with mean $\mu_F + \mu_G$, $k \geq 2$. Thus we would expect that the (mean) average number of failures per unit of time is approximately equal to $1/(\mu_F + \mu_G)$ for large $t$. In the following theorem some asymptotic results are presented that give precise formulations of this idea.

**Theorem 35** *With probability one ,*

$$\lim_{t \to \infty} \frac{N_t}{t} = \frac{1}{\mu_F + \mu_G}. \tag{4.23}$$

*Furthermore,*

$$\lim_{t \to \infty} \frac{EN_t}{t} = \frac{1}{\mu_F + \mu_G}, \tag{4.24}$$

$$\lim_{u \to \infty} E[N_{u+w} - N_u] = \frac{w}{\mu_F + \mu_G}, \tag{4.25}$$

$$\lim_{t \to \infty} \left( EN_t - \frac{t}{\mu_F + \mu_G} \right) = \frac{\sigma_F^2 + \sigma_G^2}{2(\mu_F + \mu_G)^2} - \frac{1}{2}.$$

**Proof.** These results follow directly from renewal theory, see Appendix B, pp. 246–248. ∎

## 4.2.3  The Distribution of the Downtime in a Time Interval

First we formulate and prove some results related to the mean of the downtime in the interval $[0, u]$. As before (cf. Section 4.1, p. 98), we let $Y_u$ represent the downtime in the interval $[0, u]$.

**Theorem 36** *The expected downtime in $[0, u]$ is given by*

$$EY_u = \int_0^u \bar{A}(t) dt. \tag{4.26}$$

*Asymptotically, the (expected) portion of time the system is down equals the limiting unavailability, i.e.,*

$$\lim_{u \to \infty} A^D(u) = \lim_{u \to \infty} \frac{EY_u}{u} = \bar{A}. \qquad (4.27)$$

*With probability one,*

$$\lim_{u \to \infty} \frac{Y_u}{u} = \bar{A}. \qquad (4.28)$$

**Proof.** Using the definition of $Y_u$ and Fubini's theorem we find that

$$
\begin{aligned}
EY_u &= E \int_0^u (1 - \Phi_t) dt \\
&= \int_0^u E(1 - \Phi_t) dt \\
&= \int_0^u \bar{A}(t) dt.
\end{aligned}
$$

This proves (4.26). Formula (4.27) follows by using (4.26) and the limiting availability formula (4.2), p. 101. Alternatively, we can use the Renewal Reward Theorem (Theorem 122, p. 250, in Appendix B), interpreting $Y_u$ as a reward. From this theorem we can conclude that $EY_u/u$ converges to the ratio of the expected downtime in a renewal cycle and the expected length of a cycle, i.e., to the limiting unavailability $\bar{A}$. The Renewal Reward Theorem also proves (4.28). ∎

Now we look into the problem of finding formulas for the downtime distribution.

Let $N_s^{op}$ denote the number of system failures after $s$ units of operational time, i.e.,

$$N_s^{op} = \sum_{n=1}^{\infty} I\left(\sum_{k=1}^{n} T_k \leq s\right).$$

Note that

$$N_s^{op} \geq n \Leftrightarrow \sum_{k=1}^{n} T_k \leq s, \quad n \in \mathbb{N}. \qquad (4.29)$$

Let $Z_s$ denote the total downtime associated with the operating time $s$, but not including $s$, i.e.,

$$Z_s = \sum_{i=1}^{N_{s-}^{op}} R_i,$$

where

$$N_{s-}^{op} = \lim_{u \to s-} N_u^{op}.$$

Define
$$C_s = s + Z_s.$$

We see that $C_s$ represents the calendar time after an operation time of $s$ time units and the completion of the repairs associated with the failures occurred up to $s$ but not including $s$.

The following theorem gives an exact expression of the probability distribution of $Y_u$, the total downtime in $[0, u]$.

**Theorem 37** *The distribution of the downtime in a time interval $[0, u]$ is given by*

$$P(Y_u \leq y) \quad = \quad \sum_{n=0}^{\infty} G^{*n}(y) P(N_{u-y}^{op} = n) \tag{4.30}$$

$$= \quad \sum_{n=0}^{\infty} G^{*n}(y)[F^{*n}(u-y) - F^{*(n+1)}(u-y)]. \tag{4.31}$$

**Proof.** To prove the theorem we first argue that

$$P(Y_u \leq y) \quad = \quad P(C_{u-y} \leq u) = P(u - y + Z_{u-y} \leq u)$$
$$= \quad P(Z_{u-y} \leq y).$$

This first equality follows by noting that the event $Y_u \leq y$ is equivalent to the event that the uptime in the interval $[0, u]$ is equal to or longer than $u - y$. This means that the point in time when the total uptime of the system equals $u - y$ must occur before or at $u$, i.e., $C_{u-y} \leq u$. Now using a standard conditional probability argument it follows that

$$P(Z_{u-y} \leq y) \quad = \quad \sum_{n=0}^{\infty} P(Z_{u-y} \leq y | N_{(u-y)-}^{op} = n) P(N_{(u-y)-}^{op} = n)$$

$$= \quad \sum_{n=0}^{\infty} G^{*n}(y) P(N_{(u-y)-}^{op} = n)$$

$$= \quad \sum_{n=0}^{\infty} G^{*n}(y) P(N_{u-y}^{op} = n).$$

We have used that the repair times are independent of the process $N_s^{op}$ and that $F$ is continuous. This proves (4.30). Formula (4.31) follows by using (4.29). ∎

In the case that $F$ is exponential with failure rate $\lambda$ the following simple bounds apply

$$e^{-\lambda(u-y)}[1 + \lambda(u-y)G(y)] \leq P(Y_u \leq y) \leq e^{-\lambda(u-y)[1-G(y)]}.$$

The lower bound follows by including only the first two terms of the sum in (4.30), observing that $N_t^{op}$ is Poisson distributed with mean $\lambda t$, whereas the upper bound follows by using (4.30) and the inequality

$$G^{*n}(y) \leq (G(y))^n.$$

In the case that the interval is rather long, the downtime will be approximately normally distributed, as is shown in Theorem 38 below.

**Theorem 38** *The asymptotic distribution of $Y_u$ as $u \to \infty$, is given by*

$$\sqrt{u}\left(\frac{Y_u}{u} - \bar{A}\right) \overset{D}{\to} N(0, \tau^2), \qquad (4.32)$$

*where*

$$\tau^2 = \frac{\mu_F^2 \sigma_G^2 + \mu_G^2 \sigma_F^2}{(\mu_F + \mu_G)^3}.$$

**Proof.** The result follows by applying Theorem 123, p. 251, in Appendix B, observing that the length of the first renewal cycle equals $S_1^\circ = T_1 + R_1$, the downtime in this cycle equals $Y_{S_1^\circ} = R_1$ and

$$\frac{\mathrm{Var}[R_1 - \bar{A}S_1^\circ]}{ES_1^\circ} = \frac{\mathrm{Var}[R_1 A - T_1 \bar{A}]}{ES_1^\circ}$$

$$= \frac{A^2 \mathrm{Var}[R_1] + \bar{A}^2 \mathrm{Var}[T_1]}{\mu_F + \mu_G}$$

$$= \frac{\mu_F^2 \sigma_G^2 + \mu_G^2 \sigma_F^2}{(\mu_F + \mu_G)^3}. \qquad \blacksquare$$

## 4.2.4   Steady-State Distribution

The asymptotic results established above provide good approximations for the performance measures related to a given point in time or an interval. Based on the asymptotic values we can define a stationary (steady-state) process having these asymptotic values as their distributions and means. To define such a process in our case, we generalize the model analyzed above by allowing $X_0$ to be 0 or 1. Thus the time evolution of the process is as shown in Figure 4.2 or as shown in Figure 4.1 (p. 99) beginning with an uptime. The process is characterized by the parameters $A(0)$, $F^*(t)$, $F(t)$, $G^*(t)$, $G(t)$, where $F^*(t)$ denotes the distribution of the first uptime provided that the system starts in state 1 at time 0 (i.e., $X_0 = 1$) and $G^*(t)$ denotes the distribution of the first downtime provided that the system starts in state 0 at time 0 (i.e., $X_0 = 0$). Now assuming that $F^*(t)$ and $G^*(t)$ are equal to the asymptotic distributions of the recurrence times, i.e., $F_\infty(t)$ and $G_\infty(t)$,

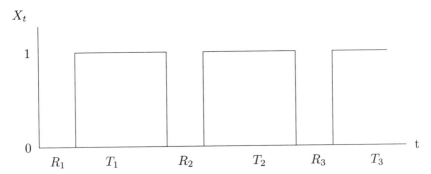

FIGURE 4.2. Time Evolution of a Failure and Repair Process for a One-Component System Starting at Time $t = 0$ in the Failure State

respectively, and $A(0) = A$, then it can be shown that the process $(X_t, \alpha_t)$ is stationary; see Birolini [46]. This means that we have, for example,

$$A(t) = A, \quad \forall t \in \mathbb{R}_+,$$

$$A[u, u + w] = \frac{\int_w^\infty \bar{F}(x)\, dx}{\mu_F + \mu_G}, \quad \forall u, w \in \mathbb{R}_+,$$

$$M(u, u + w] = \frac{w}{\mu_F + \mu_G}, \quad \forall u, w \in \mathbb{R}_+.$$

## 4.3 Point Availability and Mean Number of System Failures

Consider now a monotone system comprising $n$ independent components. For each component we define a model as in Section 4.2, indexed by "$i$." The uptimes and downtimes of component $i$ are thus denoted $T_{ik}$ and $R_{ik}$ with distributions $F_i$ and $G_i$, respectively. The lifetime distribution $F_i$ is absolutely continuous with a failure rate function $\lambda_i(t)$. The process $(N_t)$ refers now to the number of system failures, whereas $(N_t(i))$ counts the number of failures of component $i$. The counting process $(N_t(i))$ has intensity process $(\eta_t(i)) = (\lambda_i(\beta_t(i))X_t(i))$, where $(X_t(i))$ equals the state process of component $i$ and $(\beta_t(i))$ the backward recurrence time of component $i$. The mean of $(N_t(i))$ is denoted $M_i(t)$, whereas the mean of the renewal process having interarrival times $T_{ik} + R_{ik}$, $k \in \mathbb{N}$, is denoted $M_i^\circ(t)$. If the process $(\mathbf{X}_t)$ is regenerative, we denote the consecutive cycle lengths $S_1, S_2, \dots$. We write $S$ in place of $S_1$. Remember that a stochastic process $(X_t)$ is called regenerative if there exists a finite random variable $S$ such that the process beyond $S$ is a probabilistic replica of the process starting at 0. The precise definition is given in Appendix B, p. 251.

In the following we establish results similar to those obtained in the previous section. Some results are quite easy to generalize to monotone systems, others are extremely difficult. Simplifications and approximative methods are therefore sought. First we look at the point availability.

## 4.3.1   Point Availability

The following results show that the point availability (limiting availability) of a monotone system is equal to the reliability function $h$ with the component reliabilities replaced by the component availabilities $A_i(t)$ $(A_i)$.

**Theorem 39** *The system availability at time $t$, $A(t)$, and the limiting system availability, $\lim_{t \to \infty} A(t)$, are given by*

$$A(t) = h(A_1(t), A_2(t), \dots, A_n(t)) = h(\mathbf{A}(t)), \qquad (4.33)$$

$$\lim_{t \to \infty} A(t) = h(A_1, A_2, \dots, A_n) = h(\mathbf{A}). \qquad (4.34)$$

**Proof.** Formula (4.33) is simply an application of the reliability function formula (2.2), see p. 23, with $A_i(t) = P(X_t(i) = 1)$. Since the reliability function $h(\mathbf{p})$ is a linear function in each $p_i$ (see Section 2.1, p. 27), and therefore a continuous function, it follows that $A(t) \to h(A_1, A_2, \dots, A_n)$ as $t \to \infty$, which proves (4.34). ∎

The limiting system availability can also be interpreted as the expected portion of time the system is operating in the long run, or as the long run average availability, noting that

$$\lim_{t \to \infty} E\left[\frac{1}{t} \int_0^t \Phi_s ds\right] = \lim_{t \to \infty} \frac{1}{t} \int_0^t A(s)ds = \lim_{t \to \infty} A(t).$$

## 4.3.2   Mean Number of System Failures

We first state some results established in Section 3.3.2, cf. formula (3.18), p. 78. See also (4.17) and (4.18), p. 106.

**Theorem 40** *The expected number of system failures in $[0, u]$ is given by*

$$EN_u = \sum_{i=1}^{n} \int_0^u [h(1_i, \mathbf{A}(t)) - h(0_i, \mathbf{A}(t))]\, dM_i(t) \qquad (4.35)$$

$$= \sum_{i=1}^{n} \int_0^u [h(1_i, \mathbf{A}(t)) - h(0_i, \mathbf{A}(t))]\, m_i(t)\, dt$$

$$= \sum_{i=1}^{n} \int_0^u [h(1_i, \mathbf{A}(t)) - h(0_i, \mathbf{A}(t))]\, E\eta_t(i)dt,$$

*where $m_i(t)$ is the renewal density function of $M_i(t)$.*

**Corollary 41** *If component $i$ has constant failure rate $\lambda_i$, $i = 1, 2, \ldots, n$, then*

$$
\begin{aligned}
EN_u &= \sum_{i=1}^{n} \int_0^u \left[ h(1_i, \mathbf{A}(t)) - h(0_i, \mathbf{A}(t)) \right] \lambda_i A_i(t) dt, \qquad (4.36) \\
&\leq u\tilde{\lambda},
\end{aligned}
$$

*where $\tilde{\lambda} = \sum_{i=1}^{n} \lambda_i$.*

Next we will generalize the asymptotic results (4.23)–(4.25), p. 108.

**Theorem 42** *The expected number of system failures per unit of time is asymptotically given by*

$$
\lim_{u \to \infty} \frac{EN_u}{u} = \sum_{i=1}^{n} \frac{h(1_i, \mathbf{A}) - h(0_i, \mathbf{A})}{\mu_{F_i} + \mu_{G_i}}, \qquad (4.37)
$$

$$
\lim_{u \to \infty} \frac{EN_{(u, u+w]}}{w} = \sum_{i=1}^{n} \frac{h(1_i, \mathbf{A}) - h(0_i, \mathbf{A})}{\mu_{F_i} + \mu_{G_i}}. \qquad (4.38)
$$

*Furthermore, if the process $\mathbf{X}$ is a regenerative process having finite expected cycle length, i.e., $ES < \infty$, then with probability one,*

$$
\lim_{u \to \infty} \frac{N_u}{u} = \sum_{i=1}^{n} \frac{h(1_i, \mathbf{A}) - h(0_i, \mathbf{A})}{\mu_{F_i} + \mu_{G_i}}. \qquad (4.39)
$$

**Proof.** To prove these results, we make use of formula (4.35). Dividing this formula by $u$ and using the Elementary Renewal Theorem (see Appendix B, p. 247), formula (4.37) can be shown to hold noting that $E[\Phi(1_i, \mathbf{X}_t) - \Phi(0_i, \mathbf{X}_t)] \to [h(1_i, \mathbf{A}) - h(0_i, \mathbf{A})]$ as $t \to \infty$. Let $h_i^*(t) = E[\Phi(1_i, \mathbf{X}_t) - \Phi(0_i, \mathbf{X}_t)]$ and $h_i^*$ its limit as $t \to \infty$. Then we can write formula (4.35) divided by $u$ in the following form:

$$
\sum_{i=1}^{n} \left\{ h_i^* \frac{M_i(u)}{u} + \frac{1}{u} \int_0^u [h_i^*(t) - h_i^*] dM_i(t) \right\}.
$$

Hence in view of the Elementary Renewal Theorem, formula (4.37) follows if

$$
\lim_{u \to \infty} \frac{1}{u} \int_0^u [h_i^*(t) - h_i^*] dM_i(t) = 0. \qquad (4.40)
$$

But (4.40) is seen to hold true by Proposition 121, p. 249, in Appendix B.

The formula (4.38) is shown by writing

$$
E[N_{u+w} - N_u] = \sum_{i=1}^{n} \int_u^{u+w} E[\Phi(1_i, \mathbf{X}_t) - \Phi(0_i, \mathbf{X}_t)] dM_i(t)
$$

and using Blackwell's Theorem, see Theorem 117, p. 248, in Appendix B.

If we assume that the process $\mathbf{X}$ is regenerative with $ES < \infty$, it follows from the theory of renewal reward processes (see Appendix B, p. 250) that with probability one, $\lim_{u \to \infty} N_u/u$ exists and equals

$$\lim_{u \to \infty} \frac{EN_u}{u} = \frac{EN_S}{ES}.$$

Combining this with (4.37), we can conclude that (4.39) holds true, and the proof of the theorem is complete. ∎

**Definition 26** *The limit of $EN_u/u$, given by formula (4.37), is referred to as the system failure rate and is denoted $\lambda_\Phi$, i.e.,*

$$\lambda_\Phi = \lim_{u \to \infty} \frac{EN_u}{u} = \sum_{i=1}^n \frac{h(1_i, \mathbf{A}) - h(0_i, \mathbf{A})}{\mu_{F_i} + \mu_{G_i}}. \qquad (4.41)$$

**Remark 11** 1. Heuristically, the limit (4.37) can easily be established: In the interval $(t, t+w)$, $t$ large and $w$ small, the probability that component $i$ fails equals approximately $w/(\mu_{F_i} + \mu_{G_i})$, and this failure implies a system failure if $\Phi(1_i, \mathbf{X}_t) = 1$ and $\Phi(0_i, \mathbf{X}_t) = 0$, i.e., the system fails if component $i$ fails. But the probability that $\Phi(1_i, \mathbf{X}_t) = 1$ and $\Phi(0_i, \mathbf{X}_t) = 0$ is approximately equal to $h(1_i, \mathbf{A}) - h(0_i, \mathbf{A})$, which gives the desired result.

2. At time $t$ we can define a system failure rate $\lambda_\Phi(t)$ by

$$\lambda_\Phi(t) = \sum_{i=1}^n [\Phi(1_i, \mathbf{X}_t) - \Phi(0_i, \mathbf{X}_t)] \eta_t(i),$$

cf. Section 3.3.2, p. 77. Since

$$E\lambda_\Phi(t) = \sum_{i=1}^n [h(1_i, \mathbf{A}_t) - h(0_i, \mathbf{A}_t)] m_i(t),$$

where $m_i(t)$ denotes the renewal density of $M_i(t)$, we see that $E\lambda_\Phi(t) \to \lambda_\Phi$ as $t \to \infty$ provided that $m_i(t) \to 1/(\mu_{F_i} + \mu_{G_i})$. From renewal theory, see Theorem 118, p. 248, in Appendix B, we know that if the renewal cycle lengths $T_{ik} + R_{ik}$ have a density function $h$ with $h(t)^p$ integrable for some $p > 1$, and $h(t) \to 0$ as $t \to \infty$, then $M_i$ has a density $m_i$ such that $m_i(t) \to 1/(\mu_{F_i} + \mu_{G_i})$ as $t \to \infty$. See the remark following Theorem 118 for other sufficient conditions for $m_i(t) \to 1/(\mu_{F_i} + \mu_{G_i})$ to hold. If component $i$ has an exponential lifetime distribution with parameter $\lambda_i$, then $m_i(t) = \lambda_i A_i(t)$, (cf. (4.18), p. 107), which converges to $1/(\mu_{F_i} + \mu_{G_i})$.

It is intuitively clear that the process $\mathbf{X}$ is regenerative if the components have exponential lifetime distributions. Before we prove this formally, we formulate a result related to $EN_u^\circ$: the expected number of visits to the best state $(1, 1, \ldots, 1)$ in $[0, u]$. The result is analogous to (4.35) and (4.37).

**Lemma 43** *The expected number of visits to state* $(1, 1, \ldots, 1)$ *in* $[0, u]$ *is given by*

$$EN_u^\circ = \sum_{i=1}^{n} \int_0^u \prod_{j \neq i} A_j(t)\, dM_i^\circ(t). \tag{4.42}$$

*Furthermore,*

$$\lim_{u \to \infty} \frac{EN_u^\circ}{u} = \prod_{j=1}^{n} A_j \sum_{i=1}^{n} \frac{1}{\mu_{F_i}}. \tag{4.43}$$

**Proof.** Formula (4.42) is shown by arguing as in the proof of (4.35) (cf. Section 3.3.2, p. 77), writing

$$EN_u^\circ = E\left[ \sum_{i=1}^{n} \int_0^u \prod_{j \neq i} X_j(t)\, dN_t^\circ(i) \right].$$

To show (4.43) we can repeat the proof of (4.37) to obtain

$$\begin{aligned}
\lim_{u \to \infty} \frac{EN_u^\circ}{u} &= \sum_{i=1}^{n} \prod_{j \neq i} A_j \frac{1}{\mu_{F_i} + \mu_{G_i}} \\
&= \prod_{j=1}^{n} A_j \sum_{i=1}^{n} \frac{1}{\mu_{F_i}}.
\end{aligned}$$

This completes the proof of the lemma. ∎

The above result can be shown heuristically using the same type of arguments as in Remark 11. For highly available components we have $A_i \approx 1$, hence the limit (4.43) is approximately equal to

$$\sum_{i=1}^{n} \frac{1}{\mu_{F_i}}.$$

This is as expected noting that the number of visits to state $(1, 1, \ldots, 1)$ then should be approximately equal to the average number of component failures per unit of time. If a component fails, it will normally be repaired before any other component fails, and, consequently, the process again returns to state $(1, 1, \ldots, 1)$.

**Theorem 44** *If all the components have exponential lifetimes, then* **X** *is a regenerative process.*

**Proof.** Because of the memoryless property of the exponential distribution and the fact that all component uptimes and downtimes are independent, we can conclude that $\mathbf{X}$ is regenerative (as defined in Appendix B, p. 251) if we can prove that $P(S < \infty) = 1$, where $S = \inf\{t > S' : \mathbf{X}_t = (1, 1, \ldots, 1)\}$ and $S' = \min\{T_{i1} : i = 1, 2, ..., n\}$. It is clear that if $\mathbf{X}$ returns to the state $(1, 1, \ldots, 1)$, then the process beyond $S$ is a probabilistic replica of the process starting at 0.

Suppose that $P(S < \infty) < 1$. Then there exists an $\epsilon > 0$ such that $P(S < \infty) \leq 1 - \epsilon$. Now let $\tau_i$ be point in time of the $i$th visit of $\mathbf{X}$ to the state $(1, 1, \ldots, 1)$, i.e., $\tau_1 = S$ and for $i \geq 2$,

$$\tau_i = \inf\{t > \tau_{i-1} + S_i' : \mathbf{X}_t = (1, 1, \ldots, 1)\},$$

where $S_i'$ has the same distribution as $S'$. We define $\inf\{\emptyset\} = \infty$. Since $\tau_i < \infty$ is equivalent to $\tau_k - \tau_{k-1} < \infty$, $k = 1, 2, \ldots, i$ ($\tau_0 = 0$), we obtain

$$P(\tau_i < \infty) = [P(S < \infty)]^i \leq (1 - \epsilon)^i.$$

For all $t \in \mathbb{R}_+$,

$$P(N_t^{\circ} \geq i) \leq P(\tau_i < \infty),$$

and it follows that

$$
\begin{aligned}
EN_t^{\circ} &= \sum_{i=1}^{\infty} P(N_t^{\circ} \geq i) \\
&\leq \sum_{i=1}^{\infty} (1 - \epsilon)^i \\
&= \frac{1 - \epsilon}{1 - (1 - \epsilon)} = \frac{1 - \epsilon}{\epsilon} < \infty.
\end{aligned}
$$

Consequently, $EN_t^{\circ}/t \to 0$ as $t \to \infty$. But this result contradicts (4.43), and therefore $P(S < \infty) = 1$. ∎

Under the given set-up the regenerative property only holds true if the lifetimes of the components are exponentially distributed. However, this can be generalized by considering phase-type distributions with an enlarged state space, which also includes the phases; see Section 4.7.1, p. 156.

## 4.4    Distribution of the Number of System Failures

In general, it is difficult to calculate the distribution of the number of system failures $N_{(u,v]}$. Only in some special cases it is possible to obtain practical computation formulas, and in the following we look closer into some of these.

If the repair times are small compared to the lifetimes and the lifetimes are exponentially distributed with parameter $\lambda_i$, then clearly the number of failures of component $i$ in the time interval $(u, u + w]$, $N_{u+w}(i) - N_u(i)$, is approximately Poisson distributed with parameter $\lambda_i w$. If the system is a series system, and we make the same assumptions as above, it is also clear that the number of system failures in the interval $(u, u + w]$ is approximately Poisson distributed with parameter $\sum_{i=1}^{n} \lambda_i w$. The number of system failures in $[0, t]$, $N_t$, is approximately a Poisson process with intensity $\sum_{i=1}^{n} \lambda_i$.

If the system is highly available and the components have constant failure rates, the Poisson distribution (with the asymptotic rate $\lambda_\Phi$) will in fact also produce good approximations for more general systems. As motivation, we observe that $EN_{(u,u+w]}/w$ is approximately equal to the asymptotic system failure rate $\lambda_\Phi$, and $N_{(u,u+w]}$ is "nearly independent" of the history of $N$ up to $u$, noting that the process $\mathbf{X}$ frequently restarts itself probabilistically, i.e., $\mathbf{X}$ re-enters the state $(1, 1, \ldots, 1)$.

Refer to [24, 91] for Monte Carlo simulation studies of the accuracy of the Poisson approximation. As an illustration of the results obtained in these studies, consider a parallel system of two identical components where the failure rate $\lambda$ is equal to 0.05, the repair times are all equal to 1, and the expected number of system failures is equal to 5. This means, as shown below, that the time interval is about 1000 and the expected number of component failures is about 100. Using the definition of the system failure rate $\lambda_\Phi$ (cf. (4.41), p. 115) with $\mu_G = 1$, we obtain

$$
\begin{aligned}
\frac{EN_u}{u} &= \frac{5}{u} \approx \lambda_\Phi = 2\bar{A}_1 \frac{1}{\mu_{F_1} + \mu_{G_1}} = 2 \frac{\mu_G}{\frac{1}{\lambda} + \mu_G} \cdot \frac{1}{\frac{1}{\lambda} + \mu_G} \\
&\approx 2\lambda^2 = 0.005.
\end{aligned}
$$

Hence $u \approx 1000$ and $2\,EN_u(i) \approx 2\lambda u \approx 100$. Clearly, this is an approximate steady-state situation, and we would expect that the Poisson distribution gives an accurate approximation. The Monte Carlo simulations in [24] confirm this. The distance measure, which is defined as the maximum distance between the Poisson distribution (with mean $\lambda_\Phi u$) and the "true" distribution obtained by Monte Carlo simulation, is equal to 0.006. If we take instead $\lambda = 0.2$ and $EN_u = 0.2$, we find that the expected number of component failures is about 1. Thus, we are far away from a steady-state situation and as expected the distance measure is larger: 0.02. But still the Poisson approximation produces relatively accurate results.

In the following we look at the problem of establishing *formalized* asymptotic results for the distribution of the number of system failures. We first consider the interval reliability.

### 4.4.1 Asymptotic Analysis for the Time to the First System Failure

The above discussion indicates that the interval reliability $A[0, u]$, defined by $A[0, u] = P(N_u = 0)$, is approximately exponentially distributed for highly available systems comprising components with exponentially distributed lifetimes. This result can also be formulated as a limiting result as shown in the theorem below. It is assumed that the process $\mathbf{X}$ is a regenerative process with regenerative state $(1, 1, \ldots, 1)$. The variable $S$ denotes the length of the first renewal cycle of the process $\mathbf{X}$, i.e., the time until the process returns to state $(1, 1, \ldots, 1)$. Let $T_\Phi$ denote the time to the first system failure and $q$ the probability that a system failure occurs in a renewal cycle, i.e.,

$$q = P(N_S \geq 1) = P(T_\Phi < S).$$

For $q \in (0, 1)$, let $P_0$ and $P_1$ denote the conditional probability given $N_S = 0$ and $N_S \geq 1$, i.e., $P_0(\cdot) = P(\cdot | N_S = 0)$ and $P_1(\cdot) = P(\cdot | N_S \geq 1)$. The corresponding expectations are denoted $E_0$ and $E_1$. Furthermore, let $c_{0S}^2 = [E_0 S^2 / (E_0 S)^2] - 1$ denote the squared coefficient of variation of $S$ under $P_0$.

The notation $\overset{P}{\to}$ is used for convergence in probability and $\overset{D}{\to}$ for convergence in distribution, cf. Appendix A, p. 216. We write $\text{Exp}(t)$ for the exponential distribution with parameter $t$, $\text{Poisson}(t)$ for the Poisson distribution with mean $t$ and $\text{N}(\mu, \sigma^2)$ for the normal distribution with mean $\mu$ and variance $\sigma^2$.

For each component $i$ ($i \in \{1, 2, ..., n\}$) we assume that there is a sequence of uptime and downtime distributions $(F_{ij}, G_{ij})$, $j = 1, 2, \ldots$.

To simplify notation, we normally omit the index $j$. When assuming in the following that $\mathbf{X}$ is a regenerative process, it is tacitly understood for all $j \in \mathbb{N}$. We shall formulate conditions which guarantee that $\alpha T_\Phi$ is asymptotically exponentially distributed with parameter 1, where $\alpha$ is a suitable normalizing "factor" (more precisely, a normalizing sequence depending on $j$). The following factors will be studied: $q/E_0 S$, $q/ES$, $1/ET_\Phi$, and $\lambda_\Phi$. These factors are asymptotically equivalent under the conditions stated in the theorem below, i.e., the ratio of any two of these factors converges to one as $j \to \infty$. To motivate this, note that for a highly available system we have $ET_\Phi \approx E_0 S(1/q) \approx ES(1/q)$, observing that $E_0 S$ equals the length of a cycle having no system failures and $1/q$ equals the expected number of cycles until a system failure occurs (the number of such cycles is geometrically distributed with parameter $q$). We have $E_0 S \approx ES$ when $q$ is small. Note also that

$$\lambda_\Phi = \frac{E N_S}{ES} \tag{4.44}$$

by the Renewal Reward Theorem (Theorem 122, p. 250, in Appendix B). For a highly available system we have $E N_S \approx q$ and hence $\lambda_\Phi \approx q/ES$. Results from Monte Carlo simulations presented in [24] show that the factors $q/E_0 S$, $q/ES$, and $1/ET_\Phi$ typically give slightly better results (i.e., better fit to the exponential distribution) than the system failure rate $\lambda_\Phi$. From a computational point of view, however, $\lambda_\Phi$ is much more attractive than the other factors, which are in most cases quite difficult to compute. We therefore normally use $\lambda_\Phi$ as the normalizing factor.

The basic idea of the proof of the asymptotic exponentiality of $\alpha T_\Phi$ is as follows: If we assume that $\mathbf{X}$ is a regenerative process and the probability that a system failure occurs in a renewal cycle, i.e., $q$, is small (converges to zero), then the time to the first system failure will be approximately equal to the sum of a number of renewal cycles having no system failures; and this number of cycles is geometrically distributed with parameter $q$. Now if $q \to 0$ as $j \to \infty$, the desired result follows by using Laplace transformations. The result can be formulated in general terms as shown in the lemma below.

Note that series systems are excluded since such systems have $q = 1$. We will analyze series systems later in this section; see Theorem 54, p. 135.

**Lemma 45** *Let $S, S_i, i = 1, 2, \ldots$, be a sequence of non-negative i.i.d. random variables with distribution function $F(t)$ having finite mean $a$, $a > 0$ and finite variance, and let $\nu$ be a random variable independent of $(S_i)$, geometrically distributed with parameter $q$ $(0 < q \le 1)$, i.e., $P(\nu = k) = qp^{k-1}$, $k = 1, 2, \ldots$, $p = 1 - q$. Furthermore, let*

$$S^* = \sum_{i=1}^{\nu-1} S_i.$$

*Consider now a sequence $F_j$, $q_j$ $(j = 1, 2, \ldots)$ satisfying the above conditions for each $j$. Then if (as $j \to \infty$)*

$$q \to 0 \tag{4.45}$$

*and*

$$qc_S^2 \to 0, \tag{4.46}$$

*where $c_S^2$ denotes the squared coefficient of variation of $S$, we have (as $j \to \infty$)*

$$\frac{qS^*}{a} \xrightarrow{D} \mathrm{Exp}(1). \tag{4.47}$$

**Proof.** Let $\tilde{S}^* = qS^*/a$. By conditioning on the value of $\nu$, it is seen that the Laplace transform of $S^*$, $L_{S^*}(x) = Ee^{-xS^*}$, equals $q/[1 - pL(x)]$, where $L(x)$ is the Laplace transform of $S_i$. Let $\psi(x) = [L(x) - 1 + ax]/x$. Then

$$L_{S^*}(x) = \frac{q}{1 - p(1 - ax + x\psi(x))}.$$

We need to show that

$$L_{\tilde{S}^*}(x) = Ee^{-(qx/a)S^*} \to \frac{1}{1+x},$$

since the convergence theorem for Laplace transforms then give the desired result. Noting that

$$Ee^{-(qx/a)S^*} = \frac{1}{1 + px - (px/a)\psi(qx/a)},$$

we must require that

$$(x/a)\psi(qx/a) \to 0,$$

i.e.,

$$[L(qx/a) - 1 + qx]/q \to 0.$$

Using $ES = a$ and the inequalities $0 \le e^{-t} - 1 + t \le t^2/2$, we find that

$$0 \le [L(qx/a) - 1 + qx]/q \quad = \quad E[e^{-(qx/a)S} - 1 + (qx/a)S]/q$$

$$\le \quad E[(qx/a)S]^2/2q$$

$$= \quad \frac{x^2}{2}\frac{q}{a^2}ES^2$$

$$= \quad \frac{x^2}{2}q(1 + c_S^2).$$

The desired conclusion (4.47) follows now since $q \to 0$ and $qc_S^2 \to 0$ (assumptions (4.45) and (4.46)).    ∎

**Theorem 46** *Assume that $X$ is a regenerative process, and that $F_{ij}$ and $G_{ij}$ change in such a way that the following conditions hold (as $j \to \infty$):*

$$q \quad \to \quad 0, \tag{4.48}$$

$$qc_{0S}^2 \quad \to \quad 0, \tag{4.49}$$

$$\frac{qE_1S}{ES} \quad \to \quad 0, \tag{4.50}$$

$$E_1(N_S - 1) \quad \to \quad 0. \tag{4.51}$$

*Then*

$$A[0, u/\lambda_\Phi] \to e^{-u}, \ i.e., \ \lambda_\Phi T_\Phi \xrightarrow{D} \mathrm{Exp}(1). \tag{4.52}$$

**Proof.** Using Lemma 45, we first prove that under conditions (4.48)−(4.50) we have

$$\frac{T_\Phi q}{E_0 S} \xrightarrow{D} \mathrm{Exp}(1). \tag{4.53}$$

Let $\nu$ denote the renewal cycle index associated with the time of the first system failure, $T_\Phi$. Then it is seen that $T_\Phi$ has the same distribution as

$$\sum_{k=1}^{\nu-1} S_{0k} + W_\nu,$$

where $(S_{0k})$ and $(W_k)$ are independent sequences of i.i.d. random variables with

$$P(S_{0k} \le s) = P_0(S \le s)$$

and

$$P(W_k \le w) = P_1(T_\Phi \le w).$$

Both sequences are independent of $\nu$, which has a geometrical distribution with parameter $q = P(N_S \ge 1)$. Hence, (4.53) follows from Lemma 45 provided that

$$\frac{W_\nu q}{E_0 S} \xrightarrow{P} 0. \tag{4.54}$$

By a standard conditional probability argument it follows that

$$ES = (1-q)E_0 S + q E_1 S,$$

and by noting that

$$\frac{q E_1 T_\Phi}{E_0 S} = \frac{q E W}{E_0 S} \le \frac{q E_1 S}{E_0 S} = \frac{q E_1 S(1-q)}{ES - q E_1 S}$$

$$= \frac{\frac{q E_1 S}{ES}(1-q)}{1 - \frac{q E_1 S}{ES}} \to 0, \tag{4.55}$$

we see that (4.54) holds.

Using (4.44) we obtain

$$\frac{\lambda_\phi}{q/E_0 S} = \frac{\lambda_\phi}{q/ES}\frac{E_0 S}{ES}$$

$$= \frac{E N_S/ES}{q/ES}\frac{E_0 S}{ES}$$

$$= \frac{E N_S}{q}\frac{E_0 S}{ES}.$$

Now $E N_S/q = 1 + E_1(N_S - 1) \to 1$ in view of (4.51), and

$$\frac{E_0 S}{ES} = \frac{1 - q\frac{E_1 S}{ES}}{1 - q} \to 1$$

by (4.48) and (4.50). Hence the ratio of $\lambda_\phi$ and $q/E_0S$ converges to 1. Combining this with (4.53), the conclusion of the theorem follows. ■

**Remark 12** The above theorem shows that

$$\alpha T_\phi \xrightarrow{D} \text{Exp}(1)$$

for $\alpha$ equal to $\lambda_\phi$. But the result also holds for the normalizing factors $q/E_0S, q/ES$, and $1/ET_\phi$. For $q/E_0S$ and $q/ES$ this is seen from the proof of the theorem. To establish the result for $1/ET_\phi$, let

$$S^* = \sum_{i=1}^{\nu-1} S_{0i}.$$

Then $ES^* = E_0S(1-q)/q$, observing that the mean of $\nu$ equals $1/q$. It follows that

$$ET_\phi = E_0S(1-q)/q + E_1T_\phi,$$

which can be rewritten as

$$qET_\phi/E_0S = 1 - q + qE_1T_\phi/E_0S.$$

We see that the right-hand side of this expression converges to 1, remembering (4.48), (4.50), and (4.55). Hence, $1/ET_\phi$ is also a normalizing factor. Note that the condition (4.51) is not required if the normalizing factor equals either $q/E_0S, q/ES$, or $1/ET_\phi$.

We can conclude that the ratio between any of these normalizing factors converges to one if the conditions of the theorem hold true.

### 4.4.2  Some Sufficient Conditions

It is intuitively clear that if the components have constant failure rates, and the component unavailabilities converge to zero, then the conditions of Theorem 46 would hold. In Theorems 47 and 49 below this result will be formally established. We assume, for the sake of simplicity, that no single component is in series with the rest of the system. If there are one or more components in series with the rest of the system, we know that the time to failure of these components has an exact exponential distribution, and by independence it is straightforward to establish the limiting distribution of the total system.

Define

$$d = \sum_{i=1}^{n} \lambda_i \mu_{G_i}, \quad \tilde{\lambda} = \sum_{i=1}^{n} \lambda_i.$$

**Theorem 47** *Assume that the system has no components in series with the rest of the system, i.e., $\Phi(0_i, \mathbf{1}) = 1$ for $i = 1, 2, \ldots, n$. Furthermore, assume that component $i$ has an exponential lifetime distribution with failure rate $\lambda_i > 0$, $i = 1, 2, \ldots, n$.*

*If $d \to 0$ and there exist constants $c_1$ and $c_2$ such that $\lambda_i \le c_1 < \infty$ and $ER_i^2 \le c_2 < \infty$ for all $i$, then the conditions (4.48), (4.49), and (4.50) of Theorem 46 are met, and, consequently, $\alpha T_\Phi \xrightarrow{D} \mathrm{Exp}(1)$ for $\alpha$ equal to $q/E_0 S, q/ES,$ or $1/ET_\phi$.*

**Proof.** As will be shown below, it is sufficient to show that $q \to 0$ holds (condition (4.48)) and that there exists a finite constant $c$ such that

$$\tilde{\lambda}^2 E(S'')^2 \le c, \tag{4.56}$$

where $S''$ represents the "busy" period of the renewal cycle, which equals the time from the first component failure to the next regenerative point, i.e., to the time when the process again visits state $(1, 1, \ldots, 1)$. (The term "busy" period is taken from queueing theory. In the busy period at least one component is under repair.) Let $S'$ be an exponentially distributed random variable with parameter $\tilde{\lambda}$ representing the time to the first component failure. This means that we can write

$$S = S' + S''.$$

Assume that we have already proved (4.56). Then this condition and (4.48) imply (4.50), noting that

$$
\begin{aligned}
\frac{qE_1 S}{ES} &\le \tilde{\lambda} q E_1 S \\
&= \tilde{\lambda}(q E_1 S' + q E_1 S'') \\
&= q + \tilde{\lambda} q E[S'' | N_S \ge 1] \\
&= q + \tilde{\lambda} E[S'' I(N_S \ge 1)] \\
&\le q + \tilde{\lambda} q^{1/2} [E(S'')^2]^{1/2} \\
&= q + q^{1/2} [\tilde{\lambda}^2 E(S'')^2]^{1/2},
\end{aligned}
$$

where the last inequality follows from Schwartz's inequality. Furthermore, condition (4.56) together with (4.48) imply (4.49), noting that

$$
\begin{aligned}
c_{0S}^2 &\le \frac{E_0 S^2}{(E_0 S)^2} \\
&\le \tilde{\lambda}^2 E_0 S^2 \\
&= \tilde{\lambda}^2 E[S^2 I(N_S = 0)]/(1 - q) \\
&\le \tilde{\lambda}^2 E S^2 / (1 - q) \\
&= \tilde{\lambda}^2 \{E(S')^2 + E(S'')^2 + 2E[S'S'']\}/(1 - q) \\
&\le \tilde{\lambda}^2 \{(2/\tilde{\lambda}^2) + E(S'')^2 + 2(E(S')^2 E(S'')^2)^{1/2}\}/(1 - q) \\
&= \{2 + \tilde{\lambda}^2 E(S'')^2 + 2(2^{1/2})(\tilde{\lambda}^2 E(S'')^2)^{1/2}\}/(1 - q),
\end{aligned}
$$

where we again have used Schwartz's inequality. Alternatively, an upper bound on $E[S'S'']$ can be established using that $S'$ and $S''$ are independent:

$$E[S'S''] = ES'ES'' = (1/\tilde{\lambda})ES'' \leq (1/\tilde{\lambda})\{E(S'')^2\}^{1/2}.$$

Now, to establish (4.48), we note that with probability $\tilde{\lambda}_i = \lambda_i/\tilde{\lambda}$, the busy period begins at the time of the failure of component $i$. If, in the interval of repair of this component, none of the remaining components fails, then the busy period comes to an end when the repair is completed. Therefore, since there are no components in series with the rest of the system,

$$1 - q \geq \sum_{i=1}^{n} \tilde{\lambda}_i \int_0^\infty e^{-t(\tilde{\lambda}-\lambda_i)} dG_i(t),$$

where $G_i$ is the distribution of the repair time of component $i$. Hence,

$$q \leq \sum_{i=1}^{n} \tilde{\lambda}_i \int_0^\infty [1 - e^{-t(\tilde{\lambda}-\lambda_i)}] dG_i(t)$$

$$\leq \sum_{i=1}^{n} \lambda_i \int_0^\infty t\, dG_i(t) = d.$$

Consequently, $d \to 0$ implies $q \to 0$.

It remains to show (4.56). Clearly, the busy period will only increase if we assume that the flow of failures of component $i$ is a Poisson flow with parameter $\lambda_i$, i.e., we adjoin failures that arise according to a Poisson process on intervals of repair of component $i$, assuming that repair begins immediately for each failure. This means that the process can be regarded as an $M/G/\infty$ queueing process, where the Poisson input flow has parameter $\tilde{\lambda}$ and there are an infinite number of devices with servicing time distributed according to the law

$$G(t) = \sum_{i=1}^{n} \tilde{\lambda}_i G_i(t).$$

Note that the probability that a "failure is due to component $i$" equals $\tilde{\lambda}_i$. It is also clear that the busy period increases still more if, instead of an infinite number of servicing devices, we take only one, i.e., the process is a queueing process $M/G/1$. Thus, $E(S'')^2 \leq E(\tilde{S}'')^2$, where $\tilde{S}''$ is the busy period in a single-line system with a Poisson input flow $\tilde{\lambda}$ and servicing distribution $G(t)$. It is a well-known result from the theory of queueing processes (and branching processes) that the second-order moment of the busy period (extinction time) equals $ER_G^2/(1 - \tilde{\lambda}ER_G)^3$, where $R_G$ is the service time having distribution $G$, see, e.g., [89]. Hence, by introducing $d_2 = \sum_{i=1}^{n} \lambda_i ER_i^2$ we obtain

$$\tilde{\lambda}^2 E(S'')^2 \leq \frac{\tilde{\lambda}d_2}{(1-d)^3} \leq \frac{n^2 c_1^2 c_2}{(1-d)^3}.$$

The conclusion of the theorem follows.    ∎

We now give sufficient conditions for $E_1(N-1) \to 0$ (assumption (4.51) in Theorem 46).

We define

$$\check{\mu}_i = \sup_{0 \le t < t^*} \{E[R_{i1} - t | R_{i1} > t]\},$$

where $t^* = \sup\{t \in \mathbb{R}_+ : \bar{G}_i(t) > 0\}$. We see that $\check{\mu}_i$ expresses the maximum expected residual repair time of component $i$ We might have $\check{\mu}_i = \infty$, but we shall in the following restrict attention to the finite case. We know from Section 2.2, p. 39, that if $G_i$ has the NBUE property, then

$$\check{\mu}_i \le \mu_{G_i}.$$

If the repair times are bounded by a constant $c$, i.e., $P(R_{ik} \le c) = 1$, then $\check{\mu}_i \le c$. Let

$$\tilde{\mu} = \sum_{i=1}^{n} \check{\mu}_i.$$

**Lemma 48** *Assume that the lifetime of component $i$ is exponentially distributed with failure rate $\lambda_i$, $i = 1, 2, \ldots, n$. Then*

$$P_1(N_S \ge k) \le (\tilde{\lambda}\tilde{\mu})^{k-1}, k = 2, 3, \ldots. \tag{4.57}$$

**Proof.** The lemma will be shown by induction. We first prove that (4.57) holds true for $k = 2$. Suppose the first system failure occurs at time $t$. Let $L_t$ denote the number of component failures after $t$ until all components are again functioning for the first time. Furthermore, let $R_{it}$ denote the remaining repair time of component $i$ at time $t$ (put $R_{it} = 0$ if component $i$ is functioning at time $t$). Finally, let $V_t = \max_i R_{it}$ and let $G_{V_t}(v)$ denote the distribution function of $V_t$. Note that $L_t \ge 1$ implies that at least one component must fail in the interval $(t, t+V_t)$ and that the probability of at least one component failure in this interval increases if we replace the failed components at $t$ by functioning components. Using these observations and the inequality $1 - e^{-x} \le x$, we obtain

$$
\begin{aligned}
P(L_t \ge 1) &= \int_0^\infty P(L_t \ge 1 | V_t = v) dG_{V_t}(v) \\
&\le \int_0^\infty (1 - e^{-\tilde{\lambda}v}) dG_{V_t}(v) \\
&\le \tilde{\lambda} \int_0^\infty v dG_{V_t}(v) = \tilde{\lambda} E V_t \le \tilde{\lambda} E \sum_i R_{it} \\
&\le \tilde{\lambda}\tilde{\mu}.
\end{aligned}
$$

Since $N_S \ge 2$ implies $L_t \ge 1$, formula (4.57) is shown for $k = 2$ and $P_1$ conditional on the event that the first system failure occurs at time $t$.

Integrating over the failure time $t$, we obtain (4.57) for $k = 2$. Now assume that $P_1(N_S \geq k) \leq (\tilde{\lambda}\tilde{\mu})^{k-1}$ for a $k \geq 2$. We must show that

$$P_1(N_S \geq k + 1) \leq (\tilde{\lambda}\tilde{\mu})^k.$$

We have

$$
\begin{aligned}
P_1(N_S \geq k + 1) &= P_1(N_S \geq k + 1 | N_S \geq k) P_1(N_S \geq k) \\
&\leq P_1(N_S \geq k + 1 | N_S \geq k) \cdot (\tilde{\lambda}\tilde{\mu})^{k-1},
\end{aligned}
$$

thus it remains to show that

$$P_1(N_S \geq k + 1 | N_S \geq k) \leq \tilde{\lambda}\tilde{\mu}. \tag{4.58}$$

Suppose that the $k$th system failure in the renewal cycle occurs at time $t$. Then if at least one more system failure occurs in the renewal cycle, there must be at least one component failure before all components are again functioning, i.e., $L_t \geq 1$. Repeating the above arguments for $k = 2$, the inequality (4.58) follows.                                    ∎

**Remark 13** The inequality (4.57) states that the number of system failures in a renewal cycle when it is given that at least one system failure occurs is bounded in distribution by a geometrical random variable with parameter $\tilde{\lambda}\tilde{\mu}$ (provided this quantity is less than 1).

**Theorem 49** *Assume that the system has no components in series with the rest of the system. Furthermore, assume that component $i$ has an exponential lifetime distribution with failure rate $\lambda_i > 0$, $i = 1, 2, \ldots, n$. If $d' \to 0$, where $d' = \tilde{\lambda}\tilde{\mu}$, and there exist constants $c_1$ and $c_2$ such that $\lambda_i \leq c_1 < \infty$ and $ER_i^2 \leq c_2 < \infty$ for all $i$, then the conditions $(4.48) - (4.51)$ of Theorem 46 (p. 121) are all met, and, consequently, the limiting result (4.52) holds, i.e., $\lambda_\Phi T_\Phi \xrightarrow{D} \mathrm{Exp}(1)$.*

**Proof.** Since $d \leq d'$, it suffices to show that condition (4.51) holds under the given assumptions. But from (4.57) of Lemma 48 we have

$$E_1(N_S - 1) \leq d'/(1 - d'),$$

and the desired result follows.                                    ∎

The above results show that the time to the first system failure is approximately exponentially distributed with parameter $q/E_0 S \approx q/ES \approx 1/ET_\Phi \approx \lambda_\Phi$. For a system comprising highly available components, it is clear that $P(\mathbf{X}_t = 1)$ would be close to one, hence the above approximations for the interval reliability can also be used for an interval $(t, t + u]$.

### 4.4.3   Asymptotic Analysis of the Number of System Failures

For a highly available system, the downtimes will be small compared to the uptimes, and the time from when the system has failed until it returns to the state $(1, 1, \ldots, 1)$ will also be small. Hence, the above results also justify the Poisson process approximation for $N$. More formally, it can be shown that $N_{t/\alpha}$ converges in distribution to a Poisson distribution under the same assumptions as the first system failure time converges to the exponential distribution. Let $T_\Phi^*(k)$ denote the time between the $(k-1)$th and the $k$th system failure. From this sequence we define an associated sequence $T_\Phi(k)$ of i.i.d. variables, distributed as $T_\Phi$, by letting $T_\Phi(1) = T_\Phi^*(1)$, $T_\Phi(2)$ be equal to the time to the first system failure following the first regenerative point after the first system failure, etc. Then it is seen that

$$T_\Phi(1) + T_\Phi(2)(1 - I(N_{(1)} \geq 2)) \leq T_\Phi^*(1) + T_\Phi^*(2) \leq T_\Phi(1) + T_\Phi(2) + S_\nu,$$

where $N_{(1)} =$ equals the number of system failures in the first renewal cycle having one or more system failures, and $S_\nu$ equals the length of this cycle ($\nu$ denotes the renewal cycle index associated with the time of the first system failure). For $\alpha$ being one of the normalizing factors (i.e., $q/E_0S$, $q/ES$, $1/ET_\Phi$, or $\lambda_\Phi$), we will prove that $\alpha T_\Phi(2)I(N_{(1)} \geq 2)$ converges in probability to zero. It is sufficient to show that $P(N_{(1)} \geq 2) \to 0$ noting that

$$P(\alpha T_\Phi(2)I(N_{(1)} \geq 2) > \epsilon) \leq P(N_{(1)} \geq 2).$$

But

$$P(N_{(1)} \geq 2) = P_1(N_S \geq 2) \leq E_1(N_S - 1),$$

where the last expression converges to zero in view of (4.51), p. 121. The distribution of $S_\nu$ is the same as the conditional probability of the cycle length given a system failure occurs in the cycle, cf. Theorem 46 and its proof. Thus, if (4.48)–(4.51) hold, it follows that $\alpha(T_\Phi^*(1) + T_\Phi^*(2))$ converges in distribution to the sum of two independent exponentially distributed random variables with parameter 1, i.e.,

$$
\begin{aligned}
P(N_{t/\alpha} \geq 2) &= P(\alpha(T_\Phi^*(1) + T_\Phi^*(2)) \leq t) \\
&\to 1 - e^{-t} - te^{-t}.
\end{aligned}
$$

Similarly, we establish the general distribution. We summarize the result in the following theorem.

**Theorem 50** *Assume that* $\mathbf{X}$ *is a regenerative process, and that* $F_{ij}$ *and* $G_{ij}$ *change in such a way that (as* $j \to \infty$*) the conditions* (4.48) $-$ (4.50) *hold. Then (as* $j \to \infty$*)*

$$N_{t/\alpha} \xrightarrow{D} \text{Poisson}(t), \tag{4.59}$$

*where $\alpha$ is a normalizing factor that equals either $q/E_0 S$, $q/ES$, or $1/ET_\Phi$. If, in addition, condition (4.51) holds, then (4.59) is true with $\alpha = \lambda_\Phi$.*

Results from Monte Carlo simulations [24] indicate that the asymptotic system failure rate $\lambda_\Phi$ is normally preferable as parameter in the Poisson distribution when the expected number of system failures is not too small (less than one). When the expected number of system failures is small, the factor $1/ET_\Phi$ gives slightly better results. The system failure rate is however easier to compute.

## *Asymptotic Normality*

Now we turn to a completely different way to approximate the distribution of $N_t$. Above, the up and downtime distributions are assumed to change such that the system availability increases and after a time rescaling $N_t$ converges to a Poisson variable. Now we leave the up and downtime distribution unchanged and establish a central limit theorem as $t$ increases to infinity. The theorem generalizes (4.16), p. 106.

**Theorem 51** *If* $\mathbf{X}$ *is a regenerative process with cycle length $S$, $\mathrm{Var}[S] < \infty$ and $\mathrm{Var}[N_S] < \infty$, then as $t \to \infty$,*

$$\sqrt{t}\left(\frac{N_{u+t} - N_u}{t} - \lambda_\Phi\right) \xrightarrow{D} \mathrm{N}(0, \gamma_\Phi^2),$$

*where*

$$\gamma_\Phi^2 ES = \mathrm{Var}[N_S - \lambda_\Phi S]. \tag{4.60}$$

**Proof.** Noting that the system failure rate $\lambda_\Phi$ is given by

$$\lambda_\Phi = \frac{EN_S}{ES}, \tag{4.61}$$

the result follows from Theorem 123, p. 251, in Appendix B.    ∎

Below we argue that if the system failure rate is small, then we have

$$\gamma_\Phi^2 \approx \lambda_\Phi.$$

We obtain

$$\gamma_\Phi^2 = \frac{\mathrm{Var}[N_S - \lambda_\Phi S]}{ES} = \frac{E(N_S - \lambda_\Phi S)^2}{ES}$$

$$\approx \frac{EN_S^2}{ES} \approx \frac{EN_S}{ES} = \lambda_\Phi,$$

where the last approximation follows by observing that if the system failure rate is small, then $N_S$ with a probability close to one is equal to the indicator function $I(N_S \geq 1)$. More formally, it is possible to show that under certain conditions, $\gamma_\Phi^2/\lambda_\Phi$ converges to one. We formulate the result in the following proposition.

**Proposition 52** *Assume* **X** *is a regenerative process with cycle length $S$ and that $F_{ij}$ and $G_{ij}$ change in such a way that conditions (4.48) − (4.50) of Theorem 46 (p. 121) hold (as $j \to \infty$). Furthermore, assume that (as $j \to \infty$)*

$$E_1(N_S - 1)^2 \to 0 \tag{4.62}$$

*and*

$$qc_S^2 \to 0, \tag{4.63}$$

*where $c_S^2$ denotes the squared coefficient of variation of $S$. Then (as $j \to \infty$)*

$$\frac{\gamma_\Phi^2}{\lambda_\Phi} \to 1.$$

**Proof.** Using (4.60) and writing $N$ in place of $N_S$ we get

$$
\begin{aligned}
\frac{\gamma_\Phi^2}{\lambda_\Phi} &= \frac{E(N - \lambda_\Phi S)^2}{\lambda_\Phi ES} \\
&= \frac{q^{-1}EN^2 + q^{-1}(\lambda_\Phi)^2 ES^2 - 2q^{-1}\lambda_\Phi E[NS]}{q^{-1}\lambda_\Phi ES} \\
&= \frac{E_1 N^2 + q^{-1}(\lambda_\Phi)^2 ES^2 - 2q^{-1}\lambda_\Phi E[NS]}{q^{-1}\lambda_\Phi ES}.
\end{aligned}
$$

Since the denominator converges to 1 (the denominator equals the ratio between two normalizing factors), the result follows if we can show that $E_1 N^2$ converges to 1 and all the other terms of the numerator converge to zero. Writing

$$E_1 N^2 = E_1[1 + (N - 1)]^2 = 1 + E_1(N - 1)^2 + 2E_1(N - 1)$$

and using condition (4.62), it is seen that $E_1 N$ converges to 1. Now consider the term $q^{-1}(\lambda_\Phi)^2 ES^2$. Using that $\lambda_\Phi = EN/ES$ (formula (4.61)) we obtain

$$
\begin{aligned}
q^{-1}(\lambda_\Phi)^2 ES^2 &= q^{-1}(EN/ES)^2 ES^2 = q^{-1}(EN)^2\{ES^2/(ES)^2\} \\
&= q(E_1 N)^2(1 + c_S^2) = q[1 + E_1(N - 1)]^2(1 + c_S^2).
\end{aligned}
$$

Letting $q \to 0$ (condition (4.48)), and applying (4.62) and (4.63), we see that $q^{-1}(\lambda_\Phi)^2 ES^2$ converges to zero. It remains to show that $q^{-1}\lambda_\Phi E[NS]$ converges to zero. But this is shown in the same way as the previous term, noting that

$$E[NS] \leq (EN^2)^{1/2}(ES^2)^{1/2}$$

by Schwartz's inequality. This completes the proof of the proposition. ∎

**Proposition 53** *Under the same conditions as formulated in Theorem 49, p. 127, the following limiting result holds true (as $j \to \infty$):*

$$\frac{\gamma_\Phi^2}{\lambda_\Phi} \to 1.$$

**Proof.** It is sufficient to show that conditions (4.62) and (4.63) hold. Condition (4.62) follows by using that under $P_1$, $N$ is bounded in distribution by a geometrical distribution random variable with parameter $d' = \tilde{\lambda}\tilde{\mu}$, cf. (4.57) of Lemma 48, p. 126. Note that for a variable $N$ that has a geometrical distribution with parameter $d'$ we have

$$E(N-1)^2 = \sum_{k=1}^{\infty} (k-1)^2 (d')^{k-1}(1-d')$$

$$= \frac{d'(1+d')}{(1-d')^2}.$$

From this equality it follows that $E_1(N_S - 1)^2 \to 0$ as $d' \to 0$. To establish (4.63) we can repeat the arguments in the proof of Theorem 47, p. 124, showing (4.49), observing that

$$c_S^2 \leq \frac{ES^2}{(ES)^2} \leq \tilde{\lambda}^2 ES^2.$$ ∎

For a parallel system of two components it is possible to establish simple expressions for some of the above quantities, such as $q$ and $ET_\Phi$.

## Parallel System of Two Identical Components

Consider a parallel system comprising two identical components having exponential life lengths with failure rate $\lambda$. Suppose one of the components has failed. Then we see that a system failure occurs, i.e., the number of system failures in the cycle is at least 1 ($N_S \geq 1$), if the operating component fails before the repair is completed. Consequently,

$$q = P(N_S \geq 1) = \int_0^\infty F(t)dG(t) = \int_0^\infty (1 - e^{-\lambda t})dG(t),$$

where $F(t) = P(T \leq t) = 1 - e^{-\lambda t}$ and $G(t) = P(R \leq t)$ equal the component lifetime and repair time distribution, respectively. It follows that

$$q \leq \int_0^\infty \lambda t dG(t) = \lambda \mu_G.$$

Thus for a parallel system comprising two identical components, it is trivially verified that the convergence of $\lambda\mu_G$ to zero implies that $q \to 0$. From the Taylor formula we have $1 - e^{-x} = x - \frac{1}{2}x^2 + x^3 O(1)$, $x \to 0$, where $|O(1)| \leq 1$. Hence, if $\lambda\mu_G \to 0$ and $ER^3/\mu_G^3$ is bounded by a finite constant, we have

$$
\begin{aligned}
q &= \lambda\mu_G - \frac{\lambda^2}{2}ER^2 + \lambda^3 ER^3 O(1) \\
&= \lambda\mu_G - \frac{(\lambda\mu_G)^2}{2}(1 + c_G^2) + o((\lambda\mu_G)^2),
\end{aligned}
$$

where $c_G^2$ denotes the squared coefficient of variation of $G$ defined by $c_G^2 = \mathrm{Var}R/\mu_G^2$. We can conclude that if $\lambda\mu_G$ is small, then comparing distributions $G$ with the same mean, those with a large variance exhibit a small probability $q$.

If we instead apply the Taylor formula $1 - e^{-x} = x - x^2 O(1)$, we can write

$$
q = \lambda\mu_G + o(\lambda\mu_G), \quad \lambda\mu_G \to 0.
$$

For this example it is also possible to establish an explicit formula for $E_0 S$. It is seen that

$$
E_0 S = E\min\{T_1, T_2\} + E[R|R < T],
$$

where $T_1$ and $T_2$ are the times to failure of component 1 and 2, respectively. But

$$
E\min\{T_1, T_2\} = \frac{1}{2\lambda}
$$

and

$$
\begin{aligned}
E[R|R < T] &= E[RI(R < T)]/(1 - q) \\
&= \int_0^\infty re^{-\lambda r}\, dG(r)/(1 - q).
\end{aligned}
$$

This gives

$$
E_0 S = \frac{1}{2\lambda} + \frac{1}{1 - q}\int_0^\infty re^{-\lambda r}\, dG(r).
$$

From the Taylor formula we have $e^{-x} = 1 - xO(1)$, $x \to 0$, where $|O(1)| \leq 1$. Using this and noting that

$$
\int_0^\infty re^{-\lambda r}\, dG(r) = \mu_G[1 + \lambda\mu_G(c_G^2 + 1)O(1)],
$$

it can be shown that if the failure rate $\lambda$ and the squared coefficient of variation $c_G^2$ are bounded by a finite constant, then the normalizing factor $q/E_0 S$ is asymptotically given by

$$
\frac{q}{E_0 S} = 2\lambda^2\mu_G + o(\lambda\mu_G), \quad \lambda\mu_G \to 0.
$$

Now we will show that the system failure rate $\lambda_\Phi$, defined by (4.41), p. 115, is also approximately equal to $2\lambda^2\mu_G$. First note that the unavailability of a component, $\bar{A}$, is given by $\bar{A} = \lambda\mu_G/(1+\lambda\mu_G)$. It follows that

$$\lambda_\Phi = \frac{2\bar{A}}{\lambda^{-1} + \mu_G} = 2\lambda^2\mu_G + o(\lambda\mu_G), \quad \lambda\mu_G \to 0, \qquad (4.64)$$

provided that the failure rate $\lambda$ is bounded by a finite constant.

Next we will compute the exact distribution and mean of $T_\Phi$. Let us denote this distribution by $F_{T_\Phi}(t)$. In the following $F_X$ denotes the distribution of any random variable $X$ and $F_{iX}(t) = P_i(X \le t), i = 0, 1$, where $P_0(\cdot) = P(\cdot|N_S = 0)$ and $P_1(\cdot) = P(\cdot|N_S \ge 1)$. Observe that the length of a renewal cycle $S$ can be written as $S' + S''$, where $S'$ represents the time to the first failure of a component, and $S''$ represents the "busy" period, i.e., the time from when one component has failed until the process returns to the best state $(1,1)$. The variables $S'$ and $S''$ are independent and $S'$ is exponentially distributed with rate $\tilde{\lambda} = 2\lambda$. Now, assume a component has failed. Let $R$ denote the repair time of this component and let $T$ denote the time to failure of the operating component. Then

$$F_{1T}(t) = P(T \le t|T \le R) = \frac{1}{q}\int_0^\infty (1 - e^{-\lambda(t \wedge r)})dG(r),$$

where $a \wedge b$ denotes the minimum of $a$ and $b$. Furthermore,

$$F_{0R}(t) = P(R \le t|R < T) = \frac{1}{\bar{q}}\int_0^t e^{-\lambda r}dG(r),$$

where $\bar{q} = 1 - q$. Now, by conditioning on whether a system failure occurs in the first renewal cycle or not, we obtain

$$\begin{aligned} F_{T_\Phi}(t) &= qP(T_\Phi \le t|N_S \ge 1) + \bar{q}P(T_\Phi \le t|N_S = 0) \\ &= qF_{1T_\Phi}(t) + \bar{q}F_{0T_\Phi}(t). \end{aligned} \qquad (4.65)$$

To find an expression for $F_{1T_\Phi}(t)$ we use a standard conditional probability argument, yielding

$$\begin{aligned} F_{1T_\Phi}(t) &= \int_0^t P_1(T_\Phi \le t|S' = s)dF_{S'}(s) \\ &= \int_0^t P(T \le t - s|T \le R)dF_{S'}(s) \\ &= \int_0^t F_{1T}(t-s)dF_{S'}(s). \end{aligned}$$

Consider now $F_{0T_\Phi}(t)$. By conditioning on $S = s$, we obtain

$$\begin{aligned} F_{0T_\Phi}(t) &= \int_0^t P_0(T_\Phi \le t|S = s)dF_{0S}(s) \\ &= \int_0^t F_{T_\Phi}(t-s)dF_{0S}(s). \end{aligned}$$

Inserting the above expressions into (4.65) gives

$$F_{T_\Phi}(t) = h(t) + \bar{q} \int_0^t F_{T_\Phi}(t-s)dF_{0S}(s),$$

where

$$h(t) = q \int_0^t F_{1T}(t-s)dF_{S'}(s). \tag{4.66}$$

Hence, $F_{T_\Phi}(t)$ satisfies a renewal equation with the defective distribution $\bar{q}F_{0S}(s)$, and arguing as in the proof of Theorem 111, p. 245, in Appendix B, it follows that

$$F_{T_\Phi}(t) = h(t) + \int_0^t h(t-s)dM_0(s), \tag{4.67}$$

where the renewal function $M_0(s)$ equals

$$\sum_{j=1}^{\infty} \bar{q}^j F_{0S}^{*j}(s).$$

Noting that $F_{0S} = F_{S'} * F_{0R}$, the Laplace transform of $S'$ equals $2\lambda/(2\lambda+v)$, $\bar{q} = L_G(\lambda)$ and $L_{F_{0R}}(v) = L_G(v+\lambda)/L_G(\lambda)$, we see that the Laplace transform of $M_0$ takes the form

$$L_{M_0}(v) = \frac{\bar{q}\frac{2\lambda}{2\lambda+v}L_{F_{0R}}(v)}{1 - \bar{q}\frac{2\lambda}{2\lambda+v}L_{F_{0R}}(v)} = \frac{\frac{2\lambda}{2\lambda+v}L_G(v+\lambda)}{1 - \frac{2\lambda}{2\lambda+v}L_G(v+\lambda)}.$$

It is seen that the Laplace transform of $F_{1T}$ is given by

$$L_{F_{1T}}(v) = \frac{1}{1 - L_G(\lambda)}(1 - L_G(v+\lambda))\frac{\lambda}{\lambda+v}.$$

Now using (4.67) and (4.66) and the above expressions for the Laplace transform we obtain the following simple formula for $L_{F_{T_\Phi}}$:

$$L_{F_{T_\Phi}}(v) = \frac{2\lambda^2}{\lambda+v} \cdot \frac{1 - L_G(v+\lambda)}{v + 2\lambda(1 - L_G(v+\lambda))}.$$

The mean $ET_\Phi$ can be found from this formula, or alternatively by using a direct renewal argument. We obtain

$$
\begin{aligned}
ET_\Phi &= ES' + E(T_\Phi - S') \\
&= \frac{1}{2\lambda} + E\min\{R,T\} + (1-q)ET_\Phi,
\end{aligned}
$$

noting that the time one component is down before system failure occurs or the renewal cycle terminates equals $\min\{R,T\}$. If a system failure does not occur, the process starts over again. It follows that

$$ET_\Phi = \frac{1}{2q\lambda} + \frac{E\min\{R,T\}}{q}.$$

Note that

$$E\min\{R,T\} = \int_0^\infty \bar{F}(t)\bar{G}(t)dt = \int_0^\infty e^{-\lambda t}\bar{G}(t)dt.$$

It is also possible to write

$$ET_\Phi = \frac{3}{2\lambda}\frac{1 - \frac{2}{3}L_G(\lambda)}{1 - L_G(\lambda)}.$$

Now using the Taylor formula $e^{-x} = 1 - xO(1)$, $|O(1)| \leq 1$, we obtain

$$E\min\{R,T\} = \int_0^\infty e^{-\lambda t}\bar{G}(t)dt = \mu_G + \lambda\mu_G^2(c_G^2 + 1)O(1),$$

where $c_G^2$ is the squared coefficient of variation of $G$. From this it can be shown that the normalizing factor $1/ET_\Phi$ can be written in the same form as the other normalizing factors:

$$\frac{1}{ET_\Phi} = 2\lambda^2\mu_G + o(\lambda\mu_G), \quad \lambda\mu_G \to 0,$$

assuming that $\lambda$ and $c_G^2$ are bounded by a finite constant.

## Asymptotic Analysis for Systems having Components in Series with the Rest of the System

We now return to the general asymptotic analysis. Remember that $d = \sum \lambda_i\mu_{G_i}$ and $\tilde{\lambda} = \sum \lambda_i$. So far we have focused on nonseries systems (series system have $q = 1$). Below we show that a series system also has a Poisson limit under the assumption that the lifetimes are exponentially distributed. We also formulate and prove a general asymptotic result for the situation that we have some components in series with the rest of the system. A component is in series with the rest of the system if $\Phi(0_i, \mathbf{1}) = 0$.

**Theorem 54** *Assume that $\Phi$ is a series system and the lifetimes are exponentially distributed. Let $\lambda_i$ be the failure rate of component $i$. If $d \to 0$ (as $j \to \infty$), then (as $j \to \infty$)*

$$N_{t/\tilde{\lambda}} \xrightarrow{D} \text{Poisson}(t).$$

**Proof.** Let $N_t^P(i)$ be the Poisson process with intensity $\lambda_i$ generated by the consecutive uptimes of component $i$. Then it is seen that

$$\sum_{i=1}^n N_{t/\tilde{\lambda}}^P(i) - D = N_{t/\tilde{\lambda}} \leq \sum_{i=1}^n N_{t/\tilde{\lambda}}^P(i),$$

where

$$D = \sum_{i=1}^{n} N_{t/\tilde{\lambda}}^{P}(i) - N_{t/\tilde{\lambda}}.$$

We have $D \geq 0$ and hence the conclusion of the theorem follows if we can show that $ED \to 0$, since then $D$ converges in probability to zero. Note that $\sum_{i=1}^{n} N_{t/\tilde{\lambda}}^{P}(i)$ is Poisson distributed with mean

$$E \sum_{i=1}^{n} N_{t/\tilde{\lambda}}^{P}(i) = \sum_{i=1}^{n} (t/\tilde{\lambda})\lambda_i = t. \tag{4.68}$$

From (4.36) of Corollary 41, p. 114, we have

$$EN_{t/\tilde{\lambda}} = \sum_{i=1}^{n} \int_{0}^{t/\tilde{\lambda}} [h(1_i, \mathbf{A}(s)) - h(0_i, \mathbf{A}(s))] \lambda_i A_i(s)ds,$$

which gives

$$
\begin{aligned}
EN_{t/\tilde{\lambda}} &= \sum_{i=1}^{n} \int_{0}^{t/\tilde{\lambda}} \prod_{k \neq i} A_k(s)\lambda_i A_i(s)ds \\
&= \tilde{\lambda} \int_{0}^{t/\tilde{\lambda}} \prod_{k=1}^{n} A_k(s)ds.
\end{aligned}
$$

Using this expression together with (4.68), the inequalities $1 - \prod_i(1 - q_i) \leq \sum_i q_i$, and the component unavailability bound (4.22) of Proposition 34, p. 107, $(\bar{A}_i(t) \leq \lambda_i \mu_{G_i})$, we find that

$$
\begin{aligned}
ED &= \tilde{\lambda} \int_{0}^{t/\tilde{\lambda}} [1 - \prod_{i=1}^{n} A_i(s)]ds \\
&\leq \tilde{\lambda} \int_{0}^{t/\tilde{\lambda}} \sum_{i=1}^{n} \bar{A}_i(s)ds \\
&\leq \tilde{\lambda}(t/\tilde{\lambda}) \sum_{k=1}^{n} \lambda_i \mu_{G_i} \\
&= td.
\end{aligned}
$$

Now if $d \to 0$, we see that $ED \to 0$ and the proof is complete.    ∎

**Remark 14** Arguing as in the proof of the theorem above it can be shown that if $a_j \to a$ as $j \to \infty$, then

$$N_{a_j t/\tilde{\lambda}} \xrightarrow{D} \text{Poisson}(ta).$$

Observe that $\sum_{i=1}^{n} N_{a_j t/\tilde{\lambda}}^{P}(i)$ is Poisson distributed with parameter $a_j t$ and as $j \to \infty$ this variable converges in distribution to a Poisson variable with parameter $at$.

**Theorem 55** *Assume that the components have exponentially distributed lifetimes, and let $\lambda_i$ be the failure rate of component $i$. Let $A$ denote the set of components that are in series with the rest of the system, and let $B$ be the remaining components. Let $N^A, \tilde{\lambda}^A$, etc., denote the number of system failures, the total failure rate, etc., associated with the series system comprising the components in $A$. Similarly define $N^B, \alpha^B, d^B$, etc., for the system comprising the components in $B$. Assume that the following conditions hold (as $j \to \infty$) :*

1. *$d \to 0$*

2. *The conditions of Theorem 46, p. 121, i.e., (4.48) − (4.51), hold for system $B$*

3. *$\tilde{\lambda}^A/\alpha^B \to a$.*

*Then (as $j \to \infty$)*

$$N_{t/\alpha^B} \xrightarrow{D} \text{Poisson}(t(1+a)).$$

**Remark 15** The conditions of Theorem 46 ensure that

$$N_{t/\alpha^B}^{B} \xrightarrow{D} \text{Poisson}(t),$$

cf. Theorem 50, p. 128. Theorem 49, p. 127, gives sufficient conditions for (4.48)−(4.51).

**Proof.** First note that

$$N_{t/\alpha^B} \leq N_{t/\alpha^B}^{A} + N_{t/\alpha^B}^{B} = N_{a_j t/\tilde{\lambda}^A}^{A} + N_{t/\alpha^B}^{B},$$

where $a_j = \tilde{\lambda}^A/\alpha^B$. Now in view of Remark 14 above and the conditions of the theorem, it is sufficient to show that $D^*$, defined as the expected number of times system $A$ fails while system $B$ is down, or vice versa, converges to zero. But noting that the probability that system $A$ $(B)$ is not functioning is less than or equal to $d$ (the unreliability of a monotone system is bounded by the sum of the component unreliabilities, which in its turn is bounded by $d$, cf. (4.22), p. 107), it is seen that

$$
\begin{aligned}
D^* &\leq d[EN_{a_j t/\tilde{\lambda}^A}^{A} + EN_{t/\alpha^B}^{B}] \leq d[\tilde{\lambda}^A a_j t/\tilde{\lambda}^A + EN_{t/\alpha^B}^{B}] \\
&= d[a_j t + EN_{t/\alpha^B}^{B}].
\end{aligned}
$$

To find a suitable bound on $EN_{t/\alpha^B}^{B}$, we need to refer to the argumentation in the proof of Theorem 59, formulas (4.88) and (4.93), p. 148. Using these results we can show that $EN^B_{t/\alpha^B} \to t$. Hence, $D^* \to 0$ and the theorem is proved.  ∎

## 4.5  Downtime Distribution Given System Failure

In this section we study the downtime distribution of the system given that a failure has occurred. We investigate the downtime distribution given a failure at time $t$, the asymptotic (steady-state) distribution obtained by letting $t \to \infty$, and the distribution of the downtime following the $i$th system failure. Recall that $\Phi$ represents the structure function of the system and $N_t$ the number of system failures in $[0, t]$. Component $i$ generates an alternating renewal process with uptime distribution $F_i$ and downtime distribution $G_i$, with means $\mu_{F_i}$ and $\mu_{G_i}$, respectively. The lifetime distribution $F_i$ is absolutely continuous with a failure rate function $\lambda_i$. The $n$ component processes are independent.

Let $\Delta N_t = N_t - N_{t-}$. Define $G_\Phi(\cdot, t)$ as the downtime distribution at time $t$, i.e.,

$$G_\Phi(y, t) = P(Y \le y | \Delta N_t = 1),$$

where $Y$ is a random variable representing the downtime (we omit the dependency on $t$). The asymptotic (steady-state) downtime distribution is given by

$$G_\Phi(y) = \lim_{t \to \infty} G_\Phi(y, t),$$

assuming that the limit exists. It turns out that it is quite simple to establish the asymptotic (steady-state) downtime distribution of a parallel system, so we first consider this category of systems.

### 4.5.1  Parallel System

Consider a parallel system comprising $n$ stochastically identical components, with repair time distribution $G$. Since a system failure coincides with one and only one component failure, we have

$$P(Y > y | \Delta N_t = 1) = \bar{G}(y)[\bar{G}_{\alpha_t}(y)]^{n-1},$$

where $G_{\alpha_t}(y) = P(\alpha_t(i) > y | X_i(t) = 0)$ denotes the distribution of the forward recurrence time in state 0 of a component. But we know from (4.14) and (4.15), p. 105, that the asymptotic distribution of $G_{\alpha_t}(y)$ is given by

$$\lim_{t \to \infty} \bar{G}_{\alpha_t}(y) = \frac{\int_y^\infty \bar{G}(x)dx}{\mu_G} = \bar{G}_\infty(y). \tag{4.69}$$

Thus we have proved the following theorem.

**Theorem 56** *For a parallel system of $n$ identical components, the asymptotic (steady-state) downtime distribution given system failure, equals*

$$G_\Phi(y) = 1 - \bar{G}(y) \left[ \frac{\int_y^\infty \bar{G}(x)dx}{\mu_G} \right]^{n-1}. \tag{4.70}$$

Next we consider a parallel system of not necessarily identical components. We have the following result.

**Theorem 57** *Let $m_i(t)$ be the renewal density function of $M_i(t)$, and assume that $m_i(t)$ is right-continuous and satisfies*

$$\lim_{t \to \infty} m_i(t) = \frac{1}{\mu_{F_i} + \mu_{G_i}}. \tag{4.71}$$

*For a parallel system of not necessarily identical components, the asymptotic (steady-state) downtime distribution given system failure equals*

$$G_\Phi(y) = \sum_{i=1}^n c_i \left[ 1 - \bar{G}_i(y) \prod_{k \neq i} \frac{\int_y^\infty \bar{G}_k(x)\, dx}{\mu_{G_k}} \right],$$

*where*

$$c_i = \frac{1/\mu_{G_i}}{\sum_{k=1}^n 1/\mu_{G_k}} \tag{4.72}$$

*denotes the asymptotic (steady-state) probability that component $i$ causes a system failure.*

**Proof.** The proof follows the lines of the proof of Theorem 56, the difference being that we have to take into consideration which component causes system failure and the probability of this event given system failure. Clearly,

$$1 - \bar{G}_i(y) \prod_{k \neq i} \frac{\int_y^\infty \bar{G}_k(x)\, dx}{\mu_{G_k}}$$

equals the asymptotic downtime distribution given that component $i$ causes system failure. Hence it suffices to show (4.72). Since the system failure rate $\lambda_\Phi$ is given by $\lambda_\Phi = \sum_{i=1}^n \lambda_\Phi^{(i)}$, where

$$\lambda_\Phi^{(i)} = \prod_{k \neq i} \bar{A}_k \frac{1}{\mu_{F_i} + \mu_{G_i}}$$

represents the expected number of system failures per unit of time caused by failures of component $i$, an intuitive argument gives that the asymptotic (steady-state) probability that component $i$ causes system failure equals

$$\frac{\lambda_\Phi^{(i)}}{\lambda_\Phi} = \frac{\frac{1}{\mu_{F_i}+\mu_{G_i}}\prod_{k\neq i}\bar{A}_k}{\sum_{l=1}^n \frac{1}{\mu_{F_l}+\mu_{G_l}}\prod_{k\neq l}\bar{A}_k}$$

$$= \frac{\frac{1}{\mu_{G_i}}\prod_{k=1}^n \bar{A}_k}{\sum_{l=1}^n \frac{1}{\mu_{G_l}}\prod_{k=1}^n \bar{A}_k} = c_i.$$

To establish sufficient conditions for this result to hold, we need to carry out a somewhat more formal proof. Let $c_i(t)$ be defined as the conditional probability that component $i$ causes system failure given that the system failure occurs at time $t$. For each $h > 0$ let

$$N^c_{[t,t+h)}(i) = \int_{[t,t+h)} (\Phi(1_i, \mathbf{X}_s) - \Phi(0_i, \mathbf{X}_s))dN_s(i)$$

$$N^c_{[t,t+h)} = \sum_{i=1}^n N^c_{[t,t+h)}(i).$$

Then

$$c_i(t) = \lim_{h\to 0+} \frac{P(N^c_{[t,t+h)}(i) = 1)}{P(N^c_{[t,t+h)} = 1)}$$

$$= \lim_{h\to 0+} \frac{\frac{1}{h}EN^c_{[t,t+h)}(i) - o_i(1)}{\frac{1}{h}EN^c_{[t,t+h)} - o(1)}, \qquad (4.73)$$

where

$$o_i(1) = E[N^c_{[t,t+h)}(i))I(N^c_{[t,t+h)}(i) \geq 2)]/h$$

and

$$o(1) = E[N^c_{[t,t+h)})I(N^c_{[t,t+h)} \geq 2)]/h.$$

Hence it remains to study the limit of the ratio of the first terms of (4.73). Using that

$$EN^c_{[t,t+h)}(i) = \int_{[t,t+h)} (h(1_i, \mathbf{A}(s)) - h(0_i, \mathbf{A}(s)))m_i(s)ds,$$

where $A_i(s) = P(X_s(i) = 1)$ equals the availability of component $i$ at time $s$, it follows that

$$c_i(t) = \frac{\{h(1_i, \mathbf{A}(t)) - h(0_i, \mathbf{A}(t))\}m_i(t)}{\sum_{k=1}^n \{h(1_k, \mathbf{A}(t)) - h(0_k, \mathbf{A}(t))\}m_k(t)}.$$

From this expression, we see that $\lim_{t\to\infty} c_i(t) = c_i$ provided that

$$\lim_{t\to\infty} m_i(t) = \frac{1}{\mu_{F_i} + \mu_{G_i}}.$$

This completes the proof of the theorem. ∎

**Remark 16** 1. From renewal theory (see Theorem 118, p. 248, in Appendix $B$) sufficient conditions can be formulated for the limiting result (4.71) to hold true. For example, if the renewal cycle lengths $T_{ik} + R_{ik}$ have a density function $h$ with $h(t)^p$ integrable for some $p > 1$, and $h(t) \to 0$ as $t \to \infty$, then $M_i$ has a density $m_i$ such that $m_i(t) \to 1/(\mu_{F_i} + \mu_{G_i})$ as $t \to \infty$. If component $i$ has an exponential lifetime distribution with parameter $\lambda_i$, then we know that $m_i(t) = \lambda_i A_i(t)$ (cf. (4.18), p. 107), which converges to $1/(\mu_{F_i} + \mu_{G_i})$.

2. From the above proof it is seen that the downtime distribution at time $t$, $G_\Phi(y, t)$, is given by

$$G_\Phi(y, t) = \sum_{i=1}^{n} c_i(t) \left[ 1 - \bar{G}_i(y) \prod_{k \neq i} \bar{G}_{k\alpha_t}(y) \right].$$

### 4.5.2   General Monotone System

Consider now an arbitrary monotone system comprising the minimal cut sets $K_k$, $k = 1, 2, \ldots, k_0$. No simple formula exists for the downtime distribution in this case. But for highly available systems the following formula can be used to approximate the downtime distribution:

$$\sum_k r_k G_{K_k}(y),$$

where

$$r_k = \frac{\lambda_{K_k}}{\sum_l \lambda_{K_l}}.$$

Here $\lambda_{K_k}$ and $G_{K_k}$ denote the asymptotic (steady-state) failure rate of minimal cut set $K_k$ and the asymptotic (steady-state) downtime distribution of minimal cut set $K_k$, respectively, when this set is considered in isolation (i.e., we consider the parallel system comprising the components in $K_k$). We see that $r_k$ is approximately equal to the probability that minimal cut set $K_k$ causes system failure. Refer to [25, 80] for more detailed analyses in the general case. In [80] it is formally proved that the asymptotic downtime distribution exists and is equal to the steady-state downtime distribution.

### 4.5.3   Downtime Distribution of the $i$th System Failure

The above asymptotic (steady-state) formulas for $G_\Phi$ give in most cases good approximations to the downtime distribution of the $i$th system failure, $i \in \mathbb{N}$. Even for the first system failure observed, the asymptotic formulas produce relatively accurate approximations. This is demonstrated by Monte Carlo simulations in [25]. An example is given below. Let the distance measure $D_i(y)$ be defined by

$$D_i(y) = |G_\Phi(y) - \hat{G}_{i,\Phi}(y)|,$$

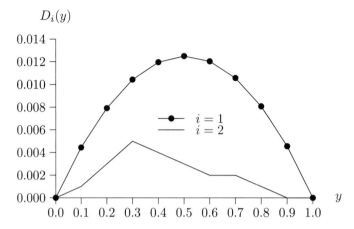

FIGURE 4.3. The Distance $D_i(y), i = 1, 2$, as a Function of $y$ for a Parallel System of Two Components with Constant Repair Times, $\mu_G = 1, \lambda = 0.1$

where $\hat{G}_{i,\Phi}(y)$ equals the "true" downtime distribution of the $i$th system failure obtained by Monte Carlo simulations. In Figure 4.3 the distance measure of the first and second system failure have been plotted as a function of $y$ for a parallel system of two identical components with constant repair times and exponential lifetimes. As we can see from the figure, the distance is quite small; the maximum distance is about 0.012 for $i = 1$ and 0.004 for $i = 2$.

Only for some special cases are explicit expressions for the downtime distribution of the $i$th system failure known. Below we present such expressions for the downtime distribution of the first failure for a two-component parallel system of identical components with exponentially distributed lifetimes.

**Theorem 58** *For a parallel system of two identical components with constant failure rate $\lambda$ and repair time distribution $G$, the downtime distribution $G_{1,p2}(y)$ of the first system failure is given by*

$$G_{1,p2}(y) \;=\; 1 - \bar{G}(y)\frac{\int_0^\infty \int_0^s \bar{G}(y+s-x)dF(x)\,dF(s)}{\int_0^\infty \int_0^s \bar{G}(s-x)dF(x)\,dF(s)} \qquad (4.74)$$

$$=\; 1 - \bar{G}(y)\frac{\int_y^\infty [1 - e^{-\lambda(r-y)}]dG(r)}{\int_0^\infty [1 - e^{-\lambda r}]dG(r)}. \qquad (4.75)$$

**Proof.** Let $T_i$ and $R_i$ have distribution function $F$ and $G$, respectively, $i = 1, 2$, and let

$$Y = \min_{1 \le i \le 2}(T_i + R_i) - \max_{1 \le i \le 2}(T_i).$$

It is seen that the downtime distribution $G_{1,p2}(y)$ equals the conditional distribution of $Y$ given that $Y > 0$. The equality (4.74) follows if we can show that

$$P(Y > y) = \bar{G}(y) \int_0^\infty \int_0^s 2\bar{G}(y + s - x) dF(x) \, dF(s). \tag{4.76}$$

Consider the event that $T_i = s$, $T_j = x$, $R_i > y$, and $T_j + R_j > y + s$ for $x < s$ and $j \neq i$. For this event it holds that $Y$ is greater than $y$. The probability of this event, integrated over all $s$ and $x$, is given by

$$\int_0^\infty \int_0^s \bar{G}(y + s - x) \bar{G}(y) dF(x) dF(s).$$

By taking the union over $i = 1, 2$, we find that (4.76) holds.

But the double integral in (4.76) can be written as

$$\int_0^\infty 2 \int_0^s \bar{G}(y + s - x) d(1 - e^{-\lambda x}) d(1 - e^{-\lambda s})$$

$$= 1 - \int_0^\infty \int_0^s G(y + s - x) 2\lambda^2 e^{-\lambda(x+s)} dx ds$$

$$= 1 - \int_0^\infty \int_x^\infty G(y + s - x)\lambda e^{-\lambda(s-x)} 2\lambda e^{-2\lambda x} ds dx.$$

Introducing $r = y + s - x$ gives

$$1 - \int_0^\infty 2\lambda e^{-2\lambda x} \int_y^\infty G(r)\lambda e^{-\lambda(r-y)} dr dx$$

$$= 1 - \int_y^\infty G(r)\lambda e^{-\lambda(r-y)} dr$$

$$= \int_y^\infty (1 - e^{-\lambda(r-y)}) dG(r).$$

Thus the formulas (4.75) and (4.74) in the theorem are identical. This completes the proof of the theorem. ∎

Now what can we say about the limiting downtime distribution of the first system failure as the failure rate converges to 0? Is it equal to the steady-state downtime distribution $G_\Phi$? Yes, for the above example we can show that if the failure rate converges to 0, the distribution $G_{1,p2}(y)$ converges to the steady-state formula, i.e.,

$$\lim_{\lambda \to 0} G_{1,p2}(y) = 1 - \bar{G}(y) \frac{\int_y^\infty \bar{G}(r) dr}{\mu_G} = G_\Phi(y).$$

This is seen by noting that

$$\lim_{\lambda \to 0} \frac{\int_y^\infty [1 - e^{-\lambda(r-y)}] dG(r)}{\int_0^\infty [1 - e^{-\lambda r}] dG(r)} = \frac{\int_y^\infty (r - y) dG(r)}{\int_0^\infty r \, dG(r)}$$

$$= \frac{\int_y^\infty \bar{G}(r)dr}{\int_0^\infty \bar{G}(r)dr}.$$

This result can be extended to general monotone systems, and it is not necessary to establish an exact expression for the distribution of the first downtime; see [80]. Consider the asymptotic set-up introduced in Section 4.4, to study highly available components, with exponential lifetime distributions $F_{ij}(t) = 1 - e^{-\lambda_{ij}t}$ and fixed repair time distributions $G_i$, and where we assume $\lambda_{ij} \to 0$ as $j \to \infty$. Then for a parallel system it can be shown that the distribution of the $i$th system downtime converges as $j \to \infty$ to the steady-state downtime distribution $G_\Phi$. For a general system it is more complicated. Assuming that the steady-state downtime distribution converges as $j \to \infty$ to $G_\Phi^*$ (say), it follows that the distribution of the $i$th system downtime converges to the same limit. See [80] for details.

## 4.6  Distribution of the System Downtime in an Interval

In this section we study the distribution of the system downtime in a time interval. The model considered is as described in Section 4.3, p. 112. The system analyzed is monotone and comprises $n$ independent components. Component $i$ generates an alternating renewal process with uptime distribution $F_i$ and downtime distribution $G_i$.

We immediately observe that the asymptotic expression for the expected average downtime presented in Theorem 36, p. 109, also holds for monotone systems, with $A = h(\mathbf{A})$. Formula (4.28) of Theorem 36 requires that the process $\mathbf{X}$ is a regenerative process with finite expected cycle length.

The rest of this section is organized as follows. First we present some approximative methods for computing the distribution of $Y_u$ (the downtime in the time interval $[0, u]$) in the case that the components are highly available, utilizing that $(Y_u)$ is approximately a compound Poisson process, denoted $(CP_u)$, and the exact one-unit formula (4.30), p. 110, for the downtime distribution. Then we formulate some sufficient conditions for when the distribution of $CP_u$ is an asymptotic limit. The framework is the same as described in Section 4.4.1, p. 119. Finally, we study the convergence to the normal distribution.

### 4.6.1  Compound Poisson Process Approximation

We assume that the components have constant failure rate and that the components are highly available, i.e., the products $\lambda_i \mu_{G_i}$ are small. Then it can be heuristically argued that the process $(Y_u), u \in \mathbb{R}_+$, is approximately a compound Poisson process,

$$Y_u \approx \sum_{i=1}^{N_u} Y_i \approx \mathrm{CP}_u. \tag{4.77}$$

Here $N_u$ is the number of system failures in $[0, u]$ and $Y_i$ is the downtime of the $i$th system failure. The dependency between $N_u$ and the random variables $Y_i$ is not "very strong" since $N_u$ is mainly governed by the renewal cycles without system failures. We can ignore downtimes $Y_i$ being the second, third, etc., system failure in a renewal cycle of the process $\mathbf{X}$. The probability of having two or more system failures in a cycle is small since we are assuming highly available components. This means that the random variables $Y_i$ are approximately independent and identically distributed.

From this we can find an approximate expression for the distribution of $Y_u$.

A closely related approximation can be established by considering system operational time, as described in the following.

Let $N_s^{op}$ be the number of system failures in $[0, s]$ when we consider operational time. Similar to the reasoning in Section 4.4, p. 118, it can be argued that $N_s^{op}$ is approximately a homogeneous Poisson process with intensity $\lambda'_{\Phi}$, where $\lambda'_{\Phi}$ is given by

$$\lambda'_{\Phi} = \sum_{i=1}^{n} \frac{h(1_i, \mathbf{A}) - h(0_i, \mathbf{A})}{(\mu_{F_i} + \mu_{G_i}) h(\mathbf{A})}.$$

To motivate this result, we note that the expected number of system failures per unit of time when considering calendar time is approximately equal to the asymptotic (steady-state) system failure rate $\lambda_{\Phi}$, given by (cf. formula (4.41), p. 115)

$$\lambda_{\Phi} = \sum_{i=1}^{n} \frac{h(1_i, \mathbf{A}) - h(0_i, \mathbf{A})}{\mu_{F_i} + \mu_{G_i}}.$$

Then observing that the ratio between calendar time and operational time is approximately $1/h(\mathbf{A})$, we see that the expected number of system failures per unit of time when considering operational time, $EN^{op}(u, u+w]/w$, is approximately equal to $\lambda_{\Phi}/h(\mathbf{A})$Furthermore, $N_{(u,u+w]}^{op}$ is "nearly independent" of the history of $N^{op}$ up to $u$, noting that the state process $\mathbf{X}$ frequently restarts itself probabilistically, i.e., $\mathbf{X}$ re-enters the state $(1, 1, \ldots, 1)$. It can be shown by repeating the proof of the Poisson limit Theorem 50, p. 128, and using the fact that $h(\mathbf{A}) \to 1$ as $\lambda_i \mu_{G_i} \to 0$, that $N_{t/\alpha}^{op}$ has an asymptotic Poisson distribution with parameter $t$. The system downtimes given system failure are approximately identically distributed with distribution function $G(y)$, say, independent of $N^{op}$, and approximately independent observing that the state process $\mathbf{X}$ with a high probability restarts itself quickly after a system failure. The distribution

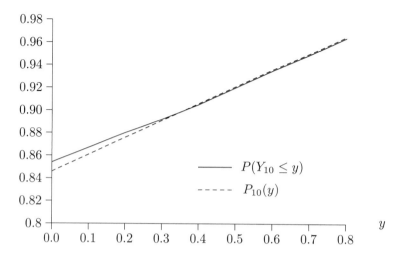

FIGURE 4.4. $P_{10}(y)$ and $P(Y_{10} \leq y)$ for a Parallel System of Two Components
with Constant Repair Times, $\mu_G = 1, \lambda = 0.1$

function $G(y)$ is normally taken as the asymptotic (steady-state) downtime
distribution given system failure or an approximation to this distribution;
see Section 4.5.

Considering the system as a one-unit system, we can now apply the
exact formula (4.30), p. 110, for the downtime distribution with the Poisson
parameter $\lambda'_\Phi$. It follows that

$$P(Y_u \leq y) \approx \sum_{n=0}^{\infty} G^{*n}(y) \frac{[\lambda'_\Phi(u-y)]^n}{n!} e^{-\lambda'_\Phi(u-y)} = P_u(y), \quad (4.78)$$

where the equality is given by definition. Formula (4.78) gives good approximations for "typical real life cases" with small component availabilities; see
[91]. Figure 4.4 presents the downtime distribution for a parallel system of
two components with the repair times identical to 1 and $\mu_F = 10$ using
the steady-state formula $G_\Phi$ for $G$ (formula (4.70), p. 139). The "true"
distribution is found using Monte Carlo simulation. We see that formula
(4.78) gives a good approximation.

## 4.6.2   Asymptotic Analysis

We argued above that $(Y_u)$ is approximately equal to a compound Poisson
process when the system comprises highly available components. In the
following theorem we formalize this result.

The set-up is the same as in Section 4.4.1, p. 119. We consider for each
component $i$ a sequence $\{F_{ij}, G_{ij}\}$, $j \in \mathbb{N}$, of distributions satisfying certain conditions. To simplify notation, we normally omit the index $j$. When

assuming in the following that $\mathbf{X}$ is a regenerative process, it is tacitly understood for all $j \in \mathbb{N}$.

We say that the renewal cycle is a "success" if no system failure occurs during the cycle and a "fiasco" if a system failure occurs.

Let $\alpha$ be a suitable normalizing factor (or more precisely, a normalizing sequence in $j$) such that $N_{t/\alpha}$ converges in distribution to a Poisson variable with mean $t$, cf. Theorem 50, p. 128. Normally we take $\alpha = \lambda_\Phi$, but we could also use $q/E_0 S, q/ES$, or $1/ET_\Phi$, where $q$ equals the probability that a system failure occurs in a cycle, $S$ equals the length of a cycle, $E_0 S$ equals the expected length of a cycle with no system failures, and $T_\Phi$ equals the time to the first system failure. Furthermore, let $Y_{i1}$ denote the length of the first downtime of the system in the $i$th "fiasco" renewal cycle, and $Y_{i2}$ the length of the remaining downtime in the same cycle. We assume that the asymptotic distribution of $Y_{i1}$ exists (as $j \to \infty$): $Y_{i1} \overset{D}{\to} G_\Phi^*$ (say).

A random variable is denoted $CP(r, G)$ if it has the same distribution as $\sum_{i=1}^N Y_i$, where $N$ is a Poisson variable with mean $r$, the variables $Y_i$ are i.i.d. with distribution function $G$, and $N$ and $Y_i$ are independent. The distribution of $CP(r, G)$ equals

$$\sum_{i=0}^\infty G^{*i} \frac{r^i}{i!} e^{-r},$$

where $G^{*i}$ denotes $i$th convolution of $G$.

**Theorem 59** *Assume that $\mathbf{X}$ is a regenerative process, and that $F_{ij}$ and $G_{ij}$ change in such a way that the following conditions hold (as $j \to \infty$):*

$$q \quad \to \quad 0, \tag{4.79}$$
$$qc_{0S}^2 \quad \to \quad 0, \tag{4.80}$$

*where $c_{0S}^2 = [E_0 S^2/(E_0 S)^2] - 1$ denotes the squared coefficient of variation of $S$ under $P_0$,*

$$\frac{qE_1 S}{ES} \quad \to \quad 0, \tag{4.81}$$
$$E_1(N_S - 1) \quad \to \quad 0, \tag{4.82}$$
$$Y_{i1} \overset{D}{\to} G_\Phi^*. \tag{4.83}$$

*Then (as $j \to \infty$)*

$$Y_{t/\alpha} \overset{D}{\to} CP(t, G_\Phi^*), \tag{4.84}$$

*where $\alpha = \lambda_\Phi, q/E_0 S, q/ES$, or $1/ET_\Phi$.*

**Proof.** First we will introduce two renewal processes, $N'$ and $N''$, having the same asymptotic properties as $N_{t/\alpha}$. From Theorem 50, p. 128, we know that

$$N_{t/\alpha} \overset{D}{\to} \text{Poisson}(t)$$

under conditions (4.79)−(4.82).

Let $\nu(1)$ equal the renewal cycle index associated with the first "fiasco" renewal cycle, and let $U_1$ denote the time to the starting point of this cycle, i.e.,

$$U_1 = \sum_{i=1}^{\nu(1)-1} S_i.$$

Note that if the first cycle is a "fiasco" cycle, then $U_1 = 0$. Starting from the beginning of the renewal cycle $\nu(1) + 1$, we define $U_2$ as the time to the starting point of the next "fiasco" renewal cycle. Similarly we define $U_3, U_4, \ldots$. The random variables $U_i$ are equal the interarrival times of the renewal process $N_t''$, i.e.,

$$N_t'' = \sum_{k=1}^{\infty} I(\sum_{i=1}^{k} U_i \leq t).$$

By repeating the proofs of Theorem 46 (p. 121) and Theorem 50 it is seen that

$$N_{t/\alpha}'' \xrightarrow{D} \text{Poisson}(t). \tag{4.85}$$

Using that the process $N_t''$ and the random variables $Y_i$ are independent, and the fact that $Y_{i1} \xrightarrow{D} G_\Phi^*$ (assumption (4.83)), it follows that

$$\sum_{i=1}^{N_{t/\alpha}''} Y_{i1} \xrightarrow{D} \text{CP}(t, G_\Phi^*). \tag{4.86}$$

A formal proof of this can be carried out using Moment Generating Functions.

Next we introduce $N_t'$ as the renewal process having interarrival times with the same distribution as $U_1 + S_{\nu(1)}$, i.e., the renewal cycle also includes the "fiasco" cycle. It follows from the proof of Theorem 46, using condition (4.81), that $N_{t/\alpha}'$ has the same asymptotic Poisson distribution as $N_{t/\alpha}$.

It is seen that

$$N_t' \leq N_t'', \tag{4.87}$$

$$N_t' \leq N_t \leq \sum_{i=1}^{N_t''} N_{(i)} = N_t'' + \sum_{i=1}^{N_t''}(N_{(i)} - 1), \tag{4.88}$$

where $N_{(i)}$ equals the number of system failures in the $i$th "fiasco" cycle. Note that $N_t''$ is at least the number of "fiasco" cycles up to time $t$, including the one that is possibly running at $t$, and $N_t'$ equals the number of finished "fiasco" cycles at time $t$ without the one possibly running at $t$.

Now to prove the result (4.84) we will make use of the following inequalities:

$$Y_{t/\alpha} \le \sum_{i=1}^{N_{t/\alpha}''} Y_{i1} + \sum_{i=1}^{N_{t/\alpha}''} Y_{i2}, \tag{4.89}$$

$$Y_{t/\alpha} \ge \sum_{i=1}^{N_{t/\alpha}''} Y_{i1} - \sum_{i=N_{t/\alpha}'+1}^{N_{t/\alpha}''} Y_{i1}. \tag{4.90}$$

In view of (4.86), and the inequalities (4.89) and (4.90), we need to show that

$$\sum_{i=1}^{N_{t/\alpha}''} Y_{i2} \xrightarrow{P} 0, \tag{4.91}$$

$$\sum_{i=N_{t/\alpha}'+1}^{N_{t/\alpha}''} Y_{i1} \xrightarrow{P} 0. \tag{4.92}$$

To establish (4.91) we first note that

$$Y_{i2} \xrightarrow{P} 0,$$

since

$$P(Y_{i2} > \epsilon) \le P_1(N_S \ge 2) \le E_1(N_S - 1) \to 0$$

by (4.82). Using Moment Generating Functions it can be shown that (4.91) holds.

The key part of the proof of (4.92) is to show that $(N_{t/\alpha}'')$ is uniformly integrable in $j$ ($t$ fixed). If this result is established, then since $N_{t/\alpha}'' \xrightarrow{D}$ Poisson($t$) by (4.85) it follows that

$$EN_{t/\alpha}'' \to t. \tag{4.93}$$

And because of the inequality (4.87), $(N_{t/\alpha}')$ is also uniformly integrable so that $EN_{t/\alpha}' \to t$, and we can conclude that (4.92) holds noting that

$$P(N_{t/\alpha}'' - N_{t/\alpha}' \ge 1) \le EN_{t/\alpha}'' - EN_{t/\alpha}' \to 0.$$

Thus it remains to show that $(N_{t/\alpha}'')$ is uniformly integrable.

Let $F_U$ denote the probability distribution of $U$ and let $V_l = \sum_{i=1}^{l} U_i$. Then we obtain

$$E[N_{t/\alpha}'' I(N_{t/\alpha}'' \ge k)] = \sum_{l=k}^{\infty} P(N_{t/\alpha}'' \ge l) + (k-1)P(N_{t/\alpha}'' \ge k)$$

$$
\begin{aligned}
&= \sum_{l=k}^{\infty} P(V_l \le t/\alpha) + (k-1)P(V_k \le t/\alpha) \\
&= \sum_{l=k}^{\infty} F_U^{*l}(t/\alpha) + (k-1)F_U^{*k}(t/\alpha) \\
&\le \sum_{l=k}^{\infty} (F_U(t/\alpha))^l + (k-1)(F_U(t/\alpha))^k \\
&= \frac{(F_U(t/\alpha))^k}{1 - F_U(t/\alpha)} + (k-1)(F_U(t/\alpha))^k.
\end{aligned}
$$

Since $F_U(t/\alpha) \to 1-e^{-t}$, as $j \to \infty$, it follows that for any sequence $F_{ij}, G_{ij}$ satisfying the conditions $(4.79)-(4.82)$, $(N''_{t/\alpha})$ is uniformly integrable. To see this, let $\epsilon$ be given such that $0 < \epsilon < e^{-t}$. Then for $j \ge j_0$ (say) we have

$$
\sup_{j \ge j_0} E[N''_{t/\alpha} I(N''_{t/\alpha} \ge k)] \le \frac{(1 - e^{-t} + \epsilon)^k}{e^{-t} - \epsilon} + (k-1)(1 - e^{-t} + \epsilon)^k.
$$

Consequently,

$$
\lim_{k \to \infty} \sup_j E[N''_{t/\alpha} I(N''_{t/\alpha} \ge k)] = 0,
$$

i.e., $(N''_{t/\alpha})$ is uniformly integrable, and the proof is complete. ■

**Remark 17** The conditions $(4.79)-(4.82)$ of Theorem 59 ensures the asymptotic Poisson distribution of $N_{t/\alpha}$, cf. Theorem 50, p. 128. Sufficient conditions for $(4.79)-(4.82)$ are given in Theorem 47, p. 124.

## Asymptotic Normality

We now study convergence to the normal distribution. The theorem below is "a time average result" — it is not required that the system is highly available. The result generalizes $(4.32)$, p. 111.

**Theorem 60** If $\mathbf{X}$ is a regenerative process with cycle length $S$ and associated downtime $Y = Y_S$, $\mathrm{Var}[S] < \infty$, and $\mathrm{Var}[Y] < \infty$, then as $t \to \infty$,

$$
\sqrt{t}\left[\frac{Y_t}{t} - \bar{A}\right] \xrightarrow{D} N(0, \tau_\Phi^2), \tag{4.94}
$$

where

$$
\tau_\Phi^2 = \frac{\mathrm{Var}[Y - \bar{A}S]}{ES} \tag{4.95}
$$

$$
\bar{A} = \frac{EY}{ES}. \tag{4.96}
$$

**Proof.** The result (4.94) follows by applying the Central Limit Theorem for renewal reward processes, Theorem 123, p. 251, in Appendix B.    ∎

In the case that the system is highly available, we have

$$\tau_\Phi^2 \approx \lambda_\Phi EY_1^2, \tag{4.97}$$

where $Y_1$ is the downtime of the first system failure (note that $Y_1 = Y_{11}$). The idea used to establish (4.97) is the following: As before, let $S$ be equal to the time of the first return to the best state $(1, 1, \ldots, 1)$. Then (4.97) follows by using (4.95), (4.96), $\bar{A} \approx 0$, the fact that $Y \approx Y_1$ if a system failure occurs in the renewal cycle, the probability of two or more failures occurring in the renewal cycle is negligible, and $\lambda_\Phi = EN_S/ES$ (by the Renewal Reward Theorem, p. 250). We obtain

$$
\begin{aligned}
\tau_\Phi^2 &= \frac{\mathrm{Var}[Y - \bar{A}S]}{ES} = \frac{E(Y - \bar{A}S)^2}{ES} \\
&\approx \frac{EY^2}{ES} = \frac{E_1 Y^2 q}{ES} \approx \frac{EY_1^2 q}{ES} \\
&\approx EY_1^2 \frac{EN_S}{ES} = EY_1^2 \lambda_\Phi,
\end{aligned}
$$

which gives (4.97).

More formally, it is possible to show that under certain conditions, the ratio $\tau_\Phi^2 / \lambda_\Phi EY_1^2$ converges to 1, see [29].

# 4.7   Generalizations and Related Models

## 4.7.1   Multistate Monotone Systems

We consider a multistate monotone system $\Phi$ as described in Section 2.1.2, p. 33, observed in a time interval $J$, with the following extensions of the model: We assume that there exists a reference level $D_t$ at time $t$, $t \in J$, which expresses a desirable level of system performance at time $t$. The reference level $D_t$ at time $t$ is a positive random variable, taking values in $\{d_0, d_1, \ldots, d_r\}$. For a flow network system we interpret $D_t$ as the demand rate at time $t$. In the following we will use the word "demand rate" also in the general case. The state of the system at time $t$, which we in the following refer to as the throughput rate, is assumed to be a function of the states of the components and the demand rate, i.e.,

$$\Phi_t = \Phi(\mathbf{X}_t, D_t).$$

If $D_t$ is a constant, we write $\Phi(\mathbf{X}_t)$. The process $(\Phi_t)$ takes values in $\{\Phi_0, \Phi_1, \ldots, \Phi_M\}$.

## Performance Measures

The performance measures introduced in Section 4.1, p. 98, can now be generalized to the above model.

a) For a fixed time $t$ we define point availabilities

$$P(\Phi_t \geq \Phi_k | D_t = d),$$
$$E[\Phi_t | D_t = d],$$
$$P(\Phi_t \geq D_t).$$

b) Let $N_J$ be defined as the number of times the system state is below demand in $J$. The following performance measures related to $N_J$ are considered

$$P(N_J \leq k),\ k \in \mathbb{N}_0,$$
$$EN_J,$$
$$P(\Phi_t \geq D_t,\ \forall t \in J) = P(N_J = 0).$$

Some closely related measures are obtained by replacing $D_t$ by $\Phi_k$ and $N_J$ by $N_J^k$, where $N_J^k$ is equal to the number of times the process $\Phi$ is below state $\Phi_k$ during the interval $J$.

c) Let

$$
\begin{aligned}
Y_J &= \int_J (D_t - \Phi_t)\, dt \\
&= \int_J D_t\, dt - \int_J \Phi_t\, dt.
\end{aligned}
$$

We see that $Y_J$ represents the lost throughput (volume) in $J$, i.e., the difference between the accumulated demand (volume) and the actual throughput (volume) in $J$. The following performance measures related to $Y_J$ are considered

$$P(Y_J \leq y),\ y \in \mathbb{R}_+,$$

$$\frac{EY_J}{|J|},$$

$$\frac{E \int_J \Phi_t\, dt}{E \int_J D_t\, dt}, \tag{4.98}$$

where $|J|$ denotes the length of the interval $J$. The measure (4.98) is called *throughput availability*.

d) Let

$$Z_J = \frac{1}{|J|} \int_J I(\Phi_t \geq D_t)\, dt.$$

The random variable $Z$ represents the portion of time the throughput rate equals (or exceeds) the demand rate. We consider the following performance measures related to $Z_J$

$$P(Z_J \leq y),\ y \in \mathbb{R}_+,$$

$$EZ_J.$$

The measure $EZ_J$ is called *demand availability*.

As in the binary case we will often use in practice the limiting values of these performance measures.

The above performance measures are the most common measures used in reliability studies of offshore oil and gas production and transport systems, see, e.g., Aven [16]. In particular, the throughput availability is very much used when predicting the performance of various design options. For economic analysis and as a basis for decision-making, however, it is essential to be able to compute the total distribution of the throughput loss, and not only the mean. The measures related to the number of times the system is below a certain demand level is also useful, but more from an operational and safety point of view.

## Computation

We now briefly look into the computation problem for some of the measures defined above. To simplify the analysis we shall make the following assumptions:

*Assumptions*

1. $J = [0, u]$.

2. The demand rate $D_t$ equals the maximum throughput rate $\Phi_M$ for all $t$.

3. The $n$ component processes $(X_t(i))$ are independent. Furthermore, with probability one, the $n$ component processes $(X_t(i))$ make no transitions ("jumps") at the same time.

4. The process $(X_t(i))$ generates an alternating renewal process $T_{i1}, R_{i1}, T_{i2}, R_{i2}, \ldots$, as described in Section 4.2, p. 99, where $T_{im}$ represents the time spent in the state $x_{iM_i}$ during the $m$th visit to this state, and $R_{im}$ represents the time spent in the states $\{x_{i0}, x_{i1}, \ldots, x_{i,M_i-1}\}$ during the $m$th visit to these states.

For all $i$ and $r$,

$$a_{ir} = \lim_{t \to \infty} P(X_t(i) = x_{ir})$$

exist.

Arguing as in the binary case we can use results from regenerative and renewal reward processes to generalize the results obtained in the previous sections. To illustrate this, we formulate some of these extensions below. The proofs are omitted. We will focus here on the asymptotic results. Refer to Theorems 39, p. 113, and 42, p. 114, for the analogous results in the binary case. We need the following notation:

$$
\begin{aligned}
\mu_i &= ET_{im} + ER_{im} \\
N_t &= N_{[0,t]}^k \quad (k \text{ is fixed}) \\
p_{ir}(t) &= P(X_t(i) = x_{ir}); \text{ if } t \text{ is fixed, we write } p_{ir} \text{ and } X(i) \\
\mathbf{p} &= (p_{10}, \ldots, p_{nM_n}) \\
\mathbf{a} &= (a_{10}, \ldots, a_{nM_n}) \\
\Phi_k(\mathbf{X}) &= I(\Phi(\mathbf{X}) \geq \Phi_k) \\
h_k(\mathbf{p}) &= E\Phi_k(\mathbf{X}) \\
h(\mathbf{p}) &= E\Phi(\mathbf{X}) \\
(1_{ir}, \mathbf{p}) &= \mathbf{p} \text{ with } p_{ir} \text{ replaced by 1 and } p_{il} = 0 \text{ for } l \neq r.
\end{aligned}
$$

We see that $\mu_i$ is equal to the expected cycle length for component $i$, $N_t$ represents the number of times the process $\Phi$ is below state $\Phi_k$ during the interval $[0, t]$, and $\Phi_k(\mathbf{X})$ equals 1 if the system is in state $\Phi_k$ or better, and 0 otherwise.

**Theorem 61** *The limiting availabilities are given by*

$$
\begin{aligned}
\lim_{t \to \infty} E\Phi(\mathbf{X}_t) &= h(\mathbf{a}), \\
\lim_{t \to \infty} P(\Phi(\mathbf{X}_t) \geq \Phi_k) &= h_k(\mathbf{a}).
\end{aligned}
$$

**Theorem 62** *Let*

$$\gamma_{ilr} = h_k(1_{il}, \mathbf{a}) - h_k(1_{ir}, \mathbf{a})$$

*and let $f_{ilr}$ denote the expected number of times component $i$ makes a transition from state $x_{il}$ to state $x_{ir}$ during a cycle of component $i$. Assume $f_{ilr} < \infty$. Then the expected number of times the system state is below $\Phi_k$ per unit of time in the long run equals*

$$\lim_{u \to \infty} \frac{EN_u}{u} = \lim_{u \to \infty} \frac{E[N_{u+s} - N_u]}{s} = \sum_{i=1}^{n} \sum_{r < l} \frac{f_{ilr} \gamma_{ilr}}{\mu_i}. \tag{4.99}$$

*If* **X** *is a regenerative process having finite expected cycle length, then with probability one,*

$$\lim_{u \to \infty} \frac{N_u}{u} = \sum_{i=1}^{n} \sum_{r < l} \frac{f_{ilr}\, \gamma_{ilr}}{\mu_i}.$$

The limit (4.99) is denoted $\lambda_\Phi$. If the random variables $T_{im}$ are exponentially distributed, then **X** is regenerative, cf. Theorem 44, p. 116.

It is also possible to extend the asymptotic results related to the distribution of the number of system failures at level $k$, and the distribution of the lost volume (downtime). We can view the system as a binary system of binary components, and the asymptotic results of Sections 4.4–4.6 apply.

## Gas Compression System

Consider the gas compression system example in Section 1.3.2, p. 14. Two design alternatives were studied:

i) One gas train with a maximum throughput capacity of 100%.

ii) Two trains in parallel, each with a maximum throughput capacity of 50%.

Normal production is 100%. Each train comprises compressor–turbine, cooler and scrubber. To analyze the performance of the system it was considered sufficient to use approximate methods developed for highly available systems, as presented in this chapter. In the system analysis, each train was treated as one component, having exponential lifetime distribution with a failure rate of 13 per year, and mean repair time equal to

$$(10/13) \cdot 12 + (2/13) \cdot 50 + (1/13) \cdot 20 \approx 18.5 \text{ (hours)}.$$

From this we find that the asymptotic unavailability $\bar{A}$, given by formula (4.2), p. 101, for a train equals 0.027, assuming 8760 hours per year. The number of system failures per unit of time is given by the system failure rate $\lambda_\Phi$. For alternative i) there is only one failure level and $\lambda_\Phi = 13$. For alternative ii) we must distinguish between failures resulting in production below 100% and below 50%. The system in these two cases can be viewed as a series system of the two trains and a parallel system of the two trains, respectively. Hence the system failure rate for these levels is approximately equal to 26 and 0.7, respectively. Note that for the latter case (cf. (4.64), p. 133),

$$\lambda_\Phi \approx 2 \cdot \bar{A} \cdot 13.$$

Using that the number of system failures is approximately Poisson distributed, we can compute the probability that a certain number of failures occurs during a specific period of time. For example, we find that for alternative ii) there is a probability of about $e^{-0.7} = 0.50$ of having no complete shutdowns during a year.

Let $EY$ denote the asymptotic mean lost production relative to the demand. For alternative i) it is clear that $EY$ equals 0.027, observing that a failure results in 100% loss and the unavailability equals 0.027. For alternative ii), we obtain the same value for the asymptotic mean lost production, as is seen from the following calculation

$$EY = 0.5 \cdot 2 \cdot 0.027 \cdot 0.973 + 1 \cdot 0.027^2 = 0.027.$$

The first term in the sum represents the contribution from failures leading to 50% loss, whereas the second term represents the contribution from failures leading to 100% loss. The latter contribution is in practice negligible compared to the former one. To compute the distribution of the lost production, we need to know more about the distribution of the repair time $R$ of the train. It was assumed in this application that $ER^2 = 1000$, which corresponds to a squared coefficient of variation equal to 1.9 and a standard deviation equal to 25.7. The unit of time is hours. This assumption makes it possible to approximate the distribution of the lost production during a year, using the normal approximation. We know the mean ($EY = 0.027$) and need to estimate the variance of $Y$. To do this we make use of (4.97), p. 151, stating that the variance in the binary case is approximately equal to $\lambda_\Phi EY_1^2/t$, where $t$ is the length of the time period considered and $Y_1$ is the downtime of the first system failure. For alternative i) we find that the variance equals approximately

$$(13/8760) \cdot 1000/8760 = 1.7 \cdot 10^{-4},$$

and for alternative ii) (we ignore situations with both components down so that the lost production is approximately 50% of the downtime)

$$(50/100)^2 \cdot (26/8760) \cdot 1000/8760 = 0.85 \cdot 10^{-4}.$$

From this we estimate, for example, that the probability that the lost production during one year is more than 4% of demand, to be 0.16 for alternative i) and 0.08 for alternative ii).

## Special Case. Phase-Type Distributions

In the asymptotic analysis in Sections 4.4–4.6 main emphasis has been placed on the situation that the lifetimes are exponentially distributed. Using the so-called phase-type approach, we can show that the multistate model also provides a framework for covering other types of distributions. The phase-type approach makes use of the fact that a distribution function can be approximated by a mixture of Erlang distributions (with the same scale parameter), cf., e.g., Asmussen [9] and Tijms [168]. It is common to use a mixture of two Erlang distributions with the first two moments matching the distribution considered. Now assume that the lifetime of component $i$, $F_i$, can be described by the sum of $M_i$ random variables, each of which is exponentially distributed with rate $\lambda_{i0}$, i.e., the lifetime of component $i$ is Erlangian distributed with parameters $\lambda_{i0}$ and $M_i$. Then we have a situation that fits into the above multistate framework and the asymptotic results can be applied. The state space for component $i$ is $\{0, 1, ..., M_i\}$. The component process $(X_t(i))$ starts in state $M_i$, it stays there a time governed by an exponential random variable with rate $\lambda_{i0}$ and jumps to state $M_i - 1$, it stays there a time governed by an exponential random variable with rate $\lambda_{i0}$ and jumps to state $M_i - 2$, and this continues until the process reaches state 0. After a duration having distribution $G_i$ in state 0 it returns to state $M_i$. We see that $f_{ilr} = 1$ if $l = r + 1$ and $f_{ilr} = 0$ otherwise (for $r < l$). Furthermore,

$$\mu_i = M_i \frac{1}{\lambda_{i0}} + \mu_{G_i} = \mu_{F_i} + \mu_{G_i},$$
$$\lambda_\Phi = \sum_{i=1}^n [h_1(1_{i1}, \mathbf{a}) - h_1(1_{i0}, \mathbf{a})] \frac{1}{\mu_i} = \sum_{i=1}^n [h(1_i, \mathbf{A}) - h(0_i, \mathbf{A})] \frac{1}{\mu_i},$$
$$a_{i0} = \frac{\mu_{G_i}}{\mu_i} = \bar{A}_i,$$

using the terminology from the binary theory. Remember that the formulas established in Sections 4.2 and 4.3 for the expected cycle length and the steady-state (un)availability of component $i$, and the system failure rate, are applicable also for nonexponential distributions.

Thus by modifying the state space, we have been able to extend the results, i.e., the Theorems 46 (p. 121), 50 (p. 128), and 59 (p. 147), in the previous sections to Erlang distributions.

Now assume that the lifetime distribution of component $i$ is a mixture of Erlang distributions, i.e., with probability $p_{ir} > 0$ the distribution equals an Erlang distribution with parameters $\lambda_{i0}$ and $M_{ir}$, $r = 1, 2, \ldots, r_i$. This situation can be analyzed as above with the state space for component $i$ given by $\{0, 1, ..., M_i\}$, where $M_i = \max_r\{M_{ir}\}$. If the component state process $(X_t(i))$ is in state 0, it will go to state $M_{ir}$ with probability $p_{ir}$. Then the component stays in this state for a time governed by an exponential distribution with parameter $\lambda_{i0}$, before it jumps to state $M_{ir} - 1$, etc. As above we can use the formulas for the binary case to compute the expected cycle length and steady-state (un)availability of component $i$, and the system failure rate. It is seen that

$$\mu_{F_i} = \sum_{r=1}^{r_i} p_{ir} M_{ir} \frac{1}{\lambda_{i0}}.$$

We can conclude that the set-up also covers mixtures of Erlang distributions, and Theorems 46, 50, and 59 apply.

Note that we have not proved that the limiting results obtained in the previous sections hold true for general lifetime distributions $F_{ij}$. We have shown that if the distributions $F_{ij}$ all belong to a certain class of mixtures of Erlang distributions, then the results hold. Starting from general distributions $F_{ij}$, we can write $F_{ij}$ as a limit of $F_{ijr}$, $r \to \infty$, where $F_{ijr}$ are mixtures of Erlang distributions. But interchanging the limits as $j \to \infty$ and as $r \to \infty$ is not justified in general. Refer also to Bibliographic Notes, p. 166, for some comments related to the non-exponential case.

## 4.7.2  Parallel System with Repair Constraints

Consider the model as described in Section 4.3, p. 112, but assume now that there are repair constraints, i.e., a maximum of $r$ ($r < n$) components can be repaired at the same time. Hence if $i$, $i > r$, components are down, the remaining $i - r$ components are waiting in a repair queue. We shall restrict attention to the case $r = 1$, i.e., there is only one repair facility (channel) available. The repair policy is first come first served. We assume exponentially distributed lifetimes.

Consider first a parallel system of two components, and the set-up of Section 4.4, p. 119. It is not difficult to see that $ET_\Phi$, $q$, and $E_0 S$ are identical to the corresponding quantities when there are no repair constraints; see section on parallel system of two identical components p. 131. We can also find explicit expressions for $ES$ and $\lambda_\Phi$. Since the time to the first component failure is exponentially distributed with parameter $2\lambda$, $ES = 1/2\lambda + ES''$, where $S''$ equals the time from the first component failure until the process again returns to $(1, 1)$. Denoting the repair time of the failed component by $R$, we see that

$$ES'' = \mu_G + qE[S'' - R|N_S \geq 1].$$

But $E[S'' - R|N_S \geq 1] = ES''$, and it follows that

$$ES = \frac{1}{2\lambda} + \frac{\mu_G}{1 - q}.$$

Hence

$$
\begin{aligned}
\lambda_\Phi &= \frac{EN_S}{ES} = \frac{q/(1 - q)}{ES} \\
&= \frac{2\lambda q}{1 - q + 2\lambda \mu_G}.
\end{aligned}
$$

Alternatively, and easier, we could have found $\lambda_\Phi$ by defining a cycle $S$ as the time between two consecutive visits to a state with just one component functioning. Then it is seen that $ES = \mu_G + (1 - q)/2\lambda$ and $EN_S = q$, resulting in the same $\lambda_\Phi$ as above.

Now suppose we have $n \geq 2$, and let $\Phi_t$ be defined as the number of components functioning at time $t$. To analyze the system, we can utilize that the state process $\Phi_t$ is a semi-Markov process with jump times at the completion of repairs. In state $0, 1, \ldots, n - 1$ the time between transitions has distribution $G(t)$ and the transition probability $P_{ij}$ is given by

$$
P_{ij} = \begin{cases}
\int_0^\infty \begin{pmatrix} i \\ i - j + 1 \end{pmatrix} F(s)^{i-j+1}(1 - F(s))^{j-1} \, dG(s), \\
\qquad\qquad\qquad\qquad\qquad\qquad\qquad 1 \leq j \leq i \leq n - 1 \\[2mm]
\int_0^\infty (1 - F(s))^i \, dG(s), \qquad\qquad\qquad j = i + 1 \\[2mm]
0, \qquad\qquad\qquad\qquad\qquad\qquad\qquad 1 \leq i < j - 1,
\end{cases}
$$

observing that if the state is $i$ and the repair is completed at time $s$, then the probability that the process jumps to state $j$, where $j \leq i \leq n - 1$, equals the probability that $i - j + 1$ components fail before $s$ and $j - 1$ components survive $s$; and, furthermore, if the state is $i$ and the repair is completed at time $s$, then the probability that the process jumps to state $i + 1$ equals the probability that $i$ components survive $s$. Now if the process is in state $n$, it stays there for an exponential time with rate $n\lambda$, and jumps to state $n - 1$.

Having established the transition probabilities, we can compute a number of interesting performance measures for the system using results from semi-Markov theory. For example, we have an explicit formula for the asymptotic probability that $P(\Phi_t = k)$ as $t \to \infty$, which depends on the mean time spent in each state and the limiting probabilities of the embedded discrete-time Markov chain; see Ross [145], p. 104.

### 4.7.3   Standby Systems

In this section we study the performance of standby systems comprising $n$ identical components of which $n - 1$ are normally operating and one is in (cold) standby. Emphasis is placed on the case that the components have constant failure rates, and the mean repair time is relatively small compared to the mean time to failure.

Standby systems as analyzed here are used in many situations in real life. As an example we return to the gas compression system in Section 1.3, p. 14 and Section 4.7.1, p. 155. To increase the availability for the alternatives considered, we may add a standby train such that when a failure of a train occurs, the standby train can be put into operation and a production loss is avoided.

## Model

The following assumptions are made:

- Normally $n - 1$ components are running and one is in standby.

- Failed components are repaired. The repair regime is characterized by

    R1 Only one component can be repaired at a time (one repair facility/channel), the repair policy is "first come first served," or

    R2 Up to $n$ repairs can be carried out at a time ($n$ repair facilities/channels).

- Switchover to the standby component is perfect, i.e., instantaneous and failure-free.

- A standby component that has completed its repair is functioning at demand, i.e., the failure rate is zero in the standby state.

- All failure times and repair times are independent with probability distributions $F(t)$ and $G(t)$, respectively. $F$ is absolutely continuous and has finite mean, and $G$ has finite third-order moment. We assume

$$\int_0^\infty F(t)dG(t) > 0.$$

In the following $T$ refers to a failure time of a component and $R$ refers to a repair time.

The squared coefficient of variation of the repair time distribution is denoted $c_G^2$.

Let $\Phi_t$ denote the state of the system at time $t$, i.e., the number of components functioning at time $t$ ($\Phi_t \in \{n, n-1, \ldots, 0\}$). For repair regime R1, $\Phi$ is generally a regenerative process, or a modified regenerative process. For a two-component system it is seen that the time points when $\Phi$ jumps to state 1 are regenerative points, i.e., the time points when i) the operating component fails and the second component is not under repair (the process jumps from state 2 to 1) or ii) both components are failed and the repair of the component being repaired is completed (the process jumps from state 0 to 1). For $n > 2$, the points in time when the process jumps from state 0 to 1 are regenerative points, noting that the situation then is characterized by one "new" component, and $n - 1$ in a repair queue. Assuming exponential lifetimes, we can define other regenerative points, e.g., consecutive visits to the best state $n$, or consecutive visits to state $n - 1$.

Also for a two-component system under repair regime R2, the process generally generates a (modified) regenerative process. The regenerative points are given by the points when the process jumps from state 2 to 1 (case i) above). If the system has more than two components $(n > 2)$, the regenerative property is not true for a general failure time distribution. However, under the assumption of an exponential time to failure, the process is regenerative. Regenerative points are given by consecutive visits to state $n$, or points when the process jumps from state $n$ to state $n - 1$. In the following, when considering a system of more than two components, we assume an exponential lifetime distribution. Remember that a cycle refers to the length between two consecutive regenerative points.

*Performance Measures*

The system can be considered as a special case of a multistate monotone system, with the demand rate $D_t$ set to $n - 1$. Hence the performance measures defined in Section 4.7.1, p. 151, also apply to the system analyzed in this section. Availability refers to the probability that at least $n - 1$ components are functioning, and system failure refers to the event that the state process $\Phi$ is below $n - 1$. Note that we cannot apply the computation results of Section 4.7.1 since the state processes of the components are not stochastically independent. The general asymptotic results obtained in Sections 4.4–4.6 for regenerative processes are however applicable.

Of the performance measures we will put emphasis on the limiting availability, and the limiting mean of the number of system failures in a time interval.

We need the following notation for $i = n, n - 1, \dots, 0$:
$$p_i(t) = P(\Phi_t = i),$$
$$p_i = \lim_{t \to \infty} p_i(t),$$
provided the limits exist. Clearly, the availability at time $t$, $A(t)$, is given by
$$A(t) = p_n(t) + p_{n-1}(t)$$
and the limiting availability, $A$, is given by
$$A = p_n + p_{n-1}.$$

*Computation*

First, we focus on the limiting unavailability $\bar{A}$, i.e., the expected portion of time in the long run that at least two components are not functioning. Under the assumption of constant failure and repair rates this unavailability can easily be computed using Markov theory, noting that $\Phi$ is a birth and death process. The probability $\tilde{p}_i$ of having $i$ components down is given by (cf. [16], p. 303)
$$\tilde{p}_i = p_{n-i} = \frac{z_i}{1 + \sum_{j=1}^n z_j}, \tag{4.100}$$

where

$$z_i = \begin{cases} \frac{(n-1)(n-1)!}{(n-i)!} \frac{1}{\prod_{l=1}^{i} u_l} \delta^i & i = 1, 2, \ldots, n \\ 1 & i = 0 \end{cases}$$

$$\delta = \mu_G/\mu_F$$

$u_l = 1$ under repair regime R1 and $l$ under repair regime R2.

Note that if $\delta$ is small, then $\tilde{p}_i \approx z_i$ for $i \geq 1$. Hence

$$\bar{A} \approx \tilde{p}_2 \approx \frac{(n-1)^2}{u_2} \delta^2. \tag{4.101}$$

We can also write

$$\bar{A} = \frac{(n-1)^2}{u_2} \delta^2 + o(\delta^2), \quad \delta \to 0.$$

In general we can find expressions for the limiting unavailability by using the regenerative property of the process $\Phi$. Defining $Y$ and $S$ as the system downtime in a cycle and the length of a cycle, respectively, it follows from the Renewal Reward Theorem (Theorem 122, p. 250, in Appendix B) that

$$\bar{A} = \frac{EY}{ES}. \tag{4.102}$$

Here system downtime corresponds to the time two or more of the components are not functioning. Let us now look closer into the problem of computing $\bar{A}$, given by (4.102), under repair regime R1.

**Repair Regime R1.** In general, semi-Markov theory can be used to establish formulas for the unavailability, cf. [30]. In practice, we usually have $\mu_G$ relatively small compared to $\mu_F$. Typically, $\delta = \mu_G/\mu_F$ is less than 0.1. In this case we can establish simple approximation formulas as shown below.

First we consider the case with two components, i.e., $n = 2$. The regenerative points for the process $\Phi$ are generated by the jumps from state 2 to 1. In view of (4.102) the limiting system unavailability $\bar{A}$ can be written as

$$\bar{A} = \frac{E[\max\{R-T,0\}]}{ET + E[\max\{R-T,0\}]} \tag{4.103}$$

$$= \frac{(\mu_G - w)}{\mu_F + (\mu_G - w)}, \tag{4.104}$$

where

$$w = E[\min\{R,T\}] = \int_0^\infty \bar{F}(t)\bar{G}(t)\,dt,$$

noting that $\max\{R - T, 0\} = R - \min\{R, T\}$ and the system downtime equals 0 if the repair of the failed component is completed before the failure of the operating component, and equals the difference between the repair time of the failed component and the time to failure of the operating component if this difference is positive. Thus we have proved the following theorem.

**Theorem 63** *If $n = 2$, then the unavailability $\bar{A}$ is given by (4.104).*

We now assume an exponential failure time distribution $F(t) = 1 - e^{-\lambda t}$. Then we have

$$\bar{A} \approx \bar{A}',\qquad(4.105)$$

where

$$\bar{A}' = \frac{\lambda^2}{2}ER^2 = \frac{\delta^2}{2}[1 + c_G^2].\qquad(4.106)$$

This gives a simple approximation formula for computing $\bar{A}$. The approximation (4.105) is established formally by the following proposition.

**Proposition 64** *If $n = 2$ and $F(t) = 1 - e^{-\lambda t}$, then*

$$0 \le \bar{A}' - \bar{A} \le (\bar{A}')^2 + \frac{\delta^3}{6}\frac{ER^3}{\mu_G^3}.\qquad(4.107)$$

**Proof.** Using that $1 - e^{-\lambda t} \le \lambda t$ and changing the order of integration, it follows that

$$\begin{aligned}
\bar{A} &= \frac{\lambda(\mu_G - w)}{1 + \lambda(\mu_G - w)} \le \lambda(\mu_G - w) &(4.108)\\
&= \lambda \int_0^\infty F(t)\bar{G}(t)dt \\
&\le \lambda \int_0^\infty (\lambda t)\bar{G}(t)dt \\
&= \lambda^2 \frac{1}{2}ER^2 = \bar{A}'. &(4.109)
\end{aligned}$$

It remains to show the right-hand inequality of (4.107). Considering

$$\begin{aligned}
\bar{A}(1 + \lambda(\mu_G - w)) &= \lambda \int_0^\infty F(t)\bar{G}(t)dt \\
&\ge \lambda \int_0^\infty (\lambda t - \frac{1}{2}(\lambda t)^2)\bar{G}(t)dt \\
&= \bar{A}' - \frac{1}{6}\lambda^3 ER^3
\end{aligned}$$

and the inequalities $\bar{A} \leq \lambda(\mu_G - w) \leq \bar{A}'$ obtained above, it is not difficult to see that

$$0 \leq \bar{A}' - \bar{A} \leq \bar{A}\lambda(\mu_G - w) + \frac{1}{6}\lambda^3 ER^3 \leq (\bar{A}')^2 + \frac{1}{6}\lambda^3 ER^3,$$

which completes the proof.    ∎

Hence $\bar{A}'$ overestimates the unavailability and the error term will be negligible provided that $\delta = \mu_G/\mu_F$ is sufficiently small.

Next, let us compare the approximation formula $\bar{A}'$ with the standard "Markov formula" $\bar{A}^M = \delta^2$, obtained by assuming exponentially distributed failure and repair times (replace $c_G^2$ by 1 in the expression (4.106) for $\bar{A}'$, or use the Markov formula (4.101), p. 162). It follows that

$$\bar{A}' = \bar{A}^M \cdot \frac{1}{2}[1 + c_G^2].    \tag{4.110}$$

From this, we see that the use of the Markov formula when the squared coefficient of variation of the repair time distribution, $c_G^2$, is not close to 1, will introduce a relatively large error. If the repair time is a constant, then $c_G^2 = 0$ and the unavailability using the Markov formula is two times $\bar{A}'$. If $c_G^2$ is large, say 2, then the unavailability using the Markov formula is $2/3$ of $\bar{A}'$.

Assume now $n > 2$. The repair regime is R1 as before. Assume that $\delta$ is relatively small. Then it is possible to generalize the approximations obtained above for $n = 2$.

Since $\delta$ is small, there will be a negligible probability of having $\Phi \leq n - 3$, i.e., three or more components not functioning at the same time. By neglecting this possibility we obtain a simplified process that is identical to the process for the two-component system analyzed above, with failure rate $(n-1)\lambda$. Hence by replacing $\lambda$ with $(n-1)\lambda$, formula (4.105) is valid for general $n$, i.e., $\bar{A} \approx \bar{A}'$, where

$$\bar{A}' = \frac{[(n-1)\delta]^2}{2}[1 + c_G^2].$$

The error bounds are, however, more difficult to obtain, see [30].

The relation between the approximation formulas $\bar{A}'$ and $\bar{A}^M$, given by (4.101), p. 162, are the same for all $n \geq 2$. Hence $\bar{A}' = \bar{A}^M \cdot \frac{1}{2}[1 + c_G^2]$ (formula (4.110)) holds for $n > 2$ too.

Next we will establish results for the long run average number of system failures. It follows from the Renewal Reward Theorem that $EN_t/t$ and $E[N_{t+s} - N_t]/s$ converge to $\lambda_\Phi = EN/ES$ as $t \to \infty$, where $N$ equals the number of system failures in one renewal cycle and $S$ equals the length of the cycle as before. With probability one, $N_t/t$ converges to the same value. Under repair regime R1, $N \in \{0, 1\}$. Hence $EN$ equals the probability that the system fails in a cycle, i.e., $EN = q$ using the terminology of Sections 4.3 and 4.4. Below we find expressions for $\lambda_\Phi$ in the case that the repair regime is R1. The regenerative points are consecutive visits to state $n - 1$.

**Theorem 65** *If $n = 2$, then*

$$\lambda_\Phi = \frac{q}{\mu_F + EY}, \tag{4.111}$$

*where*

$$q = \int_0^\infty F(t)dG(t), \tag{4.112}$$

$$EY = \int_0^\infty F(t)\bar{G}(t)\,dt.$$

**Proof.** First note that $EY$ equals the expected downtime in a cycle and is given by

$$EY = E[(R - T)I(T < R)] = E[R - \min\{R, T\}],$$

cf. (4.103)–(4.104), p. 162. We have established above that

$$\lambda_\Phi = \frac{EN}{ES} = \frac{q}{ES},$$

where $N$ equals the number of system failures in one renewal cycle, $S$ equals the length of the cycle, and $q = P(T \le R)$ equals the probability of having a system failure during a cycle. Thus it remains to show that

$$ES = \mu_F + EY. \tag{4.113}$$

Suppose the system has just jumped to state 1. We then have one component operating and one undergoing repair. Now if a system failure occurs (i.e., $T \le R$), then the cycle length equals R, and if a system failure does not occur (i.e., $T > R$), then the cycle length equals $T$. Consequently,

$$S = I(T \le R)R + I(T > R)T = T + (R - T)I(T < R).$$

Formula (4.113) follows and the proof is complete. ∎

We see from (4.111) that if $F(t)$ is exponential with rate $\lambda$ and the components are highly available, then

$$\lambda_\Phi \approx \lambda^2 \mu_G.$$

If $n > 2$ and the repair regime is R1, it is not difficult to see that $q$ is given by (4.112) with $F(t)$ replaced by $1 - e^{-(n-1)\lambda t}$. It is however more difficult to find an expression for $ES$. For highly available components,

we can approximate the system with a two-state system with failure rate
$(n - 1)\lambda$; hence,

$$\lambda_\Phi \approx [(n - 1)\lambda]^2 \mu_G,$$
$$ES \approx \frac{1}{(n - 1)\lambda}.$$

When the state process of the system jumps from state $n$ to $n - 1$, it will
return to state $n$ with a high probability and the sojourn time in state $n-1$
will be relatively short; consequently, the expected cycle length is approximately equal to the expected time in the best state $n$, i.e., $1/(n - 1)\lambda$.

**Repair regime R2.** Finally in this section we briefly comment on the
repair regime R2. We assume constant failure rates. It can be argued that
if there is ample repair facilities, i.e., the repair regime is R2, the steady-
state unavailability is invariant with respect to the repair time distribution,
cf., e.g., Smith [157] and Tijms [168], p. 175. This means that we can use
the steady-state Markov formula (4.100), p. 161, also when the repair time
distribution is not exponential. The result only depends on the repair time
distribution through its mean value. However, a strict mathematical proof
of this invariance result does not seem to have been presented yet.

**Bibliographic Notes.** Alternating renewal processes are studied in
many textbooks, e.g., Birolini [46] and Ross [145]. Different versions of the
one-component downtime distribution formula in Theorem 37 (p. 110) have
been formulated and proved in the literature, cf. [46, 47, 62, 73, 78, 166].
The first version was established by Takács. Theorem 37, which is taken
from Haukås and Aven [91], seems to be the most general formulation and
also has the simplest proof.

Some key references to the theory of point availability of monotone systems and the mean number of system failures are Barlow and Proschan [33,
34] and Ross [146]; see also Aven [16]. Parallel systems of two identical components have been studied by a number of researchers, see, e.g., [36, 81, 84].
Gaver [81] established formulas for the distribution and mean of the time
to the first system failure, identical to those presented in Section 4.4, p.
131. Our derivation of these formulas is different however from Gaver's.

Asymptotic analysis of highly available systems has been carried out by a
number of researchers. A survey is given by Gertsbakh [83], with emphasis
on results related to the convergence of the distribution of the first system
failure to the exponential distribution. See also the summary given in the
books by Gnedenko and Ushakov [84], Ushakov [169], and Kovalenko et al.
[121, 122]. Some of the earliest results go back to work done by Keilson
[115] and Solovyev [160]. A result similar to Lemma 45 (p. 120) was first
proved by Keilson [115]; see also [84, 116, 120]. Our version of this lemma is
taken from Aven and Jensen [29]. To establish the asymptotic exponential

distribution, different normalizing factors are used, e.g., $q/E_0S$, where $q$ equals the probability of having at least one system failure in a renewal cycle and $E_0S$ equals the expected cycle length given that no system failures occur in the cycle. This factor, as well as the other factors considered in the early literature in this field (cf., e.g., the references [83, 84, 169]) are generally difficult to compute. The asymptotic failure rate of the system, $\lambda_\phi$, is more attractive from a computational point of view, and is given most attention in this presentation. We find it somewhat difficult to read some of the earlier literature on availability. A large part of the research in this field has been developed outside the framework of monotone system theory. Using this framework it is possible to give a unified presentation of the results. Our set-up and results (Section 4.4) are to a large extent taken from the recent papers by Aven and Haukås [24] and Aven and Jensen [29]. These papers also cover convergence of the number of system failures to the Poisson distribution.

The literature includes a number of results proving that the exponential/Poisson distribution is the asymptotic limit of certain sums of point processes. Most of these results are related to the thinning of independent processes, see e.g., Çinlar [57], Daley and Vere-Jones [64], and Kovalenko et al. [122]. See also Lam and Lehoczky [125] and the references therein. These results are not applicable for the availability problems studied in this book.

Sections 4.5 and 4.6 are to a large extent based on Gåsemyr and Aven [80], Aven and Haukås [25], and Aven and Jensen [29]. Gåsemyr and Aven [80] and Aven and Haukås [25] study the asymptotic downtime distribution given system failure. Theorem 58 is due to Haukås (see [29, 90]) and Smith [158]. Aven and Jensen [29] gives sufficient conditions for when a compound Poisson distribution is an asymptotic limit for the distribution of the downtime of a monotone system observed in a time interval. An alternative approach for establishing the compound Poisson process limit is given by Serfozo [148]. There exist several asymptotic results in the literature linking the sums of independent point processes with integer marks to the compound Poisson process; see, e.g., [165]. It is, however, not possible to use these results for studying the asymptotic downtime distributions of monotone systems.

Section 4.7.1 generalizes results obtained in the previous sections to multistate systems. The presentation on multistate systems is based on Aven [13, 17]. For the analysis in Section 4.7.3 on standby systems, reference is given to the work by Aven and Opdal [30].

In this chapter we have primarily focused on the situation that the component lifetime distributions are exponential. In Section 4.7.1 we outlined how some of the results can be extended to phase-type distributions. A detailed analysis of the nonexponential case (nonregenerative case) is however outside the scope of this book. Further research is needed to present formally proved results for the general case. Presently, the literature covers only some particular cases. Intuitively, it seems clear that it is possible to generalize many of the results obtained in this chapter. Consider, for example, the convergence to the Poisson process for the number of system failures. As long as the components are highly available, we would expect that the number of failures are approximately Poisson distributed. But formal asymptotic results are rather difficult to establish; see, for example, [28, 113, 117, 123, 164, 174]. Strict conditions have to be imposed to establish the results, to the system structure and the component lifetime and downtime distributions. Also the general approach of showing that the compensator of the counting process converges in probability (see Daley and Vere-Jones [64], p. 552), is difficult to apply in our setting.

Of course, this chapter covers only a small number of availability models compared to the large number of models presented in the literature. We have, for example, not included models where some components remain in "suspended animation" while a component is being repaired/replaced, and models allowing preventive maintenance. For such models, and other related models, refer to the above cited references, Beichelt and Franken [38], Osaki [138], Srinivasan and Subramanian [162], Van Heijden and Schornagel [172], and Yearout et. al. [178]. See also the survey paper by Smith et al. [159].

# 5

# Maintenance Optimization

In this chapter we combine the general lifetime model of Chapter 3 with maintenance actions like repairs and replacements. Given a certain cost and reward structure an optimal repair and replacement strategy will be derived. We begin with some basic and well-known models and come then to more complex ones, which show how the general approach can be exploited to open a variety of different optimization models.

## 5.1   Basic Replacement Models

First of all we consider some basic models that are simple in both the lifetime modeling and the optimization criterion. These basic models include the age and the block replacement models that are widely used and thoroughly investigated. A technical system is considered, the lifetime of which is described by a positive random variable $T$ with distribution $F$. Upon failure the system is immediately replaced by an equivalent one and the process repeats itself. A preventive replacement can be carried out before failure. Each replacement incurs a cost of $c > 0$ and each failure adds a penalty cost $k > 0$.

### 5.1.1   Age Replacement Policy

For this policy a replacement age $s, s > 0$, is fixed for each system at which a preventive replacement takes place. If $T_i, i = 1, 2, ...$, are the successive lifetimes of the systems, then $\tau_i = T_i \wedge s$ denotes the operating time of the $i$th system and equals the $i$th cycle length. The random variables $T_i$ are assumed to form an i.i.d. sequence with common distribution $F$, i.e., $F(t) = P(T_i \leq t)$. The costs for one cycle are described by the stochastic process $Z = (Z_t), t \in \mathbb{R}_+$, $Z_t = c + kI(T \leq t)$. Clearly, the average cost after $n$ cycles is

$$\frac{\sum_{i=1}^{n} Z_{\tau_i}}{\sum_{i=1}^{n} \tau_i}$$

and the total cost per unit time up to time $t$ is given by

$$C_t = \frac{1}{t} \sum_{i=1}^{N_t} Z_{\tau_i},$$

where $(N_t), t \in \mathbb{R}_+$, is the renewal counting process generated by $(\tau_i)$ and $Z_\tau = c + kI(T \leq \tau)$ describes the incurred costs in one cycle. It is well known from renewal theory (see Appendix B, p. 250) that the limits of the expectations of these ratios, $K_s$, coincide and are equal to the ratio of the expected costs for one cycle and the expected cycle length:

$$K_s = \lim_{n \to \infty} E\left[\frac{\sum_{i=1}^{n} Z_{\tau_i}}{\sum_{i=1}^{n} \tau_i}\right] = \lim_{t \to \infty} EC_t = \frac{EZ_\tau}{E\tau}.$$

The objective is to find the replacement age that minimizes this long run average cost per unit time. Inserting the cost function $Z_t = c + kI(T \leq t)$ we get

$$K_s = \frac{c + kF(s)}{\int_0^s (1 - F(x))dx}. \tag{5.1}$$

Now elementary analysis can be used to find the optimal replacement age $s$, i.e., to find $s^*$ with

$$K_{s^*} = \inf\{K_s : s \in \mathbb{R}_+ \cup \{\infty\}\}.$$

Here $s^* = \infty$ means that preventive replacements do not pay and it is optimal to replace only at failures. As can be easily seen this case occurs if the lifetimes are exponentially distributed, i.e., if $F(t) = 1 - \exp\{-\lambda t\}$, $t \geq 0, \lambda > 0$, then $K_\infty = \lambda(c + k) \leq K_s$ for all $s > 0$.

**Example 40** Using rudimentary calculus we see that in the case of an increasing failure rate $\lambda(t) = f(t)/\bar{F}(t)$, the optimal replacement age is given by

$$s^* = \inf\{t \in \mathbb{R}_+ : \lambda(t) \int_0^t \bar{F}(x)dx - F(t) \geq \frac{c}{k}\},$$

where $\inf \emptyset = \infty$. By differentiating it is not hard to show that the left-hand side of the inequality is increasing in the IFR case so that $s^*$ can easily be determined. As an example consider the Weibull distribution $F(t) = 1 - \exp\{-(\lambda t)^\beta\}, t \geq 0$ with $\lambda > 0$ and $\beta > 1$. The corresponding failure rate is $\lambda(t) = \lambda\beta(\lambda t)^{\beta-1}$ and the optimal replacement age is the unique solution of

$$\lambda(t)\int_0^t \exp\{-(\lambda x)^\beta\}dx - 1 + \exp\{-(\lambda t)^\beta\} = \frac{c}{k}.$$

The cost minimum is then given by $K_{s^*} = k\lambda(s^*)$.

The age replacement policy allows for planning of a preventive replacement only when a new item is installed. If one wants to fix the time points for preventive replacements in advance for a longer period, one is led to the block replacement policy.

## 5.1.2 Block Replacement Policy

Under this policy the item is replaced at times $is, i = 1, 2, \ldots$ and $s > 0$, and at failures. The preventive replacements occur at regular predetermined intervals at a cost of $c$, whereas failures within the intervals incur a cost of $c + k$.

The advantage of this policy is the simple structure and administration because the time points of preventive replacements are fixed and determined in advance. On the other hand, preventive replacements are carried out, irrespective of the age of the processing unit, so that this policy is usually applied to several units at the same time and only if the replacement costs $c$ are comparatively low.

For a fixed time interval $s$ the long run average cost per unit time is

$$K_s = \frac{(c+k)M(s) + c}{s}, \tag{5.2}$$

where $M$ is the renewal function $M(t) = \sum_{j=1}^\infty F^{*j}(t)$ (see Appendix B, p. 244). If the renewal function is known explicitly, we can again use elementary analysis to find the optimal $s$, i.e., to find $s^*$ with

$$K_{s^*} = \inf\{K_s : s \in \mathbb{R}_+ \cup \{\infty\}\}.$$

In most cases the renewal function is not known explicitly. In such a case asymptotic expansions like Theorem 114, p. 247 in Appendix B or numerical methods have to be used. As is to be expected in the case of an $\text{Exp}(\lambda)$ distribution, preventive replacements do not pay: $M(s) = \lambda s$ and $s^* = \infty$.

**Example 41** Let $F$ be the Gamma distribution function with parameters $\lambda > 0$ and $n = 2$. The corresponding renewal function is

$$M(s) = \frac{\lambda s}{2} - \frac{1}{4}(1 - e^{-2\lambda s})$$

(cf. [1], p. 244) and $s^*$ can be determined as the solution of

$$\frac{d}{ds}M(s) = \frac{M(s)}{s} + \frac{c}{s(c+k)}.$$

The solution $s^*$ is finite if and only if $c/(c+k) < 1/4$, i.e., if failure replacements are at least four times more expensive than preventive replacements.

The age and block replacement policies will result in a finite optimal value of $s$ only if there is some aging and wear-out of the units, i.e., in probabilistic terms the lifetime distribution $F$ fulfills some aging condition like IFR, NBU, or NBUE (see Chapter 2 for these notions). To judge whether it pays to follow a certain policy and in order to compare the policies it is useful to consider the number of failures and the number of planned preventive replacements in a time interval $[0, t]$.

## 5.1.3  Comparisons and Generalizations

Let $F$ be the underlying lifetime distribution that generates the renewal counting process $(N_t), t \in \mathbb{R}_+$, so that $N_t$ describes the number of failures or completed replacements in $[0, t]$ following the basic policy replace at failure only. Let $N_t^A(s)$ and $N_t^B(s)$ denote the number of failures up to time $t$ following policy $A$ (age replacement) or $B$ (block replacement), respectively, and $R_t^A(s)$ and $R_t^B(s)$ the corresponding total number of removals in $[0, t]$ including failures and preventive replacements. We now want to summarize some early comparison results that can be found, including the proofs, in the monographs of Barlow and Proschan [33, 34]. We remind the reader of the notion of stochastic comparison of two positive random variables $X$ and $Y$: $X \leq_{st} Y$ means $P(X > t) \leq P(Y > t)$ for all $t \in \mathbb{R}_+$.

**Theorem 66**  *The following four assertions hold true:*
  *(i) $N_t \geq_{st} N_t^B(s)$ for all $t \geq 0, s \geq 0 \Longleftrightarrow F$ is NBU;*
  *(ii) $N_t \geq_{st} N_t^A(s)$ for all $t \geq 0, s \geq 0 \Longleftrightarrow F$ is NBU;*
  *(iii) $F$ IFR $\Rightarrow N_t \geq_{st} N_t^A(s) \geq_{st} N_t^B(s)$ for all $t \geq 0, s \geq 0$;*
  *(iv) $R_t^A(s) \leq_{st} R_t^B(s)$ for all $t \geq 0, s \geq 0$.*

Part (i) and (ii) say that under the weak aging notion NBU it is useful to apply a replacement strategy, since the number of failures is (stochastically) decreased under such a strategy. If, in addition, $F$ has an increasing failure rate, block replacement results in stochastically less failures than age replacement, and it follows that $EN_t^A(s) \geq EN_t^B(s)$. On the other hand, for any lifetime distribution $F$ (irrespective of aging notions) block policies have more removals than age policies.

**Theorem 67**  *$N_t^A(s)$ is stochastically increasing in $s$ for each $t \geq 0$ if and only if $F$ is IFR.*

This result says that IFR is characterized by the reasonable aging condition that the number of failures is growing with increasing replacement age. Somewhat weaker results hold true for the block policy (see Shaked and Zhu [154] for proofs):

**Theorem 68** *If $N_t^B(s)$ is stochastically increasing in $s$ for each $t \geq 0$, then $F$ is IFR.*

**Theorem 69** *The expected value $EN_t^B(s)$ is increasing in $s$ for each $t \geq 0$ if and only if the renewal function $M(t)$ is convex.*

Since the monographs of Barlow and Proschan appeared, many possible generalizations have been investigated concerning a) the comparison methods, b) the lifetime models and replacement policies and the cost structures. It is beyond the scope of this book to describe all of these models and refinements. Some hints for further reading can be found in the Bibliographic Notes at the end of the chapter.

Berg [39] and Dekker [69] among others use a *marginal cost analysis* for studying the optimal replacements problem. Let us, for example, consider this approach for block-type policies. In this model it is assumed that the long run average cost per unit time is given by

$$K_s = \frac{c + R(s)}{s}, \tag{5.3}$$

where $c$ is the cost of a preventive replacement and $R(s) = \int_0^s r(x)dx$ denotes the total expected costs due to deterioration over an interval of length $s$. The derivative $r$, called the (*marginal*) deterioration cost rate, is assumed to be continuous and piecewise differentiable. If in the block replacement model of the preceding Section 5.1.2 the lifetime distribution function $F$ has a bounded density $f$, then it is known (see Appendix B, p. 248) that also the corresponding renewal function $M$ admits a density $m$ and we have $R(s) = \int_0^s (c + k)m(x)dx$, which shows that this is a special case of this block-type model. Now certain properties of the marginal cost rate can be carried over to the cost function $K$. The proof of the following theorem is straightforward and can be found in [69].

**Theorem 70** *(i) If $r(t)$ is nonincreasing on $[t_0, t_1]$ for some $0 \leq t_0 < t_1$ and $r(t_0) < K_{t_0}$, then $K_s$ is also nonincreasing in $s$ on $[t_0, t_1]$;*
*(ii) if $r(t)$ increases strictly for $t > t_0$ and some $t_0 \geq 0$, where $r(t_0) < K_{t_0}$, and if either*

$$(a) \lim_{t \to \infty} r(t) = \infty \quad or \quad (b) \lim_{t \to \infty} r(t) = a \ and \ \lim_{t \to \infty} (at - R(t)) > c,$$

*then $K_s$ has a minimum, say $K^*$ at $s^*$, which is unique on $[t_0, \infty)$; moreover, $K^* = K_{s^*} = r(s^*)$.*

Thus a myopic policy, in which at every moment we consider whether to defer the replacement or not, is optimal. That is, the expected cost of deferring the replacement to level $t + \Delta t$, being $r(t)\Delta t$, should be compared with the minimum average cost over an interval of the same length, being $K^*\Delta t$. Hence if $r(t)$ is larger than $K^*$, the deferment costs are larger and we should replace. This is the idea of marginal cost analysis as described for example in [39] and [69].

The above framework can be extended to age-type policies if we consider the following long run average cost per unit time

$$K_s = \frac{c + \int_0^s r(x)\bar{F}(x)dx}{\int_0^s \bar{F}(x)dx}, \tag{5.4}$$

where $c$ is the cost of a preventive replacement and $r$ denotes the marginal deterioration cost rate. Again it can easily be seen that the basic age replacement model (5.1) is a special case setting $r(x) = k\lambda(x)$, where $\lambda(x) = f(x)/\bar{F}(x)$ is the failure rate. Now a very similar analysis can be carried out (see [69]) and the same theorem holds true for this cost criterion except that condition (ii) (b) has to be replaced by

$$\lim_{t\to\infty} r(t) = a \text{ and } a > \lim_{s\to\infty} K_s \text{ for some } a > 0.$$

This shows that behind these two quite different models the same optimizations mechanism works. This has been exploited by Aven and Bergman in [22] (see also [23]). They recognized that for many replacement models the optimization criterion can be written in the form

$$\frac{E\left[\int_0^\tau a_t h_t dt + c_0\right]}{E\left[\int_0^\tau h_t dt + p_0\right]}, \tag{5.5}$$

where $\tau$ is a stopping time based on the information about the condition of the system, $(a_t)$ is a nondecreasing stochastic process, $(h_t)$ is a nonnegative stochastic process, and $c_0$ and $p_0$ are nonnegative random variables; all variables are adapted to the information about the condition of the system. Both, the block-type model (5.3) and the age-type model (5.4) are included. Take, for example, for all random quantities deterministic values, especially $\tau = t$, $h_t = \bar{F}(t)$, $a_t = r(t)$, $p_0 = 0$, and $c_0 = c$. This leads to the age-type model. In (5.5) the stopping time $\tau$ is the control variable which should be determined in a way that (5.5) is minimized. This problem of choosing a minimizing stopping time is known as an optimal stopping problem and will be further developed in the next section.

## 5.2   A General Replacement Model

In this section we want to develop the tools that allow certain maintenance problems to be solved in a fairly general way, also considering the possibility of taking different levels of information into account.

### 5.2.1   An Optimal Stopping Problem

In connection with maintenance models as described above, we will have to solve optimization problems. Often an optimal point in time has to be determined that maximizes some reward functional. In terms of the theory of stochastic processes, this optimal point in time will be a stopping time $\tau$ that maximizes the expectation $EZ_\tau$ of some stochastic process $Z$. We will see that the smooth semimartingale (SSM) representation of $Z$, as introduced in detail in Section 3.1, is an excellent tool to carry out this optimization. Therefore, we want to solve the stopping problem and to characterize optimal stopping times for the case in which $Z$ is an SSM and $\tau$ ranges in a suitable class of stopping times, say

$$C^{\mathbb{F}} = \{\tau : \tau \text{ is an } \mathbb{F}\text{-stopping time, } \tau < \infty, EZ_\tau > -\infty\}.$$

Without any conditions on the structure of the process $Z$ one cannot hope to find an explicit solution of the stopping problem. A condition called *monotone case* in the discrete time setting can be transferred to continuous time as follows.

**Definition 27 (MON)** *Let* $Z = (f, M)$ *be an SSM. Then the following condition*

$$\{f_t \leq 0\} \subset \{f_{t+h} \leq 0\} \ \forall \ t, h \in \mathbb{R}_+, \ \bigcup_{t \in \mathbb{R}_+} \{f_t \leq 0\} = \Omega \qquad (5.6)$$

*is said to be the monotone case and the stopping time*

$$\zeta = \inf\{t \in \mathbb{R}_+ : f_t \leq 0\}$$

*is called the ILA-stopping rule (infinitesimal-look-ahead).*

Obviously in the monotone case the process $f$ driving the SSM $Z_t = \int_0^t f_s ds + M_t$ remains negative (nonpositive) if it once crosses zero from above and the ILA-stopping rule $\zeta$ is a natural candidate to solve the maximization problem.

**Theorem 71** *Let* $Z = (f, M)$ *be an* $\mathbb{F}$*-SSM and* $\zeta$ *the ILA-stopping rule. If the martingale $M$ is uniformly integrable, then in the monotone case (5.6)*

$$EZ_\zeta = \sup\{EZ_\tau : \tau \in C^{\mathbb{F}}\}.$$

**Remark 18** The condition that the martingale is uniformly integrable can be relaxed; in [109] it is shown that the condition may be replaced by

$$M_\zeta \in L^1, \zeta \in C^{\mathbb{F}}, \lim_{t \to \infty} \int_{\{\tau > t\}} M_t^- \, dP = 0 \; \forall \; \tau \in C^{\mathbb{F}},$$

where as usual $a^-$ denotes the negative part of $a \in \mathbb{R} : a^- = \max\{-a, 0\}$. But in most cases such a generalization will not be used in what follows.

**Proof.** Since $M$ is uniformly integrable we have $EM_\tau = 0$ for all $\tau \in C^{\mathbb{F}}$ as a consequence of the optional sampling theorem (cf. Appendix A, p. 231). Also $\zeta$ is an element of $C^{\mathbb{F}}$ because $\zeta < \infty$ per definition and $EZ_\zeta^- \leq E|Z_0| + E|M_\zeta| < \infty$. It remains to show that

$$E \int_0^\zeta f_s ds \geq E \int_0^\tau f_s ds$$

for all $\tau \in C^{\mathbb{F}}$. But this is an immediate consequence of $f_s > 0$ on $\{\zeta > s\}$ and $f_s \leq 0$ on $\{\zeta \leq s\}$. ∎

The following example demonstrates how this optimization technique can be applied.

**Example 42** Let $\rho$ be an exponentially distributed random variable with parameter $\lambda > 0$ on the basic probability space $(\Omega, \mathcal{F}, \mathbb{F}, P)$ equipped with the filtration $\mathbb{F}$ generated by $\rho$ :

$$\mathcal{F}_t = \sigma(\{\rho > s\}, 0 \leq s \leq t) = \sigma(I(\rho > s), 0 \leq s \leq t) = \sigma(\rho \wedge t).$$

For the latter equality we make use of our agreement that $\sigma(\cdot)$ denotes the completion of the generated $\sigma$-algebra so that, for instance, the event $\{\rho = t\} = \bigcap_{n \in \mathbb{N}}\{t - \frac{1}{n} < \rho \leq t\}$ is also included in $\sigma(\rho \wedge t)$. Then we define

$$Z_t = e^t I(\rho > t), \; t \in \mathbb{R}_+.$$

This process $Z$ can be interpreted as the potential gain in a harvesting problem (in a wider sense): There is an exponentially growing potential gain and at any time $t$ the decision-maker has to decide whether to realize this gain or to continue observations with the chance of earning a higher gain. But the gain can only be realized up to a random time $\rho$, which is unknown in advance. So there is a risk to loose all potential gains and the problem is to find an optimal harvesting time.

The process $Z$ is adapted, right-continuous and integrable with

$$E[Z_{t+h}|\mathcal{F}_t] = e^{t+h} E[I(\rho > t + h)|\mathcal{F}_t] = e^{(1-\lambda)h} Z_t, \; h, t \in \mathbb{R}_+.$$

Thus $Z$ is a submartingale (martingale, supermartingale), if $\lambda < 1$ ($\lambda = 1, \lambda > 1$). Obviously we have

$$\lim_{h \to 0+} \frac{1}{h} E[Z_{t+h} - Z_t | \mathcal{F}_t] = Z_t(1 - \lambda) = f_t.$$

Theorem 9, p. 51, states that $Z$ is an SSM with representation:

$$Z_t = 1 + \int_0^t Z_s(1 - \lambda)ds + M_t.$$

Three cases will be discussed separately:

1. $\lambda < 1$. The monotone case (5.6) holds true with the ILA stopping time $\zeta = \rho$. But $\zeta$ is not optimal, because $EZ_\zeta = 0$ and $Z$ is a submartingale with unbounded expectation function: $\sup\{EZ_\tau : \tau \in C^{\mathbb{F}}\} = \infty$.

2. $\lambda > 1$. The monotone case holds true with the ILA stopping time $\zeta = 0$. It is not hard to show that in this case the martingale

$$M_t = Z_t - 1 - \int_0^t Z_s(1 - \lambda)ds$$

is uniformly integrable. Theorem 71 ensures that $\zeta$ is optimal with $EZ_\zeta = 1$.

3. $\lambda = 1$. Again the monotone case (5.6) holds true with the ILA stopping time $\zeta = 0$. However, the martingale $M_t = e^t I(\rho > t) - 1$ is not uniformly integrable. But for all $\tau \in C^{\mathbb{F}}$ we have $EM_\tau^- \le 1$ and

$$\lim_{t \to \infty} \int_{\{\tau > t\}} M_t^- dP \le \lim_{t \to \infty} \int_{\{\tau > t\}} dP = 0,$$

so that the more general conditions mentioned in the above remark are fulfilled with $M_\zeta = 0$. This yields

$$EZ_\zeta = 1 = \sup\{EZ_\tau : \tau \in C^{\mathbb{F}}\}.$$

### 5.2.2  A Related Stopping Problem

As was described in Section 5.1, replacement policies of age and block type are strongly connected to the following stopping problem: Minimize

$$K_\tau = \frac{EZ_\tau}{EX_\tau}, \tag{5.7}$$

in a suitable class of stopping times, where $Z$ and $X$ are real stochastic processes. For a precise formulation and solution of this problem we use the set-up given in Chapter 3. On the basic complete probability space $(\Omega, \mathcal{F}, P)$ a filtration $\mathbb{F} = (\mathcal{F}_t)$, $t \in \mathbb{R}_+$, is given, which is assumed to fulfill the usual conditions concerning right continuity and completeness. Furthermore, let $Z = (Z_t)$ and $X = (X_t)$, $t \in \mathbb{R}_+$, be real right-continuous stochastic processes adapted to the filtration $\mathbb{F}$. Let $T > 0$ be a finite $\mathbb{F}$-stopping time with $EZ_T > -\infty, E|X_T| < \infty$ and

$$C_T^\mathbb{F} = \{\tau : \tau \text{ is an } \mathbb{F}\text{-stopping time}, \tau \leq T, EZ_\tau > -\infty, E|X_\tau| < \infty\}.$$

For $\tau \in C_T^\mathbb{F}$ we consider the ratio $K_\tau$ in (5.7). The stopping problem is then to find a stopping time $\sigma \in C_T^\mathbb{F}$, with

$$K^* = K_\sigma = \inf\{K_\tau : \tau \in C_T^\mathbb{F}\}. \tag{5.8}$$

In this model $T$ describes the random lifetime of some technical system. The index $t$ can be regarded as a time point and $\mathcal{F}_t$ as the $\sigma$-algebra which contains all gathered information up to time $t$. The stochastic processes $Z$ and $X$ are adapted to the stream of information $\mathbb{F}$, i.e., $Z$ and $X$ are observable with respect to the given information or in mathematical terms, $Z_t$ and $X_t$ are $\mathcal{F}_t$-measurable for all $t \in \mathbb{R}_+$. The replacement times can then be identified with stopping times not greater than the system lifetime $T$.

**Example 43** In the case of block-type models no random information is to be considered so that the filtration reduces to the trivial one and all stopping times are constants, i.e., $C_T^\mathbb{F} = \mathbb{R}_+ \cup \{\infty\}$. In this case elementary analysis manipulations yield the optimum and no additional efforts are necessary.

**Example 44** Let $Z_t = c + kI(T \leq t)$, $X_t = t$, and $\mathcal{F}_t = \sigma(Z_s, 0 \leq s \leq t) = \sigma(I(T \leq s), 0 \leq s \leq t)$ be the $\sigma$-algebra generated by $Z$, i.e., at any time $t \geq 0$ it is known whether the system works or not. The $\mathbb{F}$-stopping times $\tau \in C_T^\mathbb{F}$ are of the form $\tau = t^* \wedge T$ for some $t^* > 0$. Then we have $EZ_\tau = c + kEI(T \leq \tau) = c + kP(T \leq t^*)$ and $EX_\tau = E\tau$, which leads to the basic age replacement policy.

To solve the above-mentioned stopping problem, we will make use of semimartingale representations of the processes $Z$ and $X$. It is assumed that $Z$ and $X$ are smooth semimartingales (SSMs) as introduced in Section 3.1 with representations

$$Z_t = Z_0 + \int_0^t f_s ds + M_t,$$

$$X_t = X_0 + \int_0^t g_s ds + L_t.$$

As in Section 3.1 we use the short notation $Z = (f, M)$ and $X = (g, L)$. Almost all of the stochastic processes used in applications without predictable jumps admit such SSM representations. The following general assumption is made throughout this section:

**Assumption (A).** $Z = (f, M)$ and $X = (g, L)$ are SSMs with $EZ_0 > 0, EX_0 \geq 0, g_s > 0$ for all $s \in \mathbb{R}_+$ and $M^T, L^T \in \mathcal{M}_0$ are uniformly integrable martingales, where $M_t^T = M_{t \wedge T}, L_t^T = L_{t \wedge T}$.

Remember that all relations between real random variables hold (only) $P$-almost surely. The first step to solve the optimization problem is to establish bounds for $K^*$ in (5.8)

**Lemma 72** *Assume that $(A)$ is fulfilled and*

$$q = \inf\{\frac{f_t(\omega)}{g_t(\omega)} : 0 \leq t < T(\omega), \omega \in \Omega\} > -\infty.$$

*Then*

$$b_l \leq K^* \leq b_u$$

*holds true, where the bounds are given by*

$$b_u = \frac{EZ_T}{EX_T},$$

$$b_l = \begin{cases} \frac{E[Z_0 - qX_0]}{EX_T} + q & \text{if } E[Z_0 - qX_0] > 0 \\ \frac{EZ_0}{EX_0} & \text{if } E[Z_0 - qX_0] \leq 0. \end{cases}$$

**Proof.** Because $T \in C_T^{\mathbb{F}}$ only the lower bound has to be shown. Since the martingales $M^T$ and $L^T$ are uniformly integrable, the optional sampling theorem (see Appendix A, p. 231) yields $EM_\tau = EL_\tau = 0$ for all $\tau \in C_T^{\mathbb{F}}$ and therefore

$$K_\tau \geq \frac{EZ_0 + qE[X_\tau - X_0]}{EX_\tau} = \frac{EZ_0 - qEX_0}{EX_\tau} + q \geq b_l.$$

The lower bound is derived observing that $EX_0 \leq EX_\tau \leq EX_T$, which completes the proof. ∎

The following example gives these bounds for the basic age replacement policy.

**Example 45** (Continuation of Example 44). Let us return to the simple cost process $Z_t = c + kI(T \leq t)$ with the natural filtration as before. Then $I(T \leq t)$ has the SSM representation

$$I(T \leq t) = \int_0^t I(T > s)\lambda(s)ds + M_t',$$

where $\lambda$ is the usual failure rate of the lifetime $T$. It follows that the processes $Z$ and $X$ have representations

$$Z_t = c + \int_0^t I(T > s)k\lambda(s)ds + M_t, \quad M_t = kM_t'$$

and

$$X_t = t = \int_0^t ds.$$

Assuming the IFR property, we obtain with $\lambda(0) = \inf\{\lambda(t) : t \in \mathbb{R}_+\}$ and $q = k\lambda(0)$ the following bounds for $K^*$ in the basic age replacement model:

$$b_u = \frac{EZ_T}{EX_T} = \frac{c+k}{ET},$$

$$b_l = \frac{c}{ET} + k\lambda(0).$$

These bounds could also be established directly by using (5.1), p. 170. The benefit of Lemma 72 lies in its generality, which also allows the bounds to be found in more complex models as the following example shows.

**Example 46** (Shock Model). Consider now a compound point process model in which shocks arrive according to a marked point process $(T_n, V_n)$ as was outlined in Section 3.3.3. Here we assume that $(T_n)$ is a nonhomogeneous Poisson process with a deterministic intensity $\lambda(s)$ integrating to $\Lambda(t) = \int_0^t \lambda(s)ds$ and that $(V_n)$ forms an i.i.d. sequence of nonnegative random variables independent of $(T_n)$ with $V_n \sim F$. The accumulated damage up to time $t$ is then described by

$$R_t = \sum_{n=1}^{N_t} V_n,$$

where $N_t = \sum_{n=1}^{\infty} I(T_n \leq t)$ is the number of shocks arrived until $t$. The lifetime of the system is modeled as the first time $R_t$ reaches a fixed threshold $S > 0$:

$$T = \inf\{t \in \mathbb{R}_+ : R_t \geq S\}.$$

We stick to the simple cost structure of the basic age replacement model, i.e.,

$$Z_t = c + kI(T \leq t).$$

But now we want to minimize the expected costs per number of arrived shocks in the long run, i.e.,

$$X_t = N_t.$$

This cost criterion is appropriate if we think, for example, of systems which are used by customers at times $T_n$. Each usage causes some random damage (shock). If the customers arrive with varying intensities governed by external circumstances, e.g., different intensities at different periods of a day, it makes no sense to relate the costs to time, and it is more reasonable to relate the costs to the number of customers served.

The semimartingale representations with respect to the internal filtration generated by the marked point process are (cf. Section 3.3.5, p. 81)

$$Z_t = c + \int_0^t I(T > s)k\lambda(s)\bar{F}((S - R_s)-)ds + M_t,$$

$$X_t = \int_0^t \lambda(s)ds + L_t.$$

The martingale $M$ is uniformly integrable and so is $L^T = (L_{t \wedge T})$ if we assume that $E\int_0^T \lambda(s)ds = E\Lambda(T) < \infty$. Lemma 72 yields, with

$$q = \inf\{k\bar{F}((S - R_t)-) : 0 \le t < T(\omega), \omega \in \Omega\} = k\bar{F}(S-),$$

the following bounds for $K^* = \inf\{K_\tau : \tau \in C_T^{\mathbb{F}}\}$ :

$$b_u = \frac{c + k}{EX_T},$$

$$b_l = \frac{c}{EX_T} + k\bar{F}(S-),$$

where $EX_T = E\Lambda(T)$. Observe that $X_T = \inf\{n \in \mathbb{N} : \sum_{i=1}^n V_i \ge S\}$ and $\{X_T > k\} = \{\sum_{i=1}^k V_i < S\}$. This yields

$$EX_T = \sum_{k=0}^\infty P(\sum_{i=1}^k V_i < S) \le \sum_{k=0}^\infty F^k(S-) = \frac{1}{\bar{F}(S-)},$$

if $F(S-) < 1$. In addition, using Wald's equation $E\sum_{n=1}^{X_T} V_n = EX_T EV_1 \ge S$, we can derive the following alternative bounds

$$b_u' = (c + k)\frac{EV_1}{S},$$

$$b_l' = (c + k)\bar{F}(S-),$$

which can easily be computed.

To solve the stopping problem (5.8) for a ratio of expectations, we use the solution of the simpler case in which we look for the maximum of the expectations $EZ_\tau$, where $Z$ is an SSM and $\tau$ ranges in a suitable class of stopping times, which has been considered in detail in Section 5.2. It is a well-known technique to replace the minimization problem (5.8) by an equivalent maximization problem. Observing that $K_\tau = EZ_\tau/EX_\tau \ge K^*$ is equivalent to $K^*EX_\tau - EZ_\tau \le 0$ for all $\tau \in C_T^{\mathbb{F}}$, where equality holds for an optimal stopping time, one has the maximization problem:

Find $\sigma \in C_T^{\mathbb{F}}$ with $EY_\sigma = \sup\{EY_\tau : \tau \in C_T^{\mathbb{F}}\} = 0$, where  (5.9)
$Y_t = K^*X_t - Z_t$ and $K^* = \inf\{K_\tau : \tau \in C_T^{\mathbb{F}}\}$.

This new stopping problem can be solved by means of the semimartingale representation of the process $Y = (Y_t)$ for $t \in [0, T)$

$$Y_t = K^* X_0 - Z_0 + \int_0^t (K^* g_s - f_s) ds + R_t, \qquad (5.10)$$

where the martingale $R = (R_t), t \in \mathbb{R}_+$, is given by

$$R_t = K^* L_t - M_t.$$

Now the procedure is as follows. If the integrand $k_s = K^* g_s - f_s$ fulfills the monotone case (MON), then Theorem 71, p. 175, of Section 5.2 yields that the ILA-stopping rule $\sigma = \inf\{t \in \mathbb{R}_+ : k_t \leq 0\}$ is optimal, provided the martingale part $R$ is uniformly integrable. Note, however, that this stopping time $\sigma$ depends on the unknown value $K^*$, which can be determined from the equality $EY_\sigma = 0$.

Next we want to define monotonicity conditions that ensure (MON). Obviously under assumption (A), p. 179, the monotone case holds true if the ratio $f_s/g_s$ is increasing ($P$-a.s.) with $f_0/g_0 < K^*$ and $\lim_{s \to \infty} f_s/g_s > K^*$. The value $K^*$ is unknown so that we need to use the bounds derived, and it seems too restrictive to demand that the ratio is increasing. Especially bath-tub-shaped functions, which decrease first up to some $s_0$ and increase for $s > s_0$, should be covered by the monotonicity condition. This results in the following definition.

**Definition 28** *Let $a, b \in \mathbb{R} \cup \{-\infty, \infty\}$ be constants with $a \leq b$. Then a function $r : \mathbb{R}_+ \to \mathbb{R}$ is called*
*(i) $(a, b)$-increasing, if for all $t, h \in \mathbb{R}_+$*

$$r(t) \geq a \ \text{implies} \ r(t + h) \geq r(t) \wedge b;$$

*(ii) $(a, b)$-decreasing, if for all $t, h \in \mathbb{R}_+$*

$$r(t) \leq b \ \text{implies} \ r(t + h) \leq r(t) \vee a.$$

Roughly spoken, an $(a, b)$-increasing function $r(t)$ passes with increasing $t$ the levels $a, b$ from below and never falls back below such a level. Between $a$ and $b$ the increase is monotone. Obviously a $(0, 0)$-decreasing function fulfills (MON) if $r(\infty) \leq 0$. A $(-\infty, \infty)$-increasing (decreasing) function is monotone in the ordinary sense.

The main idea for solving the stopping problem is that, if the ratio $f_s/g_s$ satisfies such a monotonicity condition, instead of considering all stopping times $\tau \in C_T^{\mathbb{F}}$ one may restrict the search for an optimal stopping time to the class of indexed stopping times

$$\rho_x = \inf\{t \in \mathbb{R}_+ : x g_t - f_t \leq 0\} \wedge T, \ \inf \emptyset = \infty, \ x \in \mathbb{R}. \qquad (5.11)$$

The optimal stopping level $x^*$ for the ratio $f_s/g_s$ can be determined from $EY_\sigma = 0$ and coincides with $K^*$ as is shown in the following theorem.

**Theorem 73** *Assume (A)(see p. 179) and let $\rho_x, x \in \mathbb{R}$, and the bounds $b_u, b_l$ be defined as above in (5.11) and in Lemma 72, p. 179, respectively. If the process $(r_t), t \in \mathbb{R}_+$, with $r_t = f_t/g_t$ has $(b_l, b_u)$-increasing paths on $[0, T)$, then*

$$\sigma = \rho_{x^*}, \text{ with } x^* = \inf\{x \in \mathbb{R} : xEX_{\rho_x} - EZ_{\rho_x} \geq 0\}$$

*is an optimal stopping time and $x^* = K^*$.*

**Proof.** Since $r$ is $(b_l, b_u)$-increasing with $b_l \leq K^* \leq b_u$, it follows that $r$ is also $(K^*, K^*)$-increasing, i.e., passes $K^*$ at most once from below. Thus the monotone case holds true for the SSM $Y$. From the general assumption (A) on p. 179 we deduce that the martingale part of $Y$ is uniformly integrable so that

$$\sigma = \inf\{t \in \mathbb{R}_+ : K^* g_t - f_t \leq 0\} \wedge T = \rho_{K^*}$$

is optimal with $EY_\sigma = \sup\{EY_\tau : \tau \in C_T^{\mathbb{F}}\} = 0$.

It remains to show that $x^* = K^*$. Define

$$v(x) = xEX_{\rho_x} - EZ_{\rho_x} = xEX_0 - EZ_0 + E \int_0^{\rho_x} (xg_s - f_s)ds.$$

Now $v(x)$ is obviously nondecreasing in $x$ and by the definition of $\rho_x$ and (A) we have $v(x) \geq -EZ_0$. For $x < K^*$ and $v(x) > -EZ_0$ the following strict inequality holds, since in this case we have $\rho_x > 0$ and $g_s > 0$ on $[0, \rho_x]$ :

$$v(x) < K^* EX_0 - EZ_0 + E \int_0^{\rho_x} (K^* g_s - f_s)ds \leq v(K^*) = 0.$$

Equally for $x < K^*$ and $v(x) = -EZ_0$ we have $v(x) < v(K^*) = 0$ because of $EZ_0 > 0$. Therefore,

$$x^* = \inf\{x \in \mathbb{R} : v(x) \geq v(K^*) = 0\} = K^*,$$

which proves the assertion. ∎

**Remark 19** 1. If $E[Z_0 - qX_0] < 0$, then the lower bound $b_l$ in Lemma 72 is attained for $\sigma = 0$. So in this case $K^* = EZ_0/EX_0$ is the minimum without any further monotonicity assumptions.

2. If no monotonicity conditions hold at all, then $x^* = \inf\{x \in \mathbb{R} : xEX_{\rho_x} - EZ_{\rho_x} \geq 0\}$ is the cost minimum if only stopping times of type $\rho_x$ are considered. But $T = \rho_\infty$ is among this restricted class of stopping times so that $x^*$ is at least an improved upper bound for $K^*$, i.e., $b_u \geq x^*$. From the definition of $x^*$ we obtain $x^* \geq K_{\rho_{x^*}}$, which is obviously bounded below by the overall minimum $K^* : b_u \geq x^* \geq K_{\rho_{x^*}} \geq K^*$.

3. Processes $r$ with $(b_l, b_u)$-increasing paths include especially unimodal or bath-tub-shaped processes provided that $r_0 < b_l$.

The case of a deterministic process $r$ is of special interest and is stated as a corollary under the assumptions of the last theorem.

**Corollary 74** *If $(f_t)$ and $(g_t)$ are deterministic with inverse of the ratio $r^{-1}(x) = \inf\{t \in \mathbb{R}_+ : r_t = f_t/g_t \geq x\}, x \in \mathbb{R}$, and $X_0 \equiv 0$, then $\sigma = t^* \wedge T$ is optimal with $t^* = r^{-1}(K^*) \in \mathbb{R}_+ \cup \{\infty\}$ and*

$$K^* = \inf\{x \in \mathbb{R} : \int_0^{r^{-1}(x)} (xg_s - f_s)P(T > s)ds \geq EZ_0\}.$$

*If, in addition, $r$ is constant with $r_t \equiv r_0 \ \forall t \in \mathbb{R}_+$, then*

$$K^* = \frac{EZ_0}{EX_T} + r_0 \ and \ \sigma = T.$$

**Remark 20** The bounds for $K^*$ in Lemma 72 are sharp in the following sense. For constant $r_t \equiv r_0$ in the above corollary the upper and lower bounds coincide.

## 5.2.3   Different Information Levels

As indicated in Section 3.2.4 in the context of the general lifetime model, the semimartingale set-up has its advantage in opening new fields of applications. One of these features is the aspect of partial information. In the framework of stochastic process theory, the information is represented by a filtration, an increasing family of $\sigma$-fields. So it is natural to describe partial information by a family of smaller $\sigma$-fields. Let $\mathbb{A} = (\mathcal{A}_t)$ be a subfiltration of $\mathbb{F} = (\mathcal{F}_t)$, i.e., $\mathcal{A}_t \subset \mathcal{F}_t$ for all $t \in \mathbb{R}_+$. The $\sigma$-field $\mathcal{F}_t$ describes the complete information up to time $t$ and $\mathcal{A}_t$ can be regarded as the available partial information that allows us to observe versions of the conditional expectations $\hat{Z} = E[Z_t|\mathcal{A}_t]$ and $\hat{X} = E[X_t|\mathcal{A}_t]$, respectively. For all $\mathbb{A}$-stopping times $\tau$ it holds true that $EZ_\tau = E\hat{Z}_\tau$ and $EX_\tau = E\hat{X}_\tau$. So the problem to find a stopping time $\sigma$ in the class $C_T^{\mathbb{A}}$ of $\mathbb{A}$-stopping times that minimizes $K_\tau = EZ_\tau/EX_\tau$ can be reduced to the ordinary stopping problem by the means developed in the last subsection if $\hat{Z}$ and $\hat{X}$ admit $\mathbb{A}$-SSM representations:

$$K_\sigma = \inf\{K_\tau = \frac{EZ_\tau}{EX_\tau} : \tau \in C_\zeta^{\mathbb{A}}\} = \inf\{K_\tau = \frac{E\hat{Z}_\tau}{E\hat{X}_\tau} : \tau \in C_\zeta^{\mathbb{A}}\}.$$

The projection theorem (Theorem 12, p. 60) yields:
    If $Z$ is an $\mathbb{F}$-SSM with representation $Z = (f, M)$ and $\mathbb{A}$ is a subfiltration of $\mathbb{F}$, then $\hat{Z} = (E[Z|\mathcal{A}_t])$ is an $\mathbb{A}$-SSM with $\hat{Z} = (\hat{f}, \bar{M})$, where $\hat{f}$ is an $\mathbb{A}$-progressively measurable version of $(E[f_t|\mathcal{A}_t]), t \in \mathbb{R}_+$, and $\bar{M}$ is an $\mathbb{A}$-martingale.

Loosely spoken, if $f$ is the "density" of $Z$ we get the "density" $\hat{f}$ of $\hat{Z}$ simply as the conditional expectation with respect to the subfiltration $\mathbb{A}$. Then the idea is to use the projection $\hat{Z}$ of $Z$ to the $\mathbb{A}$-level and apply the above-described optimization technique to $\hat{Z}$. Of course, on the lower information level the cost minimum is increased,

$$\inf\{K_\tau : \tau \in C_\zeta^\mathbb{A}\} \geq \inf\{K_\tau : \tau \in C_\zeta^\mathbb{F}\},$$

since all $\mathbb{A}$-stopping times are also $\mathbb{F}$-stopping times, and the question, to what extent the information level influences the cost minimum, has to be investigated.

## 5.3   Applications

The general set-up to minimize the ratio of expectations allows for many special cases covering a variety of maintenance models. Some few of these will be presented in this section, which show how the general approach can be exploited.

### 5.3.1   The Generalized Age Replacement Model

We first focus on the age replacement model with the long run average cost per unit time criterion: Find $\sigma \in C_T^\mathbb{F}$ with

$$K^* = K_\sigma = \frac{EZ_\sigma}{EX_\sigma} = \inf\{K_\sigma : \tau \in C_T^\mathbb{F}\},$$

where we now insert $Z_t = c + I(T \leq t)$ and $X_t = t, t \in \mathbb{R}_+$. Without loss of generality the constant $k$, the penalty costs for replacements at failures, introduced in subsection 5.1.1 is set equal to 1. We will now make use of the general lifetime model described in detail in Chapter 3.2. This means that it is assumed that the indicator process $V_t = I(T \leq t)$ has an $\mathbb{F}$-SSM representation with a failure rate process $\lambda$ :

$$V_t = I(T \leq t) = \int_0^t I(T > s)\lambda_s ds + M_t.$$

We know then that $\lambda$ has nonnegative paths, $T$ is a totally inaccessible $\mathbb{F}$-stopping time, and $M$ a uniformly integrable $\mathbb{F}$-martingale (cf. Definition 21 and Lemma 13, p. 64). With $\lambda_{\min} = q = \inf\{\lambda_t : 0 \leq t < T(\omega), \omega \in \Omega\}$ we get from Lemma 72, p. 179, the bounds

$$b_l = \frac{c}{ET} + \lambda_{\min} \leq K^* \leq b_u = \frac{c+1}{ET}.$$

Note that in contrast to Example 45, p. 179, $\lambda$ may be a stochastic failure rate process. If the paths of $\lambda$ are $(b_l, b_u)$-increasing, then the SSMs $Z$ and $X$ meet the requirements of Theorem 73, p. 183, and it follows that

$$K^* = x^* = \inf\{x \in \mathbb{R} : xE\rho_x - EZ_{\rho_x} \geq 0\} \text{ and } \sigma = \rho_{x^*},$$

where $\rho_x = \inf\{t \in \mathbb{R}_+ : \lambda_t \geq x\} \wedge T$. Consequently, if $\lambda$ is nondecreasing or bath-tub-shaped starting at $\lambda_0 < b_l$, we get this solution of the stopping problem. The optimal replacement time is a control-limit rule for the failure rate process $\lambda$.

To give an idea of how partial information influences this optimal solution, we resume the example of a two-component parallel system with i.i.d. random variables $X_i \sim \text{Exp}(\alpha), i = 1, 2$, which describe the component lifetimes (cf. Examples 32, p. 70). Then the system lifetime is $T = X_1 \vee X_2$ with corresponding indicator process

$$\begin{aligned} V_t &= I(T \leq t) = \int_0^t I(T > s)\alpha(I(X_1 \leq s) + I(X_2 \leq s))ds + M_t \\ &= \int_0^t I(T > s)\lambda_s ds + M_t. \end{aligned}$$

Possible different information levels were described in Section 3.2.4 in detail. We restrict ourselves now to four levels:

a) The complete information level: $\mathbb{F} = (\mathcal{F}_t)$,

$$\mathcal{F}_t = \sigma(I(X_1 \leq s), I(X_2 \leq s), 0 \leq s \leq t)$$

with failure rate process $\lambda_t = \lambda_t^a = \alpha(I(X_1 \leq t) + I(X_2 \leq t))$.

b) Information only about $T$ until $h > 0$, after $h$ complete information: $\mathbb{A}^b = (\mathcal{A}_t^b)$

$$\mathcal{A}_t^b = \begin{cases} \sigma(I(T \leq s), 0 \leq s \leq t) & \text{if } 0 \leq t < h \\ \mathcal{F}_t & \text{if } t \geq h \end{cases}$$

and failure rate process

$$\hat{\lambda}_t^b = E[\lambda_t | \mathcal{A}_t^b] = \begin{cases} 2\alpha(1 - (2 - e^{-\alpha t})^{-1}) & \text{if } 0 \leq t < h \\ \lambda_t & \text{if } t \geq h. \end{cases}$$

c) Information about component lifetime $X_1$: $\mathbb{A}^c = (\mathcal{A}_t^c)$,

$$\mathcal{A}_t^c = \sigma(I(T \leq s), I(X_1 \leq s), 0 \leq s \leq t)$$

and failure rate process

$$\hat{\lambda}_t^c = E[\lambda_t | \mathcal{A}_t^c] = \alpha(I(X_1 \leq t) + I(X_1 > t)P(X_2 \leq t)).$$

d) Information only about $T$: $\mathbb{A}^d = (\mathcal{A}_t^d)$, $\mathcal{A}_t^d = \sigma(I(T \leq s), 0 \leq s \leq t)$, and failure rate (process) $\hat{\lambda}_t^d = E[\lambda_t | \mathcal{A}_t^d] = 2\alpha(1 - (2 - e^{-\alpha t})^{-1})$.

In all four cases the bounds remain the same with $ET = \frac{3}{2\alpha}$:

$$b_l = \frac{2\alpha}{3}c, \ b_u = \frac{2\alpha}{3}(c+1).$$

Since $\mathbb{A}^b$ and $\mathbb{A}^c$ are subfiltrations of $\mathbb{F}$ and include $\mathbb{A}^d$ as a subfiltration, we must have for the optimal stopping values

$$b_l \leq K_a^* \leq K_b^* \leq K_d^* \leq b_u, \ K_a^* \leq K_c^* \leq K_d^*,$$

i.e., on a higher information level we can achieve a lower cost minimum. Let us consider the complete information case in more detail. The failure rate process is nondecreasing and the assumptions of Theorem 73, p. 183, are met. For the stopping times $\rho_x = \inf\{t \in \mathbb{R}_+ : \lambda_t \geq x\} \wedge T$ we have to consider values of $x$ in $[b_l, b_u]$ and to distinguish between the cases $0 < x \leq \alpha$ and $x > \alpha$ :

- $0 < x \leq \alpha$. In this case we have $\rho_x = X_1 \wedge X_2$, $E\rho_x = \frac{1}{2\alpha}$, $EZ_{\rho_x} = c$, such that $xE\rho_x - EZ_{\rho_x} = 0$ leads to $x^* = 2\alpha c$, where $0 < x^* \leq \alpha$ is equivalent to $c \leq \frac{1}{2}$;

- $\alpha < x$. In this case we have $\rho_x = T$, $E\rho_x = \frac{3}{2\alpha}$, $EZ_{\rho_x} = c + 1$, such that $x^* = b_u$, $x^* > \alpha$ is equivalent to $c > \frac{1}{2}$.

The other information levels are treated in a similar way. Only case b) needs some special attention because the failure rate process $\hat{\lambda}^b$ is no longer monotone but only piecewise nondecreasing. To meet the $(b_l, b_u)$-increasing condition, we must have $\hat{\lambda}_h^b < b_l$, i.e., $2\alpha(1-(2-e^{-\alpha h})^{-1}) < \frac{2\alpha}{3}c$. This inequality holds for all $h \in \mathbb{R}_+$, if $c \geq \frac{3}{2}$ and for $h < h(\alpha, c) = -\frac{1}{\alpha}\ln\left(\frac{3-2c}{3-c}\right)$, if $0 < c < \frac{3}{2}$.

We summarize these considerations in the following proposition the proof of which follows the lines above and is elementary but not straightforward.

**Proposition 75** *For $0 < c \leq \frac{1}{2}$ the optimal stopping times and values $K^*$ are*

a) $K_a^* = 2\alpha c, \sigma_a = X_1 \wedge X_2$;

b) $K_b^* = \alpha \frac{c+(1-e^{\alpha h})^2}{0.5+(1-e^{\alpha h})^2}, \sigma_b = ((X_1 \wedge X_2) \vee h) \wedge T$, *if* $0 < h < h(\alpha, c)$;

c) $K_c^* = \alpha\sqrt{2c}, \sigma_c = X_1 \wedge \left(-\frac{1}{\alpha}\ln\left(1 - \frac{c}{2} - \sqrt{2c}\right)\right)$;

d) $K_d^* = 2\alpha \left(\sqrt{\frac{c^2}{4} + c} - \frac{c}{2}\right), \sigma_d = T \wedge \left(-\frac{1}{\alpha}\ln\left(1 - \frac{c}{2} - \sqrt{\frac{c^2}{4} + c}\right)\right)$.

*For $c > \frac{1}{2}$ we have on all levels $K^* = b_u$ and $\sigma = T$.*

For decreasing $c$ the differences between the cost minima increase. If the costs $c$ for a preventive replacement are greater than half of the penalty costs, i.e., $c > \frac{1}{2}k = \frac{1}{2}$, then extra information and preventive replacements are not profitable.

### 5.3.2  A Shock Model of Threshold Type

In the shock model of Example 46, p. 180, the shock arrivals were described by a marked point process $(T_n, V_n)$, where at time $T_n$ a shock causing damage of amount $V_n$ occurs. Here we assume that $(T_n)$ and $(V_n)$ are independent and that $(V_n)$ forms an i.i.d. sequence of nonnegative random variables with $V_n \sim F$. As usual $N_t = \sum_{n=1}^{\infty} I(T_n \leq t)$ counts the number of shocks until $t$ and

$$R_t = \sum_{n=1}^{N_t} V_n$$

describes the accumulated damage up to time $t$. In the threshold-type model, the lifetime $T$ is given by

$$T = \inf\{t \in \mathbb{R}_+ : R_t \geq S\}, S > 0.$$

Now $\mathbb{F}$ is the internal history generated by $(T_n, V_n)$ and $(\lambda_t)$ the $\mathbb{F}$-intensity of $(N_t)$. The costs of a preventive replacement are $c > 0$ and for a replacement at failure $c + k, k > 0$, which results in a cost process $Z_t = c + kI(T \leq t)$. The aim is to minimize the expected cost per arriving shock in the long run, i.e., to find $\sigma \in C_T^{\mathbb{F}}$ with

$$K^* = K_\sigma = \inf\{K_\tau = \frac{EZ_\tau}{EX_\tau}, \tau \in C_T^{\mathbb{F}}\},$$

where $X_t = N_t$. The only assumption concerning the shock arrival process is that the intensity $\lambda$ is positive: $\lambda_t > 0$ on $[0, T)$. According to Example 46 and Section 3.3.3 we have the following SSM representations:

$$Z_t = c + \int_0^t I(T > s)k\lambda_s \bar{F}((S - R_s)-)ds + M_t,$$

$$X_t = \int_0^t \lambda_s ds + L_t.$$

Then the cost rate process $r$ is given on $[0, T)$ by $r_t = k\bar{F}((S - R_t)-)$, which is obviously nondecreasing. Under the integrability assumptions of Theorem 73, p. 183, we see that the optimal stopping time is $\sigma = \rho_{x^*} = \inf\{t \in \mathbb{R}_+ : r_t \geq x^*\}$, where the limit $x^* = \inf\{x \in \mathbb{R} : xEX_{\rho_x} - EZ_{\rho_x} \geq 0\} = K^*$ has to be found numerically. Thus the optimal stopping time is a control-limit rule for the process $(R_t)$ : Replace the system the first time the accumulated damage hits a certain control limit.

**Example 47** Under the above assumptions let $(N_t)$ be a point process with positive intensity $(\lambda_s)$ and $V_n \sim \text{Exp}(\nu)$. Then we get with $\bar{F}(x) = \exp\{-\nu x\}$ and $EX_T = E[\inf\{n \in \mathbb{N} : \sum_{i=1}^n V_i \geq S\}] = \nu S + 1$ the bounds

$$b_l = \frac{c}{\nu S + 1} + ke^{-\nu S},$$
$$b_u = \frac{c + k}{\nu S + 1},$$

and the control-limit rules

$$\rho_x = \inf\{t \in \mathbb{R}_+ : k\exp\{-\nu(S - R_t)\} \geq x\} \wedge T$$
$$= \inf\{t \in \mathbb{R}_+ : R_t \geq \frac{1}{\nu}\ln(\frac{x}{k}) + S\} \wedge T.$$

We set $g(x) = \frac{1}{\nu}\ln(\frac{x}{k}) + S$ and observe that $\rho_x = \inf\{t \in \mathbb{R}_+ : R_t \geq g(x)\}$, if $0 < x \leq k$. For such values of $x$ we find

$$EX_{\rho_x} = \nu g(x) + 1,$$
$$EZ_{\rho_x} = c + kP(T = \rho_x) = c + ke^{-\nu(S-g(x))} = c + x.$$

The probability $P(T = \rho_x)$ is just the probability that a Poisson process with rate $\nu$ has no event in the interval $[g(x), S]$, which equals $e^{-\nu(g(x)-S)}$. By these quantities the optimal control limit $x^* = K^*$ is the unique solution of

$$x^* = \frac{c + x^*}{\nu g(x^*) + 1},$$

provided that $b_l \leq x^* \leq b_u$. As expected this solution does not depend on the specific intensity of the shock arrival process.

## 5.3.3 Information-Based Replacement of Complex Systems

In this section the basic lifetime model for complex systems is combined with the possibility of preventive replacements. A system with random lifetime $T > 0$ is replaced by a new equivalent one after failure. A preventive replacement can be carried out before failure. There are costs for each replacement and an additional amount has to be paid for replacements after failures. The aim is to determine an optimal replacement policy with respect to some cost criterion. Several cost criteria are known among which the long run average cost per unit time criterion is by far the most popular one. But the general optimization procedure also allows for other criteria. As an example the total expected discounted cost criterion will be applied in this section. We will also consider the possibility to take different information levels into account. This set-up will be applied to complex monotone systems for which in Section 3.2 some examples of various degrees of observation levels were given. For the special case of a two-component parallel system with dependent component lifetimes, it is shown how the optimal replacement policy depends on the different information levels and on the degree of dependence of the component lifetimes.

Consider a monotone system with random lifetime $T, T > 0$, with an $\mathbb{F}$-semimartingale representation

$$I(T \leq t) = \int_0^t I(T > s)\lambda_s ds + M_t, \qquad (5.12)$$

for some filtration $\mathbb{F}$. When the system fails it is immediately replaced by an identical one and the process repeats itself. A preventive replacement can be carried out before failure. Each replacement incurs a cost of $c > 0$ and each failure adds a penalty cost $k > 0$. The problem is to find a replacement (stopping) time that minimizes the *total expected discounted costs*.

Let $\alpha > 0$ be the discount rate and $(Z_\tau, \tau), (Z_{\tau_1}, \tau_1), (Z_{\tau_2}, \tau_2), \ldots$ a sequence of i.i.d. pairs of positive random variables, where $\tau_i$ represents the replacement age of the $i$th implemented system, i.e., the length of the $i$th cycle, and $Z_{\tau_i}$ describes the costs incurred during the $i$th cycle discounted to the beginning of the cycle. Then the total expected discounted costs are

$$K_\tau = E\left[ Z_{\tau_1} + e^{-\alpha \tau_1} Z_{\tau_2} + e^{-\alpha(\tau_1 + \tau_2)} Z_{\tau_3} + \cdots \right]$$

$$= \frac{E Z_\tau}{E[1 - e^{-\alpha \tau}]}.$$

It turns out that $K_\tau$ is the ratio of the expected discounted costs for one cycle and $E[1 - e^{-\alpha \tau}]$. Again the set of admissible stopping (replacement) times less or equal to $T$ is

$$C_T^{\mathbb{F}} = \{\tau : \tau \text{ is an } \mathbb{F}\text{-stopping time } \tau \leq T, E Z_\tau^- < \infty\}.$$

The stopping problem is to find a stopping time $\sigma \in C_T^{\mathbb{F}}$ with

$$K^* = K_\sigma = \inf\{K_\tau : \tau \in C_T^{\mathbb{F}}\}. \qquad (5.13)$$

Stopping at a fixed time $t$ leads to the following costs for one cycle discounted to the beginning of the cycle:

$$Z_t = (c + kI(T \leq t))e^{-\alpha t}, t \in \mathbb{R}_+.$$

Starting from (5.12) such a semimartingale representation can also be obtained for $Z = (Z_t), t \in \mathbb{R}_+$, by using the product rule for "differentiating" semimartingales introduced in Section 3.1.2. Then Theorem 108, p. 239, can be applied to yield for $t \in [0, T]$:

$$Z_t = c + \int_0^t I(T > s)\alpha e^{-\alpha s}\left(-c + \lambda_s \frac{k}{\alpha}\right) ds + R_t$$

$$= c + \int_0^t I(T > s)\alpha e^{-\alpha s} r_s ds + R_t, \qquad (5.14)$$

where $r_s = \alpha^{-1}(-\alpha c + \lambda_s k)$ is a cost rate and $R = (R_t), t \in \mathbb{R}_+$, is a uniformly integrable $\mathbb{F}$-martingale. Since $X_t = 1 - e^{-\alpha t} = \int_0^t \alpha e^{-\alpha s} ds$, the ratio of the "derivatives" of the two semimartingales $Z$ and $X$ is given by $(r_t)$.

We now consider a monotone system with random component lifetimes $T_i > 0$, $i = 1, 2, ..., n, n \in \mathbb{N}$, and structure function $\Phi : \{0, 1\}^n \to \{0, 1\}$ as introduced in Chapter 2. The system lifetime $T$ is given by $T = \inf\{t \in \mathbb{R}_+ : \Phi_t = 0\}$, where the vector process $(\mathbf{X}_t)$ describes the state of the components and $\Phi_t = \Phi(\mathbf{X}_t) = I(T > t)$ indicates the state of the system at time $t$. If the random variables $T_i$ are independent with (ordinary) failure rates $\lambda_t(i)$ and $\mathbb{F} = (\mathcal{F}_t)$ is the (complete information) filtration generated by $\mathbf{X}$, $\mathcal{F}_t = \sigma(\mathbf{X}_s, 0 \le s \le t)$, then Corollary 15 in Section 3.2.2 yields the following semimartingale representation for $\Phi_t$ :

$$1 - \Phi_t = \int_0^t I(T > s)\lambda_s ds + M_t,$$

$$\lambda_t = \sum_{i=1}^{n} (\Phi(1_i, \mathbf{X}_t) - \Phi(0_i, \mathbf{X}_t))\lambda_t(i).$$

To find the minimum $K^*$ we will proceed as before. First of all bounds $b_l$ and $b_u$ for $K^*$ are determined by means of $q = \inf\{r_t : 0 \le t < T(\omega), \omega \in \Omega\}$, the minimum of the cost rate with $q \ge -c$:

$$b_l = \frac{c}{E[1 - e^{-\alpha T}]} + q \le K^* \le b_u = \frac{E\left[(c+k)e^{-\alpha T}\right]}{E[1 - e^{-\alpha T}]}. \tag{5.15}$$

If all failure rates $\lambda_t(i)$ are of IFR-type, then the $\mathbb{F}$-failure rate process $\lambda$ and the ratio process $r$ are nondecreasing. Therefore, Theorem 73, p. 183, can be applied to yield $\sigma = \rho_{x^*}$. So the optimal stopping time is among the control-limit rules

$$\rho_x = \inf\{t \in \mathbb{R}_+ : r_t \ge x\} \wedge T$$
$$= \inf\{t \in \mathbb{R}_+ : \lambda_t \ge \frac{\alpha}{k}(c+x)\} \wedge T.$$

This means: Replace the system the first time the sum of the failure rates of critical components reaches a given level $x^*$. This level has to be determined as

$$x^* = \inf\{x \in \mathbb{R} : xE[1 - e^{-\alpha \rho_x}] - E[c + kI(T = \rho_x)e^{-\alpha \rho_x}] \ge 0\}.$$

The effect of partial information is in the following only considered for the case that no single component or only some of the $n$ components are observed, say those with index in a subset $\{i_1, i_2, ..., i_r\} \subset \{1, 2, ..., n\}, r \le n$. Then the subfiltration $\mathbb{A}$ is generated by $T$ or by $T$ and the corresponding component lifetimes, respectively. The projection theorem yields a representation on the corresponding observation level:

$$1 - \hat{\Phi} = E[I_{\{T \leq t\}} | \mathcal{A}_t] = I_{\{T \leq t\}} = \int_0^t I(T > s)\hat{\lambda}_s ds + \bar{M}_t.$$

If the A-failure rate process $\hat{\lambda}_t = E[\lambda_t | \mathcal{A}_t]$ is $(b_l, b_u)$-increasing, then the stopping problem can also be solved on the lower information level by means of Theorem 73. We want to carry out this in more detail in the next section, allowing also for dependencies between the component lifetimes. To keep the complexity of the calculations on a manageable level, we confine ourselves to a two-component parallel system.

## 5.3.4  A Parallel System with Two Dependent Components

A two-component parallel system is considered now to demonstrate how the optimal replacement rule can be determined explicitly. It is assumed that the component lifetimes $T_1$ and $T_2$ follow a bivariate exponential distribution. There are lots of multivariate extensions of the univariate exponential distribution. But it seems that only a few models like those of Freund [77] and Marshall and Olkin [132] are physically motivated.

The idea behind Freund's model is that after failure of one component the stress, placed on the surviving component, is changed. As long as both components work, the lifetimes follow independent exponential distributions with parameters $\beta_1$ and $\beta_2$. When one of the components fails, the parameter of the surviving component is switched to $\bar{\beta}_1$ or $\bar{\beta}_2$ respectively.

Marshall and Olkin proposed a bivariate exponential distribution for a two-component system where the components are subjected to shocks. The components may fail separately or both at the same time due to such shocks. This model includes the possibility of a common cause of failure that destroys the whole system at once.

As a combination of these two models the following bivariate distribution can be derived. Let the pair $(Y_1, Y_2)$ of random variables be distributed according to the model of Freund and let $Y_{12}$ be another positive random variable, independent of $Y_1$ and $Y_2$, exponentially distributed with parameter $\beta_{12}$. Then $(T_1, T_2)$ with $T_1 = Y_1 \wedge Y_{12}, T_2 = Y_2 \wedge Y_{12}$ is said to follow a combined exponential distribution. For brevity the notation $\gamma_i = \beta_1 + \beta_2 - \bar{\beta}_i, i \in \{1, 2\}$, and $\beta = \beta_1 + \beta_2 + \beta_{12}$ is introduced. The survival function

$$\bar{F}(x, y) = P(T_1 > x, T_2 > y) = P(Y_1 > x, Y_2 > y)P(Y_{12} > x \vee y)$$

is then given by

$$\bar{F}(x, y) = \begin{cases} \frac{\beta_1}{\gamma_2} e^{-\gamma_2 x - (\bar{\beta}_2 + \beta_{12})y} - \frac{\bar{\beta}_2 - \beta_2}{\gamma_2} e^{-\beta y} & \text{for} \quad x \leq y \\ \frac{\beta_2}{\gamma_1} e^{-\gamma_1 y - (\bar{\beta}_1 + \beta_{12})x} - \frac{\bar{\beta}_1 - \beta_1}{\gamma_1} e^{-\beta x} & \text{for} \quad x > y, \end{cases} \tag{5.16}$$

where here and in the following $\gamma_i \neq 0, i \in \{1, 2\}$, is assumed. For $\beta_i = \bar{\beta}_i$ this formula diminishes to the Marshall–Olkin distribution and for $\beta_{12} = 0$ (5.16) gives the Freund distribution. From (5.16) the distribution $H$ of the system lifetime $T = T_1 \wedge T_2$ can be obtained:

$$
\begin{aligned}
H(t) &= P(T \leq t) = P(T_1 \leq t, T_2 \leq t) & (5.17) \\
&= 1 - \frac{\beta_2}{\gamma_1} e^{-(\bar{\beta}_1 + \beta_{12})t} - \frac{\beta_1}{\gamma_2} e^{-(\bar{\beta}_2 + \beta_{12})t} + \frac{\beta_1 \bar{\beta}_2 + \beta_2 \bar{\beta}_1 - \bar{\beta}_1 \bar{\beta}_2}{\gamma_1 \gamma_2} e^{-\beta t}.
\end{aligned}
$$

The optimization problem will be solved for three different information levels:

- Complete information about $T_1, T_2$ (and $T$). The corresponding filtration $\mathbb{F}$ is generated by both component lifetimes:

$$
\mathcal{F}_t = \sigma(I(T_1 \leq s), I(T_2 \leq s), 0 \leq s \leq t), t \in \mathbb{R}_+.
$$

- Information about $T_1$ and $T$. The corresponding filtration $\mathbb{A}$ is generated by one component lifetime, say $T_1$, and the system lifetime:

$$
\mathcal{A}_t = \sigma(I(T_1 \leq s), I(T \leq s), 0 \leq s \leq t), t \in \mathbb{R}_+.
$$

- Information about $T$. The filtration generated by $T$ is denoted by $\mathbb{B}$:

$$
\mathcal{B}_t = \sigma(I(T \leq s), 0 \leq s \leq t), t \in \mathbb{R}_+.
$$

In the following it is assumed that $\beta_i \leq \bar{\beta}_i, i \in \{1, 2\}$, and $\bar{\beta}_1 \leq \bar{\beta}_2$, i.e., after failure of one component the stress placed on the surviving one is increased. Without loss of generality the penalty costs for replacements after failures are set to $k = 1$. The solution of the stopping problem will be outlined in the following. More details are contained in [93].

## 5.3.5  Complete Information about $T_1, T_2$ and $T$

The failure rate process $\lambda$ on the $\mathbb{F}$-observation level is given by (cf. Example 28, p. 65)

$$
\lambda_t = \beta_{12} + \bar{\beta}_2 I(T_1 < t < T_2) + \bar{\beta}_1 I(T_2 < t < T_1).
$$

Inserting $q = -c + \beta_{12}\alpha^{-1}$ in (5.15) we get the bounds for the stopping value $K^*$

$$
b_l = \frac{cv}{1-v} + \frac{\beta_{12}}{\alpha} \quad \text{and} \quad b_u = \frac{(c+1)v}{1-v},
$$

where $v = E[e^{-\alpha T}]$ can be determined by means of the distribution $H$. Since the failure rate process is monotone on $[0, T)$ the optimal stopping time can be found among the control limit rules $\rho_x = \inf\{t \in \mathbb{R}_+ : r_t \geq x\} \wedge T$:

$$
\rho_x = \begin{cases}
0 & \text{for} \quad x \leq \frac{\beta_{12}}{\alpha} - c \\
T_1 \wedge T_2 & \text{for} \quad \frac{\beta_{12}}{\alpha} - c < x \leq \frac{\bar{\beta}_1 + \beta_{12}}{\alpha} - c \\
T_1 & \text{for} \quad \frac{\bar{\beta}_1 + \beta_{12}}{\alpha} - c < x \leq \frac{\bar{\beta}_2 + \beta_{12}}{\alpha} - c \\
T & \text{for} \quad x > \frac{\bar{\beta}_2 + \beta_{12}}{\alpha} - c.
\end{cases}
$$

The optimal control limit $x^*$ is the solution of the equation

$$
x E[1 - e^{-\alpha \rho_x}] - E Z_{\rho_x} = 0.
$$

Since the optimal value $x^*$ lies between the bounds $b_l$ and $b_u$, the considerations can be restricted to the cases $x \geq b_l > \beta_{12}\alpha^{-1} - c$. In the first case when $\beta_{12}\alpha^{-1} - c < x \leq (\bar{\beta}_1 + \beta_{12})\alpha^{-1} - c$, one has $\rho_x = T_1 \wedge T_2$ and

$$
E[1 - e^{-\alpha \rho_x}] = \frac{\alpha}{\beta + \alpha}
$$

$$
E Z_{\rho_x} = c E[e^{-\alpha \rho_x}] + E[I(T \leq \rho_x)e^{-\alpha \rho_x}] = c\frac{\beta}{\beta + \alpha} + \frac{\beta_{12}}{\beta + \alpha}.
$$

The solution of the equation

$$
x^* \frac{\alpha}{\beta + \alpha} - \left( c\frac{\beta}{\beta + \alpha} + \frac{\beta_{12}}{\beta + \alpha} \right) = 0
$$

is given by

$$
x^* = \frac{1}{\alpha}(c\beta + \beta_{12}) \quad \text{if} \quad \frac{\beta_{12}}{\alpha} - c < x^* \leq \frac{\bar{\beta}_1 + \beta_{12}}{\alpha} - c.
$$

Inserting $x^*$ in the latter inequality we obtain the condition $0 < c \leq c_1$, where $c_1 = \bar{\beta}_1(\beta + \alpha)^{-1}$.

The remaining two cases $(\bar{\beta}_1 + \beta_{12})\alpha^{-1} - c < x \leq (\bar{\beta}_2 + \beta_{12})\alpha^{-1} - c$ and $x > (\bar{\beta}_2 + \beta_{12})\alpha^{-1} - c$ are treated in a similar manner. After some extensive calculations the following solution of the stopping problem is derived:

$$
\rho_{x^*} = \begin{cases}
T_1 \wedge T_2 & \text{for} \quad 0 < c \leq c_1 \\
T_1 & \text{for} \quad c_1 < c \leq c_2 \\
T & \text{for} \quad c_2 < c
\end{cases}
$$

$$
x^* = \begin{cases}
x_1^* & \text{for} \quad 0 < c \leq c_1 \\
x_2^* & \text{for} \quad c_1 < c \leq c_2 \\
x_3^* & \text{for} \quad c_2 < c,
\end{cases}
$$

where $c_1$ is defined as above and

$$
c_2 = \frac{\bar{\beta}_2}{(\beta + \alpha)} + \frac{\beta_2(\bar{\beta}_2 - \bar{\beta}_1)}{(\bar{\beta}_1 + \beta_{12} + \alpha)(\beta + \alpha)},
$$

$$x_1^* = \frac{1}{\alpha}(c\beta + \beta_{12}),$$

$$x_2^* = \frac{1}{\alpha}\left(c(\beta_1 + \beta_{12}) + \beta_{12} + \frac{(c+1)\beta_2\bar{\beta}_1 - c\beta_1\beta_2}{\bar{\beta}_1 + \beta_2 + \beta_{12} + \alpha}\right),$$

$$x_3^* = b_u.$$

The explicit formulas for the optimal stopping value were only presented here to show how the procedure works and that even in seemingly simple cases extensive calculations are necessary. The main conclusion can be drawn from the structure of the optimal policy. For small values of $c$ (note that the penalty costs for failures are $k = 1$) it is optimal to stop and replace the system at the first component failure. For mid-range values of $c$, the replacement should take place when the "better" component with a lower residual failure rate ($\bar{\beta}_1 \leq \bar{\beta}_2$) fails. If the "worse" component fails first, this results in an replacement after system failure. For high values of $c$, preventive replacements do not pay, and it is optimal to wait until system failure. In this case the optimal stopping value is equal to the upper bound $x^* = b_u$.

## Information about $T_1$ and $T$

The failure rate process corresponding to this observation level $\mathbb{A}$ is given by

$$\lambda_t = g(t)I(T_1 > t) + (\bar{\beta}_2 + \beta_{12})I(T_1 \leq t),$$

$$g(t) = \bar{\beta}_1 + \beta_{12} - \frac{\bar{\beta}_1\gamma_1}{\beta_2 e^{\gamma_1 t} + \beta_1 - \bar{\beta}_1},$$

where the function $g$ is derived by means of (5.16) as the limit

$$g(t) = \lim_{h \to 0+} \frac{1}{h}P(t < T_1 \leq t + h, T_2 \leq t + h|T_1 > t).$$

The paths of the failure rate process $\lambda$ depend only on the observable component lifetime $T_1$ and not on $T_2$. The paths are nondecreasing so that the same procedure as before can be applied. For $\gamma_1 = \beta_1 + \beta_2 - \bar{\beta}_1 > 0$ the following results can be obtained:

$$\rho_{x^*} = \begin{cases} T_1 \wedge b^* & \text{for} \quad 0 < c \leq c_1 \\ T_1 & \text{for} \quad c_1 < c \leq c_2 \\ T & \text{for} \quad c_2 < c \end{cases}$$

$$x^* = \begin{cases} x_1^* & \text{for} \quad 0 < c \leq c_1 \\ x_2^* & \text{for} \quad c_1 < c \leq c_2 \\ x_3^* & \text{for} \quad c_2 < c. \end{cases}$$

The constants $c_1, c_2$ and the stopping values $x_2^*, x_3^*$ are the same as in the complete information case. What is optimal on a higher information level and can be observed on a lower information level must be optimal on the latter too. So only the case $0 < c \leq c_1$ is new. In this case the optimal replacement time is $T_1 \wedge b^*$ with a constant $b^*$, which is the unique solution of the equation

$$d_1 \exp\{\gamma_1 b^*\} + d_2 \exp\{-(\bar{\beta}_1 + \beta_{12} + \alpha)b^*\} + d_3 = 0.$$

The constants $d_i, i \in \{1, 2, 3\}$, are extensive expressions in $\alpha$, the $\beta$ and $\gamma$ constants and therefore not presented here (see [93]). The values of $b^*$ and $x_1^*$ have to be determined numerically. For $\gamma_1 < 0$ a similar result can be obtained.

## Information about $T$

On this lowest level $\mathbb{B}$, no additional information about the state of the components is available up to the time of system failure. The failure rate is deterministic and can be derived from the distribution $H$:

$$\lambda_t = -\frac{d}{dt}(\ln(1 - H(t))).$$

In this case the replacement times $\rho_x = T \wedge b, b \in \mathbb{R}_+ \cup \{\infty\}$, are the well-known age replacement policies. Even if $\lambda$ is not monotone, such a policy is optimal on this $\mathbb{B}$-level. The optimal values $b^*$ and $x^*$ have to be determined by minimizing $K_{\rho_x}$ as a function of $b$.

## Numerical Examples

The following tables show the effects of changes of two parameters, the replacement cost parameter $c$ and the "dependence parameter" $\beta_{12}$. To be able to compare the cost minima $K^* = x^*$, both tables refer to the same set of parameters: $\beta_1 = 1, \beta_2 = 3, \bar{\beta}_1 = 1.5, \bar{\beta}_2 = 3.5, \alpha = 0.08$. The optimal replacement times are denoted:

a:   $\rho_{x^*} = T_1 \wedge T_2$      b:   $\rho_{x^*} = T_1$          c:   $\rho_{x^*} = T_1 \wedge b^*$
d:   $\rho_{x^*} = T \wedge b^*$       e:   $\rho_{x^*} = T = T_1 \vee T_2$.

Table 1: $\beta_1 = 1, \beta_2 = 3, \beta_{12} = 0.5, \bar{\beta}_1 = 1.5, \bar{\beta}_2 = 3.5, \alpha = 0.08$.

| c | $b_l$ | $\mathbb{F}$ | Information Level $\mathbb{A}$ | | $\mathbb{B}$ | | $b_u$ |
|---|---|---|---|---|---|---|---|
| 0.01 | 6.453 | 6.813 | a | 9.910 | c | 11.003 | d | 20.506 |
| 0.10 | 8.280 | 11.875 | a | 17.208 | c | 19.678 | d | 22.333 |
| 0.50 | 16.402 | 28.543 | b | 28.543 | b | 30.455 | e | 30.455 |
| 1.00 | 26.553 | 39.764 | b | 39.764 | b | 40.606 | e | 40.606 |
| 2.00 | 46.856 | 60.900 | e | 60.900 | e | 60.900 | e | 60.900 |

Table 1 shows the cost minima $x^*$ for different values of $c$. For small values of $c$, the influence of the information level is greater than for moderate values. For $c > 1.394$ preventive replacements do not pay, additional information concerning $T$ is not profitable.

**Table 2:** $\beta_1 = 1, \beta_2 = 3, \bar{\beta}_1 = 1.5, \bar{\beta}_2 = 3.5, c = 0.1, \alpha = 0.08.$

| $\beta_{12}$ | $b_l$ | $\mathbb{F}$ | | $\mathbb{A}$ | | $\mathbb{B}$ | | $b_u$ |
|---|---|---|---|---|---|---|---|---|
| | | Information Level | | | | | | |
| 0.00 | 1.505 | 5.000 | a | 10.739 | c | 13.231 | d | 16.552 |
| 0.10 | 2.859 | 6.375 | a | 12.032 | c | 14.520 | d | 17.698 |
| 1.00 | 15.067 | 18.750 | a | 23.688 | c | 26.132 | d | 28.235 |
| 10.00 | 138.106 | 142.500 | b | 142.500 | b | 144.168 | e | 144.168 |
| 50.00 | 687.677 | 689.448 | e | 689.448 | e | 689.448 | e | 689.448 |

Table 2 shows how the cost minimum depends on the parameter $\beta_{12}$. For increasing values of $\beta_{12}$ the difference between the cost minima on different information levels decreases, because the probability of a common failure of both components increases and therefore extra information about a single component is not profitable.

## 5.3.6  A Burn-in Model

Many manufactured items, for example, electronic components, tend either to last a relatively long time or to fail very early. A technique used to screen out the items with short lifelengths before they are delivered to the customer is the so-called *burn-in*. To burn-in an item means that before the item is released, it undergoes a test during which it is examined under factory conditions or it is exposed to extra stress. After the test phase of (random) length $\tau$, the item is put into operation.

Considering $m$ produced items, and given some cost structure such as costs for failures during and after the test and gains per unit time for released items, one problem related to burn-in is to determine the optimal burn-in duration. This optimal burn-in time may either be fixed in advance and it is therefore deterministic, or one may consider the random information given by the lifelengths of the items failing during the test and obtain a random burn-in time.

We consider a semimartingale approach for solving the optimal stopping problem. In our model, the lifelengths of the items need not be identically distributed, and the stress level during burn-in may differ from the one after burn-in. The information at time $t$ consists of whether and when components failed before $t$. Under these assumptions, we determine the optimal burn-in time $\zeta$.

Let $T_j$, $j = 1, \ldots, m$, be independent random variables representing the lifelengths of the items that are burned in. We assume that $ET_j < \infty$ for all $j$. We consider burn-in under severe conditions. That means that we assume the items to have different failure rates during and after burn-in, $\lambda_j^0(t)$ and $\lambda_j^1(t)$, respectively, where it is supposed that $\lambda_j^0(t) \geq \lambda_j^1(t)$ for all $t \geq 0$. We assume that the lifelength $T_j$ of the $j$th item admits the following representation:

$$I(T_j \leq t) = \int_0^t I(T_j > s)\lambda_j^{Y_s}(s)ds + M_t(j), \quad j = 1,\dots,m, \qquad (5.18)$$

where $Y_t = I(\tau < t)$, $\tau$ is the burn-in time and $M(j) \in \mathcal{M}$ is bounded in $L^2$.

This representation can also be obtained by modeling the lifelength of the $j$th item in the following way:

$$T_j = Z_j \wedge \tau + R_j I(Z_j > \tau), \qquad (5.19)$$

where $Z_j, R_j$, $j = 1,\dots,m$, are independent random variables and $a \wedge b$ denotes the minimum of $a$ and $b$; $Z_j$ is the lifelength of the $j$th item when it is exposed to a higher stress level and $R_j$ is the operating time of the item if it survived the burn-in phase. Let $F_j$ be the lifelength distribution, $H_j$ denote the distribution function of $Z_j$, $j = 1,\dots,m$, and let $H_j(0) = F_j(0) = 0, \bar{H}_j(t) = 1 - H_j(t)$, $\bar{F}_j(t) = 1 - F_j(t)$. Furthermore, we assume that $H_j$ and $F_j$ admit densities $h_j$ and $f_j$, respectively. It is assumed that the operating time $R_j$ follows the conditional survival distribution corresponding to $F_j$ :

$$\begin{aligned} P(T_j \leq t + s | \tau = t < Z_j) &= P(R_j \leq s | \tau = t < Z_j) \\ &= \frac{F_j(t+s) - F_j(t)}{\bar{F}_j(t)}, \quad t, s \in \mathbb{R}_+. \end{aligned}$$

In order to determine the optimal burn-in time, we introduce the following cost and reward structure: There is a reward of $c > 0$ per unit operating time of released items. In addition there are costs for failures, $c_B > 0$ for a failure during burn-in and $c_F > 0$ for a failure after the burn-in time $\tau$, where $c_F > c_B$. If we fix the burn-in time for a moment to $\tau = t$, then the net reward is given by

$$Z_t = c\sum_{j=1}^m (T_j - t)^+ - c_B \sum_{j=1}^m I(T_j \leq t) - c_F \sum_{j=1}^m I(T_j > t), t \in \mathbb{R}_+. \quad (5.20)$$

Since we assume that the failure time of any item can be observed during the burn-in phase, the observation filtration, generated by the lifelengths of the items, is given by

$$\mathbb{F} = (\mathcal{F}_t), t \in \mathbb{R}_+, \mathcal{F}_t = \sigma(I(T_j \leq s), 0 \leq s \leq t, j = 1,\dots,m).$$

In order to determine the optimal burn-in time, we are looking for an $\mathbb{F}$-stopping time $\zeta \in C^{\mathbb{F}}$ satisfying

$$EZ_\zeta = \sup\{EZ_\tau : \tau \in C^{\mathbb{F}}\}.$$

In other words, at any time $t$ the observer has to decide whether to stop or to continue with burn-in with respect to the available information up to time $t$. Since $Z$ is not adapted to $\mathbb{F}$, i.e., $Z_t$ cannot be observed directly, we consider the conditional expectation

$$\hat{Z}_t = E[Z_t|\mathcal{F}_t] = c\sum_{j=1}^{m} I(T_j > t)E[(T_j - t)^+|T_j > t] - mc_F$$

$$+(c_F - c_B)\sum_{j=1}^{m} I(T_j \le t). \qquad (5.21)$$

As an abbreviation we use

$$\mu_j(t) = E[(T_j - t)^+|T_j > t] = \frac{1}{\bar{F}_j(t)}\int_t^\infty \bar{F}_j(x)dx, \ t \in \mathbb{R}_+,$$

for the mean residual lifelength. The derivative with respect to $t$ is given by $\mu_j'(t) = -1 + \lambda_j^1(t)\mu_j(t)$. We are now in a position to apply Theorem 71, p. 175, and formulate conditions under which the monotone case holds true.

**Theorem 76** *Suppose that the functions*

$$g_j(t) = -c - c\mu_j(t)(\lambda_j^0(t) - \lambda_j^1(t)) + (c_F - c_B)\lambda_j^0(t)$$

*satisfy the following condition:*

$$\sum_{j\in\mathcal{J}} g_j(t) \le 0 \text{ implies } g_j(s) \le 0 \ \forall j \in \mathcal{J}, \ \forall \mathcal{J} \subseteq \{1,\ldots,m\}, \ \forall s \ge t. \quad (5.22)$$

*Then*

$$\zeta = \inf\{t \in \mathbb{R}_+ : \sum_{j=1}^{m} I(T_j > t)g_j(t) \le 0\}$$

*is an optimal burn-in time:*

$$EZ_\zeta = \sup\{EZ_\tau : \tau \in C^{\mathbb{F}}\}.$$

**Proof.** In order to obtain a semimartingale representation for $\hat{Z}$ in (5.21) we derive such a representation for $I(T_j > t)\mu_j(t)$. Since $\mu_j(\cdot)$ and $I(T_j > \cdot)$ are right-continuous and of bounded variation on $[0,t]$, we can use the integration by parts formula for Stieltjes integrals (pathwise) to obtain

$$\mu_j(t)I(T_j > t) = \mu_j(0)I(T_j > 0) + \int_0^t \mu_j(s-)dI(T_j > s)$$

$$+ \int_0^t I(T_j > s)d\mu_j(s).$$

Substituting

$$I(T_j > s) = 1 + \int_0^s (-I(T_j > x)\lambda_j^0(x))dx + M_j(s)$$

in this formula and using the continuity of $\mu$, we obtain

$$
\begin{aligned}
\mu_j(t)I(T_j > t) &= \mu_j(0) + \int_0^t [-\mu_j(s)I(T_j > s)\lambda_j^0(s) + I(T_j > s)\mu_j'(s)]ds \\
&\quad + \int_0^t \mu_j(s)dM_j(s) \\
&= \mu_j(0) + \int_0^t I(T_j > s)\left[-1 - \mu_j(s)(\lambda_j^0(s) - \lambda_j^1(s))\right]ds \\
&\quad + \tilde{M}_j(t),
\end{aligned}
$$

where $\tilde{M}_j$ is a martingale, which is bounded in $L^2$. This yields the following semimartingale representation for $\hat{Z}$ :

$$
\begin{aligned}
\hat{Z}_t &= -mc_F + c\sum_{j=1}^m \mu_j(0) \\
&\quad + \int_0^t \sum_{j=1}^m cI(T_j > s)[-1 - \mu_j(s)(\lambda_j^0(s) - \lambda_j^1(s))]ds \\
&\quad + (c_F - c_B)\int_0^t \sum_{j=1}^m I(T_j > s)\lambda_j^0(s)ds + L_t \\
&= -mc_F + c\sum_{j=1}^m \mu_j(0) + \int_0^t \sum_{j=1}^m I(T_j > s)g_j(s)ds + L_t
\end{aligned}
$$

with a uniformly integrable martingale

$$L = c\sum_{j=1}^m \tilde{M}_j + (c_F - c_B)\sum_{j=1}^m M_j \in \mathcal{M}.$$

Since for all $\omega \in \Omega$ and all $t \in \mathbb{R}_+$, there exists some $\mathcal{J} \subseteq \{1,\ldots,m\}$ such that $\sum_{j=1}^m I(T_j > t)g_j(t) = \sum_{j\in\mathcal{J}} g_j(t)$, condition (5.22) in the theorem ensures that the monotone case (MON), p. 175, holds true. Therefore we get the desired result by Theorem 71 and the proof is complete. ∎

**Remark 21** The structure of the optimal stopping time shows that high rewards per unit operating time lead to short burn-in times whereas great differences $c_F - c_B$ between costs for failures in different phases lead to long testing times, as expected.

Equivalent characterizations of condition (5.22) in Theorem 76 are given in the following lemma. The proof can be found in [96].

**Lemma 77** *Let $t_{\mathcal{J}} = \inf\{t \in \mathbb{R}_+ : \sum_{j \in \mathcal{J}} g_j(t) \leq 0\}$ and denote $t_j = t_{\{j\}}$ for all $j \in \{1, \ldots, m\}$. Then the following conditions are equivalent:*

*(i)* $\sum_{j \in \mathcal{J}} g_j(t) \leq 0$ *implies* $g_j(s) \leq 0 \; \forall j \in \mathcal{J}, \; \forall \mathcal{J} \subseteq \{1, \ldots, m\}$ *and* $\forall s \geq t$.

*(ii)* $t_{\mathcal{J}} = \max_{j \in \mathcal{J}} t_j \; \forall \mathcal{J} \subseteq \{1, \ldots, m\}$ *and* $g_j(s) \leq 0 \; \forall s \geq t_j, \; \forall j \in \{1, \ldots, m\}$.

*(iii)*

$$\left| \sum_{j : g_j(t) \leq 0} g_j(t) \right| < \min_{j : g_j(t) > 0} g_j(t) \quad \forall\, t < \max_{j=1,\ldots,m} t_j$$

*and* $g_j(s) \leq 0 \; \forall s \geq t_j, \; \forall j \in \{1, \ldots, m\}$.

The following special cases illustrate the result of the theorem.

1. **Burn-in forever.** If $g_j(t) > 0$ for all $t \in \mathbb{R}_+$, $j = 1, \ldots, m$, then $\zeta = \max\{T_1, \ldots, T_m\}$, i.e., burn-in until all items have failed.

2. **No burn-in.** If $g_j(0) \leq 0, j = 1, \ldots, m$, then $\zeta = 0$ and no burn-in takes place. This case occurs for instance if the costs for failures during and after burn-in are the same: $c_B = c_F$.

3. **Identical items.** If all failure rates coincide, i.e., $\lambda_1^0(t) = \ldots = \lambda_m^0(t)$ and $\lambda_1^1(t) = \ldots = \lambda_m^1(t)$ for all $t \geq 0$, then $g_j(t) = g_1(t)$ for all $j \in \{1, \ldots, m\}$ and condition (A.1) reduces to

$$g_1(s) \leq 0 \text{ for } s \geq t_1 = \inf\{t \in \mathbb{R}_+ : g_1(t) \leq 0\}.$$

   If this condition is satisfied, the optimal stopping time is of the form $\zeta = t_1 \wedge \max\{T_1, \ldots, T_m\}$, i.e., stop burn-in as soon as $g_1(s) \leq 0$ or as soon as all items have failed, whatever occurs first.

4. **The exponential case.** If all failure rates are constant, equal to $\lambda_j^0$ and $\lambda_j^1$, respectively, then $\mu_j$ and therefore $g_j$ is constant, too, and $\zeta(\omega) \in \{0, T_1(\omega), \ldots, T_m(\omega)\}$, if condition (5.22) is satisfied. If, furthermore, the items are "identical," then we have $\zeta = 0$ or $\zeta = \max\{T_1, \ldots, T_m\}$.

5. **No random information.** In some situations the lifelengths of the items cannot be observed continuously. In this case one has to maximize the expectation function

$$EZ_t = E\hat{Z}_t = -mc_F + c \sum_{j=1}^{m} \bar{H}_j(t)\mu_j(t) + (c_F - c_B) \sum_{j=1}^{m} H_j(t)$$

   in order to obtain the (deterministic) optimal burn-in time. This can be done using elementary calculus.

## 5.4   Repair Replacement Models

In this section we consider models in which repairs are carried out in negligible time up to the time of a replacement. So the observation of the system does not end with a failure, as in the first sections of this chapter, but are continued until it is decided to replace the system by a new one. Given a certain cost structure the optimal replacement time is derived with respect to the available information.

### 5.4.1   Optimal Replacement under a General Repair Strategy

We consider a system that fails at times $T_n$, according to a point process $(N_t), t \in \mathbb{R}_+$, with an intensity $(\lambda_t)$ adapted to some filtration $\mathbb{F}$. At failures a repair is carried out at cost of $c > 0$, which takes negligible time. A replacement can be carried out at any time $t$ at an additional cost $k > 0$. Following the average cost per unit time criterion, we have to find a stopping time $\sigma$, if there exists one, with

$$K^* = K_\sigma = \inf\{K_\tau = \frac{cEN_\tau + k}{E\tau} : \tau \in C^\mathbb{F}\},$$

where $C^\mathbb{F} = \{\tau : \tau \; \mathbb{F}\text{-stopping time}, E\tau < \infty\}$ is a suitable class of stopping times. To solve this problem we can adopt the procedure of Section 5.2.1 with some slight modifications.

First of all we have $K_\tau = \frac{EZ_\tau}{EX_\tau}$ with SSM representations

$$Z_t = k + \int_0^t c\lambda_s ds + M_t, \qquad (5.23)$$

$$X_t = \int_0^t ds.$$

Setting $\tau = T_1$, we derive the simple upper bound $b_u$ :

$$b_u = \frac{c + k}{ET_1} \geq K^*.$$

The process $Y$ corresponding to (5.10) on p. 182 now reads

$$Y_t = -k + \int_0^t (K^* - c\lambda_s) ds + R_t$$

and therefore we know that, if there exists an optimal finite stopping time $\sigma$, then it is among the indexed stopping times

$$\rho_x = \inf\{t \in \mathbb{R}_+ : \lambda_t \geq \frac{x}{c}\}, 0 \leq x \leq b_u,$$

provided $\lambda$ has nondecreasing paths. We summarize this in a corollary to Theorem 73, p. 183.

**Corollary 78** *Let the martingale $M$ in (5.23) be such that $(M_{t \wedge \rho_{b_u}})$ is uniformly integrable. If $\lambda$ has nondecreasing paths and $E\rho_{b_u} < \infty$, then*

$$\sigma = \rho_{x^*}, \quad \text{with } x^* = \inf\{x \in \mathbb{R}_+ : xE\rho_x - cEN_{\rho_x} \geq k\},$$

*is an optimal stopping time and $x^* = K^*$.*

**Example 48** Considering a nonhomogeneous Poisson process with a nondecreasing deterministic intensity $\lambda_t = \lambda(t)$, we observe that the stopping times $\rho_x = \lambda^{-1}(x/c)$ are constants. If $\lambda^{-1}(b_u/c) < \infty$, then the corollary can be applied and the optimal stopping time $\sigma$ is a finite constant.

The simplest case is that of a Poisson process with constant rate $\lambda > 0$. In this case we have $b_u = c\lambda + k\lambda > c\lambda$ and $\rho_{b_u} = \infty$, so that the corollary does not apply. But in this case it is easily seen that additional stopping (replacement) costs do not pay and we get that $\sigma = \infty$ is optimal with $K^* = c\lambda$.

**Example 49** Consider the shock model with state-dependent failure probability of Section 3.3.4 in which shocks arrive according to a Poisson process with rate $\nu$ (cf. Example 35, p. 81). The failure intensity is of the form

$$\lambda_t = \nu \int_0^\infty p(X_t + y)dF(y),$$

where $p(X_t + y)$ denotes the probability of a failure at the next shock if the accumulated damage is $X_t$ and the next shock has amount $y$. Here we assume that this probability function $p$ does not depend on the number of failures in the past. Obviously $\lambda_t$ is nondecreasing so that Corollary 78 applies provided that the integrability conditions are met.

A variety of point process models as described in Section 3.3 can be used in this set-up. Also more general cost structures could be applied as for example random costs $k = (k_t)$, if $k$ admits an SSM representation. Other modifications (discounted cost criterion, different information levels) can be worked out easily apart of some technical problems.

## 5.4.2 A Markov-Modulated Repair Process – Optimization with Partial Information

In this section a model with a given reward structure is investigated in which an optimal operating time of a system has to be found that balances some flow of rewards and the increasing cost rate due to (minimal) repairs. Consider a one-unit system that fails from time to time according to a point process. After failure a minimal repair is carried out that leaves the state of the system unchanged. The system can work in one of $m$ *unobservable* states. State "1" stands for new or in good condition and "$m$" is defective or in bad condition. Aging of the system is described by a link between the failure point process and the unobservable state of the system. The failure or minimal repair intensity may depend on the state of the system. There is some constant flow of income, on the one hand, and on the other hand, each minimal repair incurs a random cost amount. The question is when to stop processing the system and carrying out an inspection or a renewal in order to maximize some reward functional.

For the basic set-up we refer to Example 21, p. 56 and Section 3.3.9. Here we recapitulate the main assumptions of the model:

The basic probability space $(\Omega, \mathcal{F}, P)$ is equipped with a filtration $\mathbb{F}$, the complete information level, to which all processes are adapted, and $S = \{1, ..., m\}$ is the set of unobservable environmental states. The changes of the states are driven by a homogeneous Markov process $Y = (Y_t), t \in \mathbb{R}_+$, with values in $S$ and infinitesimal parameters $q_i$, the rate to leave state $i$, and $q_{ij}$, the rate to reach state $j$ from state $i$. The time points of failures (minimal repairs) $0 < T_1 < T_2 < ...$ form a point process and $N = (N_t), t \in \mathbb{R}_+$, is the corresponding counting process:

$$N_t = \sum_{n=1}^{\infty} I(T_n \leq t).$$

It is assumed that $N$ has a stochastic intensity $\lambda_{Y_t}$ that depends on the unobservable state, i.e., $N$ is a so-called Markov-modulated Poisson process with representation

$$N_t = \int_0^t \lambda_{Y_s} ds + M_t,$$

where $M$ is an $\mathbb{F}$-martingale and $0 < \lambda_i < \infty, i \in S$.

Furthermore, let $(X_n), n \in \mathbb{N}$, be a sequence of positive i.i.d. random variables, independent of $N$ and $Y$, with common distribution $F$ and finite mean $\mu$. The cost caused by the $n$th minimal repair at time $T_n$ is described by $X_n$.

There is an initial capital $u$ and an income of constant rate $c > 0$ per unit time.

Now the process $R$, given by

$$R_t = u + ct - \sum_{n=1}^{N_t} X_n,$$

describes the available capital at time $t$ as the difference of the income and the total amount of costs for minimal repairs up to time $t$.

The process $R$ is well-known in other branches of applied probability like queueing or collective risk theory, where the time to ruin $\tau = \inf\{t \in \mathbb{R}_+ : R_t < 0\}$ is investigated (cf. Section 3.3.9). Here the focus is on determining the optimal operating time with respect to the given reward structure. To achieve this goal one has to estimate the unobservable state of the system at time $t$, given the history of the process $R$ up to time $t$. This can be done using results in filtering theory as is shown below. Stopping at a fixed time $t$ results in the net gain

$$Z_t = R_t - \sum_{j=1}^{m} k_j U_t(j),$$

where $U_t(j) = I(Y_t = j)$ is the indicator of the state at time $t$ and $k_j \in \mathbb{R}$, $j \in S$, are stopping costs (for inspection and replacement), which may depend on the stopping state. The process $Z$ cannot be observed directly because only the failure time points and the costs for minimal repairs are known to an observer. The observation filtration $\mathbb{A} = (\mathcal{A}_t), t \in \mathbb{R}_+$, is given by

$$\mathcal{A}_t = \sigma(N_s, X_i, 0 \le s \le t, i = 1, ..., N_t).$$

Let $C^\mathbb{A} = \{\tau : \tau \text{ is a finite } \mathbb{A}\text{-stopping time}, EZ_\tau^- < \infty\}$ be the set of feasible stopping times in which the optimal one has to be found. As usual $a^- = -\min\{0, a\}$ denotes the negative part of $a \in \mathbb{R}$. So the problem is to find $\tau^* \in C^\mathbb{A}$ which maximizes the expected net gain:

$$EZ_{\tau^*} = \sup\{EZ_\tau : \tau \in C^\mathbb{A}\}.$$

For the solution of this problem an $\mathbb{F}$-semimartingale representation of the process $Z$ is needed, where it is assumed that the complete information filtration $\mathbb{F}$ is generated by $Y, N$, and $(X_n)$:

$$\mathcal{F}_t = \sigma(Y_s, N_s, X_i, 0 \le s \le t, i = 1, ..., N_t).$$

Such a representation can be obtained by means of an SSM representation for the indicator process $U_t(j)$,

$$U_t(j) = U_0(j) + \int_0^t \sum_{i=1}^{m} U_s(i) q_{ij} ds + m_t(j), m(j) \in \mathcal{M}_0, \qquad (5.24)$$

as follows (see [106] for details):

$$Z_t = u - \sum_{j=1}^{m} k_j U_0(j) + \int_0^t \sum_{j=1}^{m} U_s(j) r_j ds + M_t, t \in \mathbb{R}_+, \qquad (5.25)$$

where $M = (M_t)$ is an $\mathbb{F}$-martingale and the constants $r_j$ are defined by

$$r_j = c - \lambda_j \mu - \sum_{\nu \ne j} (k_\nu - k_j) q_{j\nu}.$$

These constants can be interpreted as net gain rates in state $j$:

- $c$ is the income rate.

- $\lambda_j$, the failure rate in state $j$, is the expected number of failures per unit of time, $\mu$ is the expected repair cost for one minimal repair. So $\lambda_j \mu$ is the repair cost rate.

- the remaining sum is the stopping cost rate by leaving state $j$.

Since the state indicators $U(j)$ and therefore $Z$ cannot be observed, a projection to the observation filtration $\mathbb{A}$ is needed. As described in Section 3.1.3 such a projection from the $\mathbb{F}$-level (5.25) to the $\mathbb{A}$-level leads to the following conditional expectations:

$$\hat{Z}_t = E[Z_t|\mathcal{A}_t] = u - \sum_{j=1}^{m} k_j \hat{U}_0(j) + \int_0^t \sum_{j=1}^{m} \hat{U}_s(j) r_j ds + \bar{M}_t, t \in \mathbb{R}_+. \quad (5.26)$$

The integrand $\sum_{j=1}^{m} \hat{U}_s(j) r_j$ with $\hat{U}_s(j) = E[U_s|\mathcal{A}_s] = P(Y_s = j|\mathcal{A}_s)$ is the conditional expectation of the net gain rate at time $s$ given the observations up to time $s$. If this integrand has nonincreasing paths, then we know that we are in the "monotone case" (cf. p. 175) and the stopping problem could be solved under some additional integrability conditions. To state monotonicity conditions for the integrand in (5.26), an explicit representation of $\hat{U}_t(j)$ is needed, which can be obtained by means of results in filtering theory (see [52] p. 98, [104]) in the form of "differential equations":

- between the jumps of $N : T_n \leq t < T_{n+1}$

$$\hat{U}_t(j) = \hat{U}_{T_n}(j) + \int_{T_n}^t \left( \sum_{i=1}^{m} \hat{U}_s(i)\{q_{ij} + \hat{U}_s(j)(\lambda_i - \lambda_j)\} \right) ds,$$

$$q_{jj} = -q_j, \quad (5.27)$$

$$\hat{U}_0(j) = P(Y_0 = j), \ j \in S.$$

- at jumps

$$\hat{U}_{T_n}(j) = \frac{\lambda_j \hat{U}_{T_n-}(j)}{\sum_{i=1}^{m} \lambda_i \hat{U}_{T_n-}(i)}, \quad (5.28)$$

where $U_{T_n-}(j)$ denotes the left limit.

The following conditions ensure that the system ages, i.e., it moves from the "good" states with high net gains and low failure rates to the "bad" states with low and possibly negative net gains and high failure rates, and it is never possible to return to a "better" state:

$$q_i > 0, i = 1, ..., m - 1, q_{ij} = 0 \text{ for } i > j, i, j \in S,$$

$$r_1 \geq r_2 \geq ... \geq r_m = c - \lambda_m \mu, r_m < 0, \quad (5.29)$$

$$0 < \lambda_1 \leq \lambda_2 \leq ... \leq \lambda_m.$$

A reasonable candidate for an optimal $\mathbb{A}$-stopping time is

$$\tau^* = \inf\{t \in \mathbb{R}_+ : \sum_{j=1}^{m} \hat{U}_t(j)r_j \leq 0\}, \qquad (5.30)$$

the first time the conditional expectation of the net gain rate falls below 0.

**Theorem 79** *Let $\tau^*$ be the $\mathbb{A}$-stopping time (5.30) and assume that conditions (5.29) hold true. If, in addition, $q_{im} > \lambda_m - \lambda_i, i = 1, ..., m-1$, then $\tau^*$ is optimal:*

$$EZ_{\tau^*} = \sup\{EZ_\tau : \tau \in C^{\mathbb{A}}\}.$$

**Proof.** Because of $EZ_\tau = E\hat{Z}_\tau$ for all $\tau \in C^{\mathbb{A}}$ we can apply Theorem 71, p. 175, of Chapter 3 taking the $\mathbb{A}$-SSM representation (5.26) of $\hat{Z}$. We will proceed in two steps:

a) First, we prove that the monotone case holds true.

b) Second, we show that the martingale part $\bar{M}$ in (5.26) is uniformly integrable.

a) We start showing that the integrand $\sum_{j=1}^{m} \hat{U}_s(j)r_j$ has nonincreasing paths. A simple rearrangement gives

$$\sum_{j=1}^{m} \hat{U}_s(j)r_j = r_m + (r_{m-1} - r_m)\sum_{j=1}^{m-1} \hat{U}_s(j) + \cdots + (r_1 - r_2)\hat{U}_s(1).$$

Since we have from (5.29) that $r_{k-1} - r_k \geq 0, k = 2, ..., m$, it remains to show that $\sum_{\nu=1}^{j} \hat{U}_s(\nu)$ is nonincreasing in $s$ for $j = 1, ..., m-1$. Denoting $\bar{\lambda}(s) = \sum_{j=1}^{m} \hat{U}_s(j)\lambda_j$ we get from (5.27) between jumps $T_n < s < T_{n+1}$, where $T_0 = 0$,

$$\begin{aligned}
\frac{d}{ds}\left(\sum_{\nu=1}^{j} \hat{U}_s(\nu)\right) &= \sum_{\nu=1}^{j}\left(\sum_{i=1}^{m} \hat{U}_s(i)\{q_{i\nu} + \hat{U}_s(\nu)(\lambda_i - \lambda_\nu)\}\right) \\
&= \sum_{i=1}^{m}\sum_{\nu=1}^{j} \hat{U}_s(i)q_{i\nu} + \sum_{\nu=1}^{j} \hat{U}_s(\nu)(\bar{\lambda}(s) - \lambda_\nu) \\
&= \sum_{i=1}^{j} \hat{U}_s(i)\left(-\sum_{k=j+1}^{m} q_{ik} + \bar{\lambda}(s) - \lambda_i\right)
\end{aligned}$$

using $q_{ij} = 0$ for $i > j$ and $q_{ii} = -\sum_{k=i+1}^{m} q_{ik}, i = 1, ..., m-1$.

From $q_{im} > \lambda_m - \lambda_i \geq \bar{\lambda}(s) - \lambda_i$ it follows that

$$\frac{d}{ds}\left(\sum_{\nu=1}^{j} \hat{U}_s(\nu)\right) \leq 0, j = 1, ..., m-1.$$

At jumps $T_n$ we have from (5.28)

$$\sum_{\nu=1}^{j}(\hat{U}_{T_n}(\nu) - \hat{U}_{T_n-}(\nu)) = \sum_{\nu=1}^{j}\hat{U}_{T_n-}(\nu)\frac{\lambda_\nu - \bar{\lambda}(T_n-)}{\bar{\lambda}(T_n-)}.$$

The condition $\lambda_1 \leq \cdots \leq \lambda_m$ ensures that the latter sum is not greater than 0. This is obvious in the case $\lambda_j \leq \bar{\lambda}(T_n-)$; otherwise, if $\lambda_j > \bar{\lambda}(T_n-)$, this follows from

$$0 = \sum_{\nu=1}^{m}\hat{U}_{T_n-}(\nu)\frac{\lambda_\nu - \bar{\lambda}(T_n-)}{\bar{\lambda}(T_n-)} \geq \sum_{\nu=1}^{j}\hat{U}_{T_n-}(\nu)\frac{\lambda_\nu - \bar{\lambda}(T_n-)}{\bar{\lambda}(T_n-)} .$$

For the monotone case to hold it is also necessary that

$$\bigcup_{t\in\mathbb{R}_+}\{\sum_{j=1}^{m}\hat{U}_t(j)r_j \leq 0\} = \Omega$$

or equivalently $\tau^* < \infty$. From (5.24) we obtain by means of the projection theorem

$$\hat{U}_t(m) = \hat{U}_0(m) + \int_0^t \sum_{i=1}^{m-1}\hat{U}_s(i)q_{im}ds + \bar{m}_t(j)$$

with a nonnegative integrand. This shows that $\hat{U}_t(m)$ is a bounded submartingale. Thus, the limit

$$\hat{U}_\infty(m) = \lim_{t\to\infty}\hat{U}_t(m) = E[U_\infty(m)|\mathcal{A}_\infty]$$

exists and is identical to 1 since $\lim_{t\to\infty}Y_t = m$ and hence $U_\infty(m) = 1$. Because $r_m < 0$, it is possible to choose some $\epsilon > 0$ such that $(1-\epsilon)r_m + \epsilon\sum_{i=1}^{m-1}r_i < 0$. Therefore, we have

$$\tau^* = \inf\{t \in \mathbb{R}_+ : \sum_{j=1}^{m}\hat{U}_t(j)r_j \leq 0\} \leq \inf\{t \in \mathbb{R}_+ : \hat{U}_t(m) \geq 1 - \epsilon\} < \infty.$$

b) To show that $\bar{M}$ is uniformly integrable we consider a decomposition of the drift term of the F-SSM representation of $Z$ :

$$\int_0^t \sum_{j=1}^{m}U_s(j)r_jds = \int_0^t \sum_{j=1}^{m}U_s(j)(r_j - r_m)ds + tr_m,$$

where $tr_m$ is obviously A-adapted. We use the projection Theorem 12, p. 60, in the extended version. To this end we have to show that
(1) $Z_0 = c - \sum_{j=1}^{m}k_jU_0(j)$ and $\int_0^\infty |\sum_{j=1}^{m}U_s(j)(r_j - r_m)|ds$ are square integrable, and that

(2) $M$ is square integrable.

The details of these parts are omitted here and can be found in [104] and [106].

To sum up, by a) the monotone case holds true for $\hat{Z}$ with a martingale part $\bar{M}$, which is by b) square integrable and hence uniformly integrable. The monotone stopping Theorem 71 can then be applied and the assertion of the theorem follows.    ∎

## 5.4.3   The Case of m=2 States

For two states the stopping problem can be reformulated as follows. At an unobservable random time, say $\sigma$, there occurs a switch from state 1 to state 2. Detect this change as well as possible (with respect to the given optimization criterion) by means of the failure process observations. The conditions (5.29) now read

$$
\begin{aligned}
& q_1 = q_{12} = q > 0, q_2 = q_{21} = 0, \\
& r_1 = c - \lambda_1\mu - q(k_2 - k_1) > 0 > r_2 = c - \lambda_2\mu, \quad\quad (5.31) \\
& 0 < \lambda_1 \le \lambda_2 = 1.
\end{aligned}
$$

The conditional distribution of $\sigma$ can be obtained explicitly as the solution of the above differential equations. To obtain this explicit solution we assume in addition $P(Y_0 = 1)$. The result of the (lengthy) calculations is

$$
\hat{U}_t(2) = P(\sigma \le t | \mathcal{A}_t) = 1 - \frac{e^{-g_n(t)}}{d_n + (\lambda_2 - \lambda_1)\int_{T_n}^t e^{-g_n(s)}ds}, T_n \le t < T_{n+1},
$$

$$
\hat{U}_{T_n}(2) = \frac{\lambda_2 \hat{U}_{T_n-}(2)}{\lambda_1 + (\lambda_2 - \lambda_1)\hat{U}_{T_n-}(2)},
$$

where $d_n = \left(1 - \hat{U}_{T_n}(2)\right)^{-1}$, $g_n(t) = (q - (\lambda_2 - \lambda_1))(t - T_n)$. The stopping time $\tau^*$ in (5.30) can now be written as

$$
\tau^* = \inf\{t \in \mathbb{R}_+ : \hat{U}_t(2) > z^*\}, \quad z^* = \frac{r_1}{r_1 - r_2}.
$$

For $0 < q < \lambda_2 - \lambda_1, \hat{U}_t(2)$ increases as long as $\hat{U}_t(2) < q/(\lambda_2 - \lambda_1) = r$. When $\hat{U}_t(2)$ jumps above this level, then between jumps $\hat{U}_t(2)$ decreases but not below the level $r$. So even in this case under conditions (5.31) the monotone case holds true if $z^* \le q/(\lambda_2 - \lambda_1)$. As a consequence of Theorem 79 we have the following corollary.

**Corollary 80** *Assume conditions (5.31) with stopping rule* $\tau^* = \inf\{t \in \mathbb{R}_+ : \hat{U}_t(2) > z^*\}$. *Then* $\tau^*$ *is optimal in* $C^{\mathbb{A}}$ *if either* $q > \lambda_2 - \lambda_1$ *or* $z^* \le q/(\lambda_2 - \lambda_1)$.

**Remark 22** If the failure rates in both states coincide, i.e., $\lambda_1 = \lambda_2$, the observation of the failure time points should give no additional information about the change time point from state 1 to state 2. Indeed, in this case the conditional distribution of $\sigma$ is deterministic,

$$P(\sigma \le t | \mathcal{A}_t) = P(\sigma \le t) = 1 - \exp\{-qt\}$$

and $\tau^*$ is a constant. As to be expected, random observations are useless in this case.

In general, the value of the stopping problem $\sup\{EZ_\tau : \tau \in C^{\mathbb{A}}\}$, the best possible expected net gain, cannot be determined explicitly. But it is possible to determine bounds for this value. For this, the semimartingale representation turns out to be useful again, because it allows, by means of the projection theorem, comparisons of different information levels. The constant stopping times are contained in $C^{\mathbb{A}}$ and $C^{\mathbb{A}} \subset C^{\mathbb{F}}$. Therefore, the following inequality applies:

$$\sup\{EZ_t : t \in \mathbb{R}_+\} \le \sup\{EZ_\tau : \tau \in C^{\mathbb{A}}\} \le \sup\{EZ_\tau : \tau \in C^{\mathbb{F}}\}.$$

At the complete information level $\mathbb{F}$ the change time point $\sigma$ can be observed, and it is obvious that under conditions (5.31) the $\mathbb{F}$-stopping time $\sigma$ is optimal in $C^{\mathbb{F}}$. Thus, we have the following upper and lower bounds $b_u$ and $b_l$:

$$b_l \le \sup\{EZ_\tau : \tau \in C^{\mathbb{A}}\} \le b_u$$

with

$$
\begin{aligned}
b_l &= \sup\{EZ_t : t \in \mathbb{R}_+\}, \\
b_u &= \sup\{EZ_\tau : \tau \in C^{\mathbb{F}}\} = EZ_\sigma.
\end{aligned}
$$

Some elementary calculations yield

$$
\begin{aligned}
b_l &= u - k_2 + \frac{1}{q}(c - \lambda_1 \mu) - \frac{r_2}{q} \ln\left(\frac{-r_2}{r_1 - r_2}\right), \\
b_u &= u - k_2 + \frac{1}{q}(c - \lambda_1 \mu).
\end{aligned}
$$

For $\lambda_1 = \lambda_2$ the optimal stopping time is deterministic so that in this case the lower bound is attained.

**Bibliographic Notes.** A fundamental reference for basic replacement models is Barlow and Proschan [33]. There is an extensive literature about preventive replacement models, which is surveyed in the overviews of Pierskalla and Voelker [140], Sherif and Smith [155], Valdez-Flores and Feldman [170], and Jensen [105]. Block and Savits [49] and Boland and Proschan [51] give overviews over comparison methods and stochastic order in reliability theory. Shaked and Szekli [153] and Last and Szekli [127] compare replacement policies via point process methods. A good source for overviews of the vastly increasing literature on replacement and maintenance optimization models is the book *Reliability and Maintenance of Complex Systems* edited by Özekici in the NATO ASI Series.

The presentation in Section 5.2 follows the lines of [107]. A general setup for cost minimizing problems is introduced in Jensen [107] similar to Bergman [40] and Aven and Bergman [22]. It allows for specialization in different directions. As an example the model presented by Aven [21] covering the total expected discounted cost criterion is included. What goes beyond the results in [22] is the possibility to take different information levels into account.

There are lots of multivariate extensions of the univariate exponential distribution, for an overview see Hutchinson and Lai [101] or Basu [35], which also cover the models of Freund [77] and Marshall and Olkin [132]. A detailed derivation, statistical properties, and methods of parameter estimation of the combined exponential distribution can be found in [92]. The optimization problem also for more general cost structures is treated in Heinrich and Jensen [94]. An alternative approach to solve optimization problems of the kind treated in this chapter is to use Markov decision processes. It has not been within the scope of this book to develop this theory here. An introduction to this theory can be found in the books of Puterman [141], Bertsekas [43], Davis [65], and Van der Duyn Schouten [171], which also contains applications in reliability.

An overview of several problems related to burn-in and the corresponding literature is given in the review articles by Block and Savits [48], Kuo and Kuo [124], and Leemis and Beneke [129]. The problem of sequential burn-in, where the failures of the items are observed and the burn-in time depends on these failures, is treated in the article of Marcus and Blumenthal [131]. In the papers of Costantini and Spizzichino [61] and Spizzichino [161] the assumption that the component lifelengths are independent is dropped and replaced by certain dependence models.

The problem of finding optimal replacement times for general repair processes has been treated by Aven in [15] and [18]. The presentation of Markov-modulated minimal repair processes follows the lines of [104] and [106] which include the technical details. A similar model considering interest rates has been investigated by Schöttl [147].

# Appendix A
## Background in Probability and Stochastic Processes

This appendix serves as background for Chapters 3–5. The focus is on stochastic processes on the positive real time axis $\mathbb{R}_+ = [0, \infty)$. Our aim is to give that basis of the measure-theoretic framework that is necessary to make the text intelligible and accessible to those who are not familiar with the general theory of stochastic processes. For detailed presentations of this framework we recommend texts like Dellacherie and Meyer [67], [68], Rogers and Williams [143], and Kallenberg [112]. The point process theory is treated in Karr [114], Daley and Vere-Jones [64], and Brémaud [52]. A "nontechnical" introduction to parts of the general theory accompanied by comprehensive historical and bibliographic remarks can be found in Chapter II of the monograph of Andersen et al. [2]. A good introduction to basic results of probability theory is Williams [176].

## A.1  Basic Definitions

We use the following notation

$\mathbb{N} = \{1, 2, ...\}$
$\mathbb{N}_0 = \{0, 1, 2, ...\}$
$\mathbb{Z} = \{0, +1, -1, +2, -2, ...\}$ set of integers
$\mathbb{Q} = \{\frac{p}{q} : p \in \mathbb{Z}, q \in \mathbb{N}\}$ set of rationals
$\mathbb{R} = (-\infty, +\infty)$ set of real numbers
$\mathbb{R}_+ = [0, \infty)$ set of nonnegative real numbers

$f \vee g$ and $f \wedge g$ denote $\max\{f, g\}$ and $\min\{f, g\}$, respectively, where $f$ and $g$ can be real-valued functions or real numbers. We denote $f^{+} = f \vee 0$ and $f^{-} = -(f \wedge 0)$.

$\inf \emptyset = \infty$, $\sup \emptyset = 0$. Ratios of the form $\frac{0}{0}$ are set equal to 0.

A function $f$ from a set $A$ to a set $B$ is denoted by $f : A \rightarrow B$ and $f(a)$ is the value of $f$ at $a \in A$. To simplify the notation we also speak of $f(a)$ as a function.

For a function $f : \mathbb{R} \rightarrow \mathbb{R}$ we denote the left and right limit at $a$ (in the case of existence) by

$$
\begin{aligned}
f(a-) &= \lim_{t \rightarrow a-} f(t) = \lim_{h \rightarrow 0, h > 0} f(a - h) , \\
f(a+) &= \lim_{t \rightarrow a+} f(t) = \lim_{h \rightarrow 0, h > 0} f(a + h) .
\end{aligned}
$$

For two functions $f, g : \mathbb{R} \rightarrow \mathbb{R}$ we write $f(h) = o(g(h)), h \rightarrow h_0$, for some $h_0 \in \mathbb{R} \cup \{\infty\}$, if

$$
\lim_{h \rightarrow h_0} \frac{f(h)}{g(h)} = 0;
$$

we write $f(h) = O(g(h)), h \rightarrow h_0$, for some $h_0 \in \mathbb{R} \cup \{\infty\}$, if

$$
\limsup_{h \rightarrow h_0} \frac{|f(h)|}{|g(h)|} < \infty.
$$

An integral $\int f(s)ds$ of a real-valued measurable function is always an integral with respect to Lebesgue-measure. Integrals over finite intervals $\int_a^b$, $a \leq b$, are always integrals $\int_{[a,b]}$ over the closed interval $[a, b]$.

The indicator function of a set $A$ taking only the values 1 and 0 is denoted $I(A)$. This notation is preferred rather than $I_A$ or $I_A(a)$ in the case of descriptions of sets $A$ by means of random variables.

In the following we always refer to a basic *probability space* $(\Omega, \mathcal{F}, P)$, where

- $\Omega$ is a fixed nonempty set.

- $\mathcal{F}$ is a $\sigma$-*algebra* or $\sigma$-*field* on $\Omega$, i.e., a collection of subsets of $\Omega$ including $\Omega$, which is closed under countable unions and finite differences.

- $P$ is a probability measure on $(\Omega, \mathcal{F})$, i.e., a $\sigma$-additive, $[0, 1]$-valued function on $\mathcal{F}$ with $P(\Omega) = 1$.

If $\mathcal{A}$ is a collection of subsets of $\Omega$, then $\sigma(\mathcal{A})$ denotes the smallest $\sigma$-algebra containing $\mathcal{A}$, the $\sigma$-algebra generated by $\mathcal{A}$.

If $S$ is some set and $\mathcal{S}$ a $\sigma$-algebra of subsets of $S$, then the pair $(S, \mathcal{S})$ is called a measurable space. Let $S$ be a metric space (usually $\mathbb{R}$ or $\mathbb{R}^n$)

and $\mathcal{O}$ the collection of its open sets. Then the $\sigma$-algebra generated by $\mathcal{O}$ is called *Borel-$\sigma$-algebra* and denoted $\mathcal{B}(S)$, especially we denote $\mathcal{B} = \mathcal{B}(\mathbb{R})$.

If $\mathcal{A}$ and $\mathcal{C}$ are two sub-$\sigma$-algebras of $\mathcal{F}$, then $\mathcal{A} \vee \mathcal{C}$ denotes the $\sigma$-algebra generated by the union of $\mathcal{A}$ and $\mathcal{C}$. The *product* $\sigma$-algebra of $\mathcal{A}$ and $\mathcal{C}$, generated by the sets $A \times C$, where $A \in \mathcal{A}$ and $C \in \mathcal{C}$, is denoted $\mathcal{A} \otimes \mathcal{C}$.

## A.2   Random Variables, Conditional Expectations

### A.2.1   Random Variables and Expectations

On the fixed probability space $(\Omega, \mathcal{F}, P)$ we consider a mapping $X$ into the measurable space $(\mathbb{R}, \mathcal{B})$. If $X$ is measurable (or more exactly $\mathcal{F}$-$\mathcal{B}$-measurable), i.e., $X^{-1}(\mathcal{B}) = \{X^{-1}(B) : B \in \mathcal{B}\} \subset \mathcal{F}$, then it is called a *random variable*. The $\sigma$-algebra $\sigma(X) = X^{-1}(\mathcal{B})$ is the smallest one with respect to which $X$ is measurable. It is called the $\sigma$-algebra generated by $X$.

**Definition 29 (Independence)** *(i) Two events $A, B \in \mathcal{F}$ are called independent, if $P(A \cap B) = P(A)P(B)$.*

*(ii) Suppose $\mathcal{A}_1$ and $\mathcal{A}_2$ are subfamilies of $\mathcal{F}$: $\mathcal{A}_1, \mathcal{A}_2 \subset \mathcal{F}$. Then $\mathcal{A}_1$ and $\mathcal{A}_2$ are called independent, if $P(A_1 \cap A_2) = P(A_1)P(A_2)$ for all $A_1 \in \mathcal{A}_1$, $A_2 \in \mathcal{A}_2$.*

*(iii) Two random variables $X$ and $Y$ on $(\Omega, \mathcal{F})$ are called independent, if $\sigma(X)$ and $\sigma(Y)$ are independent.*

The *expectation $EX$* (or $E[X]$) of a random variable is defined in the usual way as the integral $\int X dP$ with respect to the probability measure $P$. If the expectation $E|X|$ is finite, we call $X$ *integrable*. The law or distribution of $X$ on $(\mathbb{R}, \mathcal{B})$ is given by $F_X(B) = P(X \in B)$, $B \in \mathcal{B}$, and $F_X(t) = F_X((-\infty, t])$ is the distribution function. Often the index $X$ in $F_X$ is omitted when it is clear which random variable is considered. Let $g : \mathbb{R} \to \mathbb{R}$ be a measurable function and suppose that $g(X)$ is integrable. Then

$$Eg(X) = \int_\Omega g(X)dP = \int_\mathbb{R} g(t)dF_X(t).$$

If $X$ has a density $f_X : \mathbb{R} \to \mathbb{R}_+$, i.e., $P(X \in B) = \int_B f_X(t)dt$, $B \in \mathcal{B}$, then the expectation can be calculated as

$$Eg(X) = \int_\mathbb{R} g(t)f_X(t)dt.$$

The variance of a random variable $X$ with $E[X^2] < \infty$ is denoted $\mathrm{Var}[X]$ and defined by $\mathrm{Var}[X] = E[(X - EX)^2]$.

We now present some classical inequalities:

- *Markov* inequality: Suppose that $X$ is a random variable and $g :$ $\mathbb{R}_+ \to \mathbb{R}_+$ a measurable nondecreasing function such that $g(|X|)$ is integrable. Then for any real $c > 0$

$$Eg(|X|) \geq g(c)P(|X| \geq c).$$

- *Jensen's* inequality: Suppose that $g : \mathbb{R} \to \mathbb{R}$ is a convex function and that $X$ is a random variable such that $X$ and $g(X)$ are integrable. Then

$$g(EX) \leq Eg(X).$$

- *Hölder's* inequality: Let $p, q \in \mathbb{R}$ such that $p > 1$ and $1/p + 1/q = 1$. Suppose $X$ and $Y$ are random variables such that $|X|^p$ and $|Y|^q$ are integrable. Then $XY$ is integrable and

$$E|XY| \leq E[|X|^p]^{1/p} E[|Y|^q]^{1/q}.$$

Taking $p = q = 2$ this inequality reduces to *Schwarz's* inequality.

- *Minkowski's* inequality: Suppose that $X$ and $Y$ are random variables such that $|X|^p$ and $|Y|^p$ are integrable for some $p \geq 1$. Then we have the triangle law

$$E[|X + Y|^p]^{1/p} \leq E[|X|^p]^{1/p} + E[|Y|^p]^{1/p}.$$

At the end of this section we list some types of convergence of real-valued random variables. Let $X, X_n, n \in \mathbb{N}$, be random variables carried by the triple $(\Omega, \mathcal{F}, P)$ and taking values in $(\mathbb{R}, \mathcal{B})$ with distribution functions $F, F_n$. Then the following forms of convergence $X_n \to X$ are fundamental in probability theory.

- *Almost sure convergence*: We say $X_n \to X$ almost surely $(P\text{-a.s.})$ if

$$P(\lim_{n \to \infty} X_n = X) = 1.$$

- *Convergence in probability*: We say $X_n \xrightarrow{P} X$ in probability, if for every $\epsilon > 0$,

$$\lim_{n \to \infty} P(|X_n - X| > \epsilon) = 0.$$

- *Convergence in distribution*: We say $X_n \xrightarrow{D} X$ in distribution, if for every $x$ of the set of continuity points of $F$,

$$\lim_{n \to \infty} F_n(x) = F(x).$$

- *Convergence in the pth mean* or *convergence in $L^p$*: We say $X_n \to X$ in the $p$th mean, $p \geq 1$, or in $L^p$, if $|X|^p, |X_n|^p$ are integrable and

$$\lim_{n \to \infty} E|X_n - X|^p = 0.$$

The relationships between these forms of convergence are the following:

$$X_n \to X, P\text{-a.s.} \quad \Rightarrow \quad X_n \xrightarrow{P} X,$$
$$X_n \to X \text{ in } L^p \quad \Rightarrow \quad X_n \xrightarrow{P} X,$$
$$X_n \xrightarrow{P} X \quad \Rightarrow \quad X_n \xrightarrow{D} X.$$

## A.2.2 $L^p$-Spaces and Conditioning

We introduce the vector spaces $L^p = L^p(\Omega, \mathcal{F}, P), p \geq 1$, of (equivalence classes of) random variables $X$ such that $|X|^p$ is integrable, without distinguishing between random variables $X, Y$ with $P(X = Y) = 1$. With the norm $\| X \|_p = (E|X|^p)^{1/p}$ the space $L^p$ becomes a complete space in that for any Cauchy sequence $(Y_n), n \in \mathbb{N}$, there exists a $Y \in L^p$ such that $\| Y_n - Y \|_p \to 0$ for $n \to \infty$. A sequence $(Y_n)$ is called Cauchy sequence if

$$\sup_{r,s \geq k} \| Y_r - Y_s \|_p \to 0 \text{ for } k \to \infty.$$

So $L^p$ is a complete and metric vector space or *Banach* space. For $1 \leq p \leq q$ and $X \in L^q$ it follows by Jensen's inequality that

$$\| X \|_p \leq \| X \|_q.$$

So $L^q$ is a subspace of $L^p$ if $q \geq p$. For $p = 2$ we define the scalar product $\langle X, Y \rangle = E[XY]$, which makes $L^2$ a Hilbert space, i.e., a Banach space with a norm induced by a scalar product.

We have introduced $L^p$-spaces to be able to look at conditional expectations from a geometrical point of view. Before we give a formal definition of conditional expectations, we consider the orthogonal projection in Hilbert spaces.

**Theorem 81** *Let $K$ be a complete vector subspace of $L^2$ and $X \in L^2$. Then there exists $Y$ in $K$ such that*
*(i) $\| X - Y \|_2 = \inf\{\| X - Z \|_2 : Z \in K\}$,*
*(ii) $X - Y \perp Z$, i.e., $E[(X - Y)Z] = 0$, for all $Z \in K$.*
*Properties (i) and (ii) are equivalent and if $Y^*$ shares either property (i) or (ii) with $Y$, then $P(Y = Y^*) = 1$.*

The short proof of this result can be found in Williams [176]. The theorem states that there is one unique element in the subspace $K$ that has the shortest distance from a given element in $L^2$ and the projection direction is orthogonal on $K$. A similar projection can be carried out from $L^1(\Omega, \mathcal{F}, P)$ onto $L^1(\Omega, \mathcal{A}, P)$, where $\mathcal{A} \subset \mathcal{F}$ is some sub-$\sigma$-algebra of $\mathcal{F}$. Of course, any $\mathcal{A}$-measurable random variable of $L^1(\Omega, \mathcal{A}, P)$ is also in $L^1(\Omega, \mathcal{F}, P)$. Thus, for a given $X$ in $L^1(\Omega, \mathcal{F}, P)$, we are looking for the "best" approximation in $L^1(\Omega, \mathcal{A}, P)$. A solution to this problem is given by the following fundamental theorem and definition.

**Theorem 82** *Let $X$ be a random variable in $L^1(\Omega, \mathcal{F}, P)$ and let $\mathcal{A}$ be a sub-$\sigma$-algebra of $\mathcal{F}$. Then there exists a random variable $Y$ in $L^1(\Omega, \mathcal{A}, P)$ such that*

$$\int_A Y\,dP = \int_A X\,dP, \text{ for all } A \in \mathcal{A}. \tag{A.1}$$

*If $Y^*$ is another random variable in $L^1(\Omega, \mathcal{A}, P)$ with property (A.1), then $P(Y = Y^*) = 1$.*

*A random variable $Y \in L^1(\Omega, \mathcal{A}, P)$ with property (A.1) is called (a version of) the conditional expectation $E[X|\mathcal{A}]$ of $X$ given $\mathcal{A}$. We write $Y = E[X|\mathcal{A}]$ noting that equality holds $P$-a.s.*

The standard proof of this theorem uses the Radon–Nikodym theorem (cf. for example Billingsley [44]). A more constructive proof is via the Orthogonal Projection Theorem 81. In the case that $EX^2 < \infty$, i.e., $X \in L^2(\Omega, \mathcal{F}, P)$, we can use Theorem 81 directly with $K = L^2(\Omega, \mathcal{A}, P)$. Let $Y$ be the projection of $X$ in $K$. Then property (ii) of Theorem 81 yields $E[(X - Y)Z] = 0$ for all $Z \in K$. Take $Z = I_A, A \in \mathcal{A}$. Then $E[(X - Y)I_A] = 0$ is just condition (A.1), which shows that $Y$ is a version of the conditional expectation $E[X|\mathcal{A}]$. If $X$ is not in $L^2$, we split $X$ as $X^+ - X^-$ and approximate both parts by sequences $X_n^+ = X^+ \wedge n$ and $X_n^- = X^- \wedge n, n \in \mathbb{N}$, of $L^2$-random variables. A limiting argument for $n \to \infty$ yields the desired result (see [176] for a complete proof).

Conditioning with respect to a $\sigma$-algebra is in general not very concrete, so the idea of projecting onto a subspace may give some additional insight. Another point of view is to look at conditioning as an averaging operator. The sub-$\sigma$-algebra $\mathcal{A}$ lies between the extremes $\mathcal{F}$ and $\mathcal{G} = \{\emptyset, \Omega\}$, the trivial $\sigma$-field. As can be easily verified from the definition, the corresponding conditional expectations of $X$ are $X = E[X|\mathcal{F}]$ and $EX = E[X|\mathcal{G}]$. So for $\mathcal{A}$ with $\mathcal{G} \subset \mathcal{A} \subset \mathcal{F}$ the conditional expectation $E[X|\mathcal{A}]$ lies "between" $X$ (no averaging, complete information about the value of $X$) and $EX$ (overall average, no information about the value of $X$). The more events of $\mathcal{F}$ are included in $\mathcal{A}$ the more is $E[X|\mathcal{A}]$ varying and the closer is this conditional expectation to $X$ in a sense made precise in the following proposition.

**Proposition 83** *Suppose $X \in L^2(\Omega, \mathcal{F}, P)$ and let $\mathcal{A}_1$ and $\mathcal{A}_2$ be sub-$\sigma$-algebras of $\mathcal{F}$ such that $\mathcal{A}_1 \subset \mathcal{A}_2 \subset \mathcal{F}$. Then, denoting $Y_i = E[X|\mathcal{A}_i], i = 1, 2$, we have the following inequalities:*

($i$) $\| X - Y_2 \|_2 \leq \| X - Y_1 \|_2 \leq \| X - Y_2 \|_2 + \| Y_2 - Y_1 \|_2$.
($ii$) $\| Y_1 - EX \|_2 \leq \| Y_2 - EX \|_2 \leq \| Y_1 - EX \|_2 + \| Y_2 - Y_1 \|_2$.

**Proof.** The right-hand side inequalities are just special cases of the triangle law for the $L^2$-norm or Minkowski's inequality. So we need to prove the left-hand inequalities.

(i) Since $Y_2$ is the projection of $X$ on $L^2(\Omega, \mathcal{A}_2, P)$ and

$$Y_1 \in L^2(\Omega, \mathcal{A}_1, P) \subset L^2(\Omega, \mathcal{A}_2, P),$$

we can use Theorem 81 to yield

$$\| X - Y_2 \|_2 = \inf\{ \| X - Z \|_2 : Z \in L^2(\Omega, \mathcal{A}_2, P) \} \leq \| X - Y_1 \|_2.$$

(ii) Denoting $\widetilde{Y}_i = Y_i - EX$ we see that $\widetilde{Y}_1$ is the projection of $\widetilde{Y}_2$ on $L^2(\Omega, \mathcal{A}_1, P)$. Again from Theorem 81 it follows that $\widetilde{Y}_2 - \widetilde{Y}_1$ and $\widetilde{Y}_1$ are orthogonal. The Pythagoras Theorem then takes the form

$$\| \widetilde{Y}_2 \|_2^2 = \| \widetilde{Y}_2 - \widetilde{Y}_1 + \widetilde{Y}_1 \|_2^2 = \| \widetilde{Y}_2 - \widetilde{Y}_1 \|_2^2 + \| \widetilde{Y}_1 \|_2^2,$$

which gives $\| \widetilde{Y}_1 \|_2 \leq \| \widetilde{Y}_2 \|_2$. ∎

**Remark 23** 1. Using some of the properties of conditional expectations stated below, all the inequalities but the first in (i) of the proposition can be shown to hold also in $L^p$-norm, $p \geq 1$, provided that $X \in L^p$.

2. If we view $E[X|\mathcal{A}]$ as a predictor of the unknown $X$, then Proposition 83 says that the closer $\mathcal{A}$ is to $\mathcal{F}$ the better in the mean square sense is this estimate and the bigger is the variance $\mathrm{Var}[E[X|\mathcal{A}]]$ of this random variable.

In particular, if $\mathcal{A}$ is generated by a finite or countable partition of $\Omega$, then the conditional expectation can be given explicitly.

**Theorem 84** *Let $X$ be an integrable random variable, i.e., $X \in L^1$, and let $\mathcal{A}$ be a sub-$\sigma$-algebra of $\mathcal{F}$ generated by a finite or countable partition $A_1, A_2, \ldots$ of $\Omega$. Then,*

$$E[X|\mathcal{A}] = \frac{1}{P(A_i)} \int_{A_i} X\,dP = \frac{E[I_{A_i} X]}{P(A_i)}, \quad \omega \in A_i, P(A_i) > 0.$$

*If $P(A_i) = 0$, the value of $E[X|\mathcal{A}]$ over $A_i$ is set to 0.*

### A.2.3   Properties of Conditional Expectations

Here and in the following relations like $<, \leq, =$ between random variables are always assumed to hold with probability one and the term $P$-a.s. is suppressed. All random variables in this subsection are assumed to be integrable, i.e., to be elements of $L^1(\Omega, \mathcal{F}, P)$. Let $\mathcal{A}$ and $\mathcal{C}$ denote sub-$\sigma$-algebras of $\mathcal{F}$. Then the following properties for conditional expectations hold true.

1. If $Y$ is any version of $E[X|\mathcal{A}]$, then $EY = EX$.

2. If $X$ is $\mathcal{A}$-measurable $(\sigma(X) \subset \mathcal{A})$, then $E[X|\mathcal{A}] = X$.

3. **Linearity.** $E[aX + bY|\mathcal{A}] = aE[X|\mathcal{A}] + bE[Y|\mathcal{A}]$, $a, b \in \mathbb{R}$.

4. **Monotonicity.** If $X \leq Y$, then $E[X|\mathcal{A}] \leq E[Y|\mathcal{A}]$.

5. **Monotone Convergence.** If $X_n$ is an increasing sequence and $X_n \to X$ $P$-a.s., then $E[X_n|\mathcal{A}]$ converges almost surely:
$$\lim_{n\to\infty} E[X_n|\mathcal{A}] = E[X|\mathcal{A}].$$

6. **Dominated Convergence.** If $X_n$ is a sequence of random variables such that $\sup |X_n|$ is integrable and $X_n \to X$ $P$-a.s., then $E[X_n|\mathcal{A}]$ converges almost surely:
$$\lim_{n\to\infty} E[X_n|\mathcal{A}] = E[X|\mathcal{A}].$$

7. **Jensen's Inequality.** If $g : \mathbb{R} \to \mathbb{R}$ is convex and $g(X)$ is integrable, then
$$E[g(X)|\mathcal{A}] \geq g(E[X|\mathcal{A}]),$$
in particular
$$\| X \|_p \geq \| E[X|\mathcal{A}]\|_p, \text{ for } p \geq 1.$$

8. **Successive Conditioning.** If $\mathcal{H}$ is a sub-$\sigma$-algebra of $\mathcal{A}$, then
$$E[E[X|\mathcal{A}]|\mathcal{H}] = E[X|\mathcal{H}].$$

9. **Factoring.** Let the random variable $Z$ be $\mathcal{A}$-measurable and suppose that $ZX$ is integrable. Then
$$E[ZX|\mathcal{A}] = ZE[X|\mathcal{A}].$$

10. **Independent Conditioning.** Let $\mathcal{C}$ and $\mathcal{A}$ be sub-$\sigma$-algebras of $\mathcal{F}$ such that $\mathcal{C}$ is independent of $\sigma(X)\vee\mathcal{A}$. Then
$$E[X|\mathcal{C} \vee \mathcal{A}] = E[X|\mathcal{A}].$$
In particular, if $X$ is independent of $\mathcal{C}$, then $E[X|\mathcal{C}] = EX$.

The proofs of all these properties are mainly based on the definition of the conditional expectation and follow the ideas of the corresponding proofs for unconditional expectations, e.g., for monotone and dominated convergence (cf. Williams [176], pp. 89-90).

## A.2.4  Regular Conditional Probabilities

We define the conditional probability of an event $A \in \mathcal{F}$, given a sub-$\sigma$-algebra $\mathcal{A}$ as

$$P(A|\mathcal{A}) = E[I_A|\mathcal{A}].$$

Clearly, by the monotonicity, linearity, and monotone convergence properties we have

$$0 \leq P(A|\mathcal{A}) \leq 1,$$

$$P(\Omega|\mathcal{A}) = 1,$$

and

$$P(\bigcup_{n=1}^{\infty} A_n|\mathcal{A}) = \sum_{n=1}^{\infty} P(A_n|\mathcal{A})$$

for a fixed sequence $A_1, A_2, \ldots$ of disjoint events of $\mathcal{F}$. From this we cannot conclude that for almost all $\omega \in \Omega$ the map $A \longmapsto P(A|\mathcal{A})(\omega)$ defines a probability on $\mathcal{F}$. Although we often dispense with a discussion of $P$-zero sets, it is important here. For example, the last equation showing the $\sigma$-additivity of conditional probability only holds with probability 1. Except in trivial cases, there are uncountable many sequences of disjoint events and each of these sequences determines an exceptional $P$-zero set. The union of all these exceptional sets need not have probability 0 (it need not even be an element of $\mathcal{F}$). But fortunately, for most cases encountered in applications there exists a so-called regular conditional probability.

**Definition 30** *A map $Q : \Omega \times \mathcal{F} \rightarrow [0,1]$ is called* regular conditional probability *given $\mathcal{A} \subset \mathcal{F}$, if*
  *(i) for all $A \in \mathcal{F}$, $\omega \longmapsto Q(\omega, A)$ is a version of $E[I_A|\mathcal{A}]$;*
  *(ii) there exists some $N_0 \in \mathcal{F}$, $P(N_0) = 0$ such that the map $A \longmapsto Q(\omega, A)$ is a probability measure on $\mathcal{F}$ for all $\omega \notin N_0$.*

## A.2.5  Computation of Conditional Expectations

Besides the simple case of a sub-$\sigma$-algebra $\mathcal{A}$ generated by a countable partition of $\Omega$ mentioned in Theorem 84, we consider two further ways to determine conditional expectations $E[X|\mathcal{A}]$.

  1. If there exists a regular conditional probability $Q$ given $\mathcal{A}$, we can determine the conditional distribution $Q_X$ of a random variable $X$ given $\mathcal{A}$: $Q_X(\omega, B) = Q(\omega, X^{-1}(B))$. Then for any measurable function $g : \mathbb{R} \rightarrow \mathbb{R}$ such that $g(X)$ is integrable,

$$\int_{\mathbb{R}} g(x) Q_X(\omega, dx)$$

is a version of $E[g(X)|\mathcal{A}]$.

2. We consider two random variables $X$ and $Y$ and a measurable function $g$ such that $g(X)$ is integrable. We write

$$E[g(X)|Y] = E[g(X)|\sigma(Y)]$$

for the conditional expectation of $g(X)$ given $Y$. By definition $E[g(X)|Y]$ is $\sigma(Y)$-measurable and by Doob's representation theorem (cf. [67], p. 12) there exists a measurable function $h : Y(\Omega) \to \mathbb{R}$ such that

$$E[g(X)|Y] = h(Y).$$

If we know such a function $h$, we can also determine $h(y) = E[g(X)|Y = y], y \in \mathbb{R}$, the conditional expectation of $g(X)$ given that $Y$ has realization $y$. Of course, if $P(Y = y) > 0$, we have

$$h(y) = E[g(X)|Y = y] = \frac{1}{P(Y = y)} \int_{\{Y=y\}} g(X) dP.$$

But even if the set $\{Y = y\}$ has probability 0, we are now able to determine the conditional expectation of $g(X)$ given that $Y$ takes the value $y$ (provided we know $h$). Consider the case that a joint density $f_{XY}(x, y)$ of $X$ and $Y$ is known. Let $f_Y(y) = \int_{\mathbb{R}} f_{XY}(x, y) dx$ be the density of the (marginal) distribution of $Y$ and

$$f_{X|Y}(x|y) = \begin{cases} f_{XY}(x, y)/f_Y(y) & \text{if} \quad f_Y(y) \neq 0 \\ 0 & \text{otherwise} \end{cases}$$

the elementary conditional density of $X$ given $Y$. A natural choice for the function $h$ would then be

$$h(y) = \int_{\mathbb{R}} g(x) f_{X|Y}(x, y) dx.$$

We claim that $h(Y)$ is a version of the conditional expectation $E[g(X)|Y]$. To prove this note that the elements of the $\sigma$-algebra $\sigma(Y)$ are of the form $Y^{-1}(B) = \{\omega : Y(\omega) \in B\}, B \in \mathcal{B}$. Therefore, we have to show that

$$E[g(X)I_B(Y)] = \int \int g(x) I_B(y) f_{XY}(x, y) dx dy$$

equals

$$E[h(Y)I_B(Y)] = \int h(y) I_B(y) f_Y(y) dy$$

for all $B \in \mathcal{B}$. But this follows directly from Fubini's Theorem, which proves the assertion.

## A.3    Stochastic Processes on a Filtered Probability Space

**Definition 31** 1. *A stochastic process is a family $X = (X_t), t \in \mathbb{R}_+$, of random variables all defined on the same probability space $(\Omega, \mathcal{F}, P)$ with values in a measurable space $(S, \mathcal{S})$.*

2. *For $\omega \in \Omega$ the mapping $t \to X_t(\omega)$ is called path.*

3. *Two stochastic processes $X, Y$ are called indistinguishable, if $P$-almost all paths are identical: $P(X_t = Y_t, \forall t \in \mathbb{R}_+) = 1$.*

If it is claimed that a process is unique, we mean uniqueness up to indistinguishability. Also for conditional expectations no distinction will be made between one version of the conditional expectation and the equivalence class of $P$-a.s. equal versions. A real-valued process is called right- or left-continuous, nondecreasing, of bounded variation on finite intervals etc., if $P$-almost all paths have this property, i.e., if the process is indistinguishable from a process, the paths of which all have that property. In particular a process is called *cadlag* (continu à droite, limité à gauche), if almost all paths are right-continuous and left-limited.

If not otherwise mentioned, we always refer in the following to real-valued stochastic processes, i.e., to processes $X = (X_t)$ for which the $X_t$ take values in $(S, \mathcal{S}) = (\mathbb{R}, \mathcal{B})$, where $\mathcal{B} = \mathcal{B}(\mathbb{R})$ is the Borel $\sigma$-algebra on $\mathbb{R}$.

**Definition 32** *A stochastic process $X$ is called*

1. *integrable, if $E|X_t| < \infty$, $\forall t \in \mathbb{R}_+$;*
2. *square integrable, if $EX_t^2 < \infty, \forall t \in \mathbb{R}_+$;*
3. *bounded in $L^p, p \geq 1$, if $\sup_{t \in \mathbb{R}_+} E|X_t|^p < \infty$;*
4. *uniformly integrable, if $\lim_{c \to \infty} \sup_{t \in \mathbb{R}_+} E[|X_t|I(|X_t| > c)] = 0$.*

Deviating from our notation some authors call an $L^2$-bounded stochastic process square integrable.

Uniform integrability plays an important role in martingale theory. Therefore, we look for criteria for this property. A very useful one is given in the following proposition.

**Proposition 85** *A stochastic process $X$ is uniformly integrable if and only if there exists a positive increasing convex function $G : \mathbb{R}_+ \to \mathbb{R}_+$ such that*

1. *$\lim_{t \to \infty} \frac{G(t)}{t} = \infty$ and*
2. *$\sup_{t \in \mathbb{R}_+} EG(|X_t|) < \infty$.*

In particular, taking $G(t) = t^p$, we see that a process $X$, which is bounded in $L^p$ for some $p > 1$, is uniformly integrable. A process bounded in $L^1$ is not necessarily uniformly integrable. The property of uniform integrability links the convergence in probability with convergence in $L^1$.

**Theorem 86** *Let* $(X_n), n \in \mathbb{N}$*, be a sequence of integrable random variables that converges in probability to a random variable* $X$*, i.e.,* $P(|X_n - X| > \epsilon) \to 0$ *as* $n \to \infty$ $\forall \epsilon > 0$*. Then*

$$X \in L^1 \text{ and } X_n \overset{L^1}{\to} X, \text{ i.e., } E|X_n - X| \to 0 \text{ as } n \to \infty$$

*if and only if* $(X_n)$ *is uniformly integrable.*

So if $X_n \to X$ $P$-a.s. and the sequence is uniformly integrable, then it follows that $EX_n \to EX, n \to \infty$. At first sight it seems reasonable that under uniform integrability almost sure convergence can be carried over also to conditional expectations $E[X_n|\mathcal{A}]$ for some sub-$\sigma$-algebra $\mathcal{A} \subset \mathcal{F}$. But (surprisingly) this does not hold true in general, for a counterexample see Jensen [108]. The condition $\sup X_n \in L^1$ in the dominated convergence theorem for conditional expectations as stated above is necessary for the convergence result and cannot be weakened.

To describe the information that is gathered observing some stochastic phenomena in time, we introduce filtrations.

**Definition 33** *1. A family* $\mathbb{F} = (\mathcal{F}_t), t \in \mathbb{R}_+$*, of sub-$\sigma$-algebras of* $\mathcal{F}$ *is called a filtration if it is nondecreasing, i.e., if* $s \leq t$*, then* $\mathcal{F}_s \subset \mathcal{F}_t$*. We denote* $\mathcal{F}_\infty = \bigvee_{t \in \mathbb{R}_+} \mathcal{F}_t = \sigma(\bigcup_{t \in \mathbb{R}_+} \mathcal{F}_t)$*.*
*2. If* $\mathbb{F} = (\mathcal{F}_t)$ *is a filtration, then we write*

$$\mathcal{F}_{t+} = \bigcap_{h > 0} \mathcal{F}_{t+h} \text{ and } \mathcal{F}_{t-} = \sigma\left(\bigcup_{h > 0} \mathcal{F}_{t-h}\right).$$

*3. A filtration* $(\mathcal{F}_t)$ *is called right-continuous, if for all* $t \in \mathbb{R}_+$*, we have* $\mathcal{F}_{t+} = \mathcal{F}_t$*.*
*4. A probability space* $(\Omega, \mathcal{F}, P)$ *together with a filtration* $\mathbb{F}$ *is called a stochastic basis:* $(\Omega, \mathcal{F}, \mathbb{F}, P)$*.*
*5. A stochastic basis* $(\Omega, \mathcal{F}, \mathbb{F}, P)$ *is called complete, if* $\mathcal{F}$ *is complete, i.e.,* $\mathcal{F}$ *contains all subsets of* $P$*-null sets, and if each* $\mathcal{F}_t$ *contains all* $P$*-null sets of* $\mathcal{F}$*.*
*6. A filtration* $\mathbb{F}$ *is said to fulfill the* usual conditions, *if it is right-continuous and complete.*

The $\sigma$-algebra $\mathcal{F}_t$ is often interpreted as the information gathered up to time $t$, or more precisely, the set of events of $\mathcal{F}$, which can be distinguished at time $t$. If a stochastic process $X = (X_t), t \in \mathbb{R}_+$, is observed, then a natural choice for a corresponding filtration would be $\mathcal{F}_t = \mathcal{F}_t^X = \sigma(X_s, 0 \leq s \leq t)$, which is the smallest $\sigma$-algebra such that all random variables $X_s, 0 \leq s \leq t$, are $\mathcal{F}_t$-measurable. Here we assume that $\mathcal{F}_t^X$ is augmented so that the generated filtration fulfills the usual conditions. Such an augmentation is always possible (cf. Dellacherie and Meyer [67], p. 115).

**Remark 24** Sometimes it is discussed whether such an augmentation affects the filtration too strongly. Indeed, if we consider, for example, two mutually singular probability measures, say $P$ and $Q$ on the measurable space $(\Omega, \mathcal{F})$ such that $P(A) = 1 - Q(A) = 1$ for some $A \in \mathcal{F}$, then completing each $\mathcal{F}_t$ with all $P$ and $Q$ negligible sets may result in $\mathcal{F}_t = \mathcal{F}$ for all $t \in \mathbb{R}_+$, which is a rather uninteresting case destroying the modeling of the evolution in time. But in the material we cover in this book such cases are not essential and we always assume that a stochastic basis is given with a filtration meeting the usual conditions.

**Definition 34** *A stochastic process* $X = (X_t), t \in \mathbb{R}_+$, *is called* adapted *to a filtration* $\mathbb{F} = (\mathcal{F}_t)$, *if* $X_t$ *is* $\mathcal{F}_t$-*measurable for all* $t \in \mathbb{R}_+$.

**Definition 35** *A stochastic process* $X$ *is* $\mathbb{F}$-progressive *or* progressively measurable, *if for every* $t$, *the mapping* $(s, \omega) \rightarrow X_s(\omega)$ *on* $[0, t] \times \Omega$ *is measurable with respect to the product* $\sigma$-*algebra* $\mathcal{B}([0, t]) \otimes \mathcal{F}_t$, *where* $\mathcal{B}([0, t])$ *is the Borel* $\sigma$-*algebra on* $[0, t]$.

**Theorem 87** *Let* $X$ *be a real-valued stochastic process. If* $X$ *is left- or right-continuous and adapted to* $\mathbb{F}$, *then it is* $\mathbb{F}$-*progressive. If* $X$ *is* $\mathbb{F}$-*progressive, then so is* $\int_0^t X_s ds$.

A further measurability restriction is needed in connection with stochastic processes in continuous time. This is the fundamental concept of predictability.

**Definition 36** *Let* $\mathbb{F}$ *be a filtration on the basic probability space and let* $\mathcal{P}(\mathbb{F})$ *be the* $\sigma$-*algebra on* $(0, \infty) \times \Omega$ *generated by the system of sets*

$$(s, t] \times A, 0 \le s < t, A \in \mathcal{F}_s, t > 0.$$

$\mathcal{P}(\mathbb{F})$ *is called the* $\mathbb{F}$-*predictable* $\sigma$-*algebra on* $(0, \infty) \times \Omega$. *A stochastic process* $X = (X_t)$ *is called* $\mathbb{F}$-*predictable, if* $X_0$ *is* $\mathcal{F}_0$-*measurable and the mapping* $(t, \omega) \rightarrow X_t(\omega)$ *on* $(0, \infty) \times \Omega$ *into* $\mathbb{R}$ *is measurable with respect to* $\mathcal{P}(\mathbb{F})$.

**Theorem 88** *Every left-continuous process adapted to* $\mathbb{F}$ *is* $\mathbb{F}$-*predictable.*

In all applications, we will be concerned with predictable processes that are left-continuous. Note that $\mathbb{F}$-predictable processes are also $\mathbb{F}$-progressive. A property that explains the term predictable is given in the following theorem.

**Theorem 89** *Suppose the process* $X$ *is* $\mathbb{F}$-*predictable. Then for all* $t > 0$ *the variable* $X_t$ *is* $\mathcal{F}_{t-}$-*measurable.*

## A.4   Stopping Times

Suppose we want to describe a point in time at which a stochastic process first enters a given set, say when it hits a certain level. So this point in time is a random time because it depends on the random evolution of the process. Observing this stochastic process, it is possible to decide at any time $t$ whether this random time has occurred or not. Such random times, which are based on the available information not anticipating the future, are defined as follows.

**Definition 37** *Suppose* $\mathbb{F} = (\mathcal{F}_t), t \in \mathbb{R}_+$, *is a filtration on the measurable space* $(\Omega, \mathcal{F})$. *A random variable* $\tau : \Omega \to [0, \infty]$ *is said to be a stopping time if for every* $t \in \mathbb{R}_+$,

$$\{\tau \leq t\} = \{\omega : \tau(\omega) \leq t\} \in \mathcal{F}_t.$$

In particular, a constant random variable $\tau = t_0 \in \mathbb{R}_+$ is a stopping time. Since we assume that the filtration is right-continuous, we can equivalently describe stopping times by the condition $\{\tau < t\} \in \mathcal{F}_t$ : If $\{\tau < t\} \in \mathcal{F}_t$ for all $t \in \mathbb{R}_+$, then

$$\{\tau \leq t\} = \bigcap_{n \in \mathbb{N}} \{\tau < t + \frac{1}{n}\} \in \bigcap_{n \in \mathbb{N}} \mathcal{F}_{t+\frac{1}{n}} = \mathcal{F}_{t+}.$$

Conversely, if $\{\tau \leq t\} \in \mathcal{F}_t$ for all $t \in \mathbb{R}_+$, then

$$\{\tau < t\} = \bigcup_{n \in \mathbb{N}} \{\tau \leq t - \frac{1}{n}\} \in \mathcal{F}_t \text{ for } t > 0 \text{ and } \{\tau < 0\} = \emptyset \in \mathcal{F}_0.$$

**Proposition 90** *Suppose* $\sigma$ *and* $\tau$ *are stopping times. Then* $\sigma \wedge \tau$, $\sigma \vee \tau$, *and* $\sigma + \tau$ *are stopping times. Let* $(\tau_n), n \in \mathbb{N}$, *be a sequence of stopping times. Then* $\sup \tau_n$ *and* $\inf \tau_n$ *are also stopping times.*

**Proof.** First we show that $\sigma + \tau$ is a stopping time and consider the complement of the event $\{\sigma + \tau \leq t\}$ :

$$\{\sigma + \tau > t\} = \{\sigma > t\} \cup \{\tau > t\} \cup \{\sigma \geq t, \tau > 0\} \cup \{0 < \sigma < t, \sigma + \tau > t\}.$$

The first three events of this union are clearly in $\mathcal{F}_t$. The fourth event

$$\{0 < \sigma < t, \sigma + \tau > t\} = \bigcup_{r \in \mathbb{Q} \cap [0,t)} \{r < \sigma < t, \tau > t - r\}$$

is the countable union of events of $\mathcal{F}_t$ and therefore $\sigma + \tau$ is a stopping time.

The proof of the remaining assertions follows from

$$\{\sup \tau_n \le t\} = \bigcap_{n\in\mathbb{N}} \{\tau_n \le t\} \in \mathcal{F}_t,$$

$$\{\inf \tau_n < t\} = \bigcup_{n\in\mathbb{N}} \{\tau_n \le t\} \in \mathcal{F}_t,$$

using the fact that for a right-continuous filtration it suffices to show $\{\inf \tau_n < t\} \in \mathcal{F}_t$. ∎

For a sequence of stopping times $(\tau_n)$ the random variables $\sup \tau_n$, $\inf \tau_n$ are stopping times, so that $\limsup \tau_n$, $\liminf \tau_n$ and $\lim \tau_n$ (if it exists) are also stopping times.

We now define the $\sigma$-algebra of the past of a stopping time $\tau$.

**Definition 38** *Suppose $\tau$ is a stopping time with respect to the filtration $\mathbb{F}$. Then the $\sigma$-algebra $\mathcal{F}_\tau$ of events occurring up to time $\tau$ is*

$$\mathcal{F}_\tau = \{A \in \mathcal{F}_\infty : A \cap \{\tau \le t\} \in \mathcal{F}_t \text{ for all } t \in \mathbb{R}_+\}.$$

We note that $\tau$ is $\mathcal{F}_\tau$-measurable and that for a constant stopping time $\tau = t_0 \in \mathbb{R}_+$ we have $\mathcal{F}_\tau = \mathcal{F}_{t_0}$.

**Theorem 91** *Suppose $\sigma$ and $\tau$ are stopping times.*
*(i) If $\sigma \le \tau$, then $\mathcal{F}_\sigma \subset \mathcal{F}_\tau$.*
*(ii) If $A \in \mathcal{F}_\sigma$, then $A \cap \{\sigma \le \tau\} \in \mathcal{F}_\tau$.*
*(iii) $\mathcal{F}_{\sigma\wedge\tau} = \mathcal{F}_\sigma \cap \mathcal{F}_\tau$.*

**Proof.** (i) For $B \in \mathcal{F}_\sigma$ and $t \in \mathbb{R}_+$ we have

$$B \cap \{\tau \le t\} = B \cap \{\sigma \le t\} \cap \{\tau \le t\} \in \mathcal{F}_t,$$

which proves (i).
(ii) Suppose $A \in \mathcal{F}_\sigma$. Then

$$A \cap \{\sigma \le \tau\} \cap \{\tau \le t\} = A \cap \{\sigma \le t\} \cap \{\tau \le t\} \cap \{\sigma \wedge t \le \tau \wedge t\}.$$

Now $A \cap \{\sigma \le t\}$ and $\{\tau \le t\}$ are elements of $\mathcal{F}_t$ by assumption and the random variables $\sigma \wedge t$ and $\tau \wedge t$ are both $\mathcal{F}_t$-measurable. This shows that $\{\sigma \wedge t \le \tau \wedge t\} \in \mathcal{F}_t$.
(iii) Since $\sigma \wedge \tau \le \sigma$ and $\sigma \wedge \tau \le \tau$ we obtain from (i)

$$\mathcal{F}_{\sigma\wedge\tau} \subset \mathcal{F}_\sigma \cap \mathcal{F}_\tau.$$

Conversely, for $A \in \mathcal{F}_\sigma \cap \mathcal{F}_\tau$ we have

$$A \cap \{\sigma \wedge \tau \le t\} = (A \cap \{\sigma \le t\}) \cup (A \cap \{\tau \le t\}) \in \mathcal{F}_t,$$

which proves (iii).    ∎

This theorem shows that some of the properties known for fixed time points $s, t$ also hold true for stopping times $\sigma, \tau$. Next we consider the link between a stochastic process $X = (X_t), t \in \mathbb{R}_+$, and a stopping time $\sigma$. It is natural to investigate variables $X_{\sigma(\omega)}(\omega)$ with random index and the stopped process $X_t^\sigma(\omega) = X_{\sigma \wedge t}(\omega)$ on $\{\sigma < \infty\}$. To ensure that $X_\sigma$ is a random variable, we need that $X_t$ fulfills a measurability requirement in $t$.

**Theorem 92** *If $\sigma$ is a stopping time and $X = (X_t), t \in \mathbb{R}_+$, is an $\mathbb{F}$-progressive process, then $X_\sigma$ is $\mathcal{F}_\sigma$-measurable and $X^\sigma$ is $\mathbb{F}$-progressive.*

**Proof.** We must show that for any Borel set $B \in \mathcal{B}$, $\{X_\sigma \in B\} \cap \{\sigma \leq t\}$ belongs to $\mathcal{F}_t$. This intersection equals $\{X_{\sigma \wedge t} \in B\} \cap \{\sigma \leq t\}$, so we need only show that $X^\sigma$ is progressive. Now $\sigma \wedge t$ is $\mathcal{F}_t$-measurable. Hence, $(s, \omega) \to (\sigma(\omega) \wedge s, \omega)$ is $\mathcal{B}([0, t]) \otimes \mathcal{F}_t$-measurable. Therefore, the map $(s, \omega) \to X_{\sigma(\omega) \wedge s}(\omega)$ is measurable as it is the composition of two measurable maps. Hence $X^\sigma$ is progressive.    ∎

Most important for applications are those random times $\sigma$ that are defined as first entrance times of a stochastic process $X$ into a Borel set $B$: $\sigma = \inf\{t \in \mathbb{R}_+ : X_t \in B\}$. In general, it is very difficult to show that $\sigma$ is a stopping time. For a discussion of the usual conditions in this connection, see Rogers and Williams [143], pp. 183–191. For a complete proof of the following theorem we refer to Dellacherie and Meyer [67], p. 116.

**Theorem 93** *Let $X$ be an $\mathbb{F}$-progressive process with respect to the complete and right-continuous filtration $\mathbb{F}$ and $B \in \mathcal{B}$ a Borel set. Then*

$$\sigma(\omega) = \inf\{t \in \mathbb{R}_+ : X_t(\omega) \in B\}$$

*is an $\mathbb{F}$-stopping time.*

**Proof.** We only show the simple case where $X$ is right-continuous and $B$ is an open set. Then the right continuity implies that

$$\{\sigma < t\} = \bigcup_{r \in \mathbb{Q} \cap [0,t)} \{X_r \in B\} \in \mathcal{F}_t.$$

Using the right-continuity of $\mathbb{F}$ it is seen that $\sigma$ is an $\mathbb{F}$-stopping time .    ∎

Note that the right-continuity of the paths was used to express $\{\sigma < t\}$ as the union of events $\{X_r \in B\}$ and that we could restrict ourselves to a countable union because $B$ is an open set.

# A.5   Martingale Theory

An overview over the historical development of martingale theory can be found in monographs such as Andersen et al. [2], pp. 115–120, or Kallenberg [112], pp. 464–485. We fix a stochastic basis $(\Omega, \mathcal{F}, \mathbb{F}, P)$ and define stochastic processes with certain properties which are known as the stochastic analogues to constant, increasing and decreasing functions.

**Definition 39** *An integrable $\mathbb{F}$-adapted process $X = (X_t), t \in \mathbb{R}_+$, is called a martingale if*

$$X_t = E[X_s|\mathcal{F}_t] \qquad (A.2)$$

*for all $s \geq t, s, t \in \mathbb{R}_+$. A supermartingale is defined in the same way, except that (A.2) is replaced by*

$$X_t \geq E[X_s|\mathcal{F}_t],$$

*and a submartingale is defined with (A.2) being replaced by*

$$X_t \leq E[X_s|\mathcal{F}_t].$$

Forming expectations on both sides of the (in)equality we obtain $EX_t = (\geq, \leq)EX_s$, which shows that a martingale is constant on average, a supermartingale decreases, and a submartingale increases on average, respectively.

**Example 50** Let $X$ be an integrable $\mathbb{F}$-adapted process. Suppose that the increments $X_s - X_t$ are independent of $\mathcal{F}_t$ for all $s > t, s, t \in \mathbb{R}_+$. If these increments have zero expectation (thus the expectation function $EX_t$ is constant), then $X$ is a martingale:

$$E[X_s|\mathcal{F}_t] = E[X_t|\mathcal{F}_t] + E[X_s - X_t|\mathcal{F}_t] = X_t.$$

Of particular importance are the following cases.

(i) If $X$ is continuous, $X_0 = 0$, and the increments $X_s - X_t$ are normally distributed with mean 0 and variance $s - t$, then $X$ is an $\mathbb{F}$-Brownian motion. In addition to $X$, also the process $Y_t = X_t^2 - t$ is a martingale:

$$E[Y_s|\mathcal{F}_t] = E[(X_s - X_t)^2|\mathcal{F}_t] + 2X_t E[X_s - X_t|\mathcal{F}_t] + X_t^2 - s$$
$$= s - t + 0 + X_t^2 - s = Y_t.$$

(ii) If $X_0 = 0$ and the increments $X_s - X_t$ follow a Poisson distribution with mean $s - t$, for $s > t$, then $X$ is a Poisson process. Now $X$ is a submartingale because of

$$E[X_s|\mathcal{F}_t] = X_t + E[X_s - X_t|\mathcal{F}_t] = X_t + s - t \geq X_t$$

and $X_t - t$ is a martingale.

**Example 51** Let $Y$ be an integrable random variable and define $M_t = E[Y|\mathcal{F}_t]$. Then $M$ is a martingale because of the successive conditioning property:

$$E[M_s|\mathcal{F}_t] = E[E[Y|\mathcal{F}_s]|\mathcal{F}_t] = E[Y|\mathcal{F}_t] = M_t, s \geq t.$$

So $M_t$ is a predictor of $Y$ given the information $\mathcal{F}_t$ gathered up to time $t$. Furthermore, $M$ is a uniformly integrable martingale. To see this we have to show that $\lim_{c\to\infty} \sup_{t\in\mathbb{R}_+} E[|M_t|I(|M_t| > c)] \to 0$ as $c \to \infty$. By Jensen's inequality for conditional expectations we obtain

$$E[|M_t|I(|M_t| > c)] \leq E[E[|Y|I(|M_t| > c)|\mathcal{F}_t]] = E[|Y|I(|M_t| > c)].$$

Since $Y$ is integrable and $cP(|M_t| > c) \leq E|M_t| \leq E|Y|$, it follows that $P(|M_t| > c) \to 0$ uniformly in $t$, which shows that $M$ is uniformly integrable.

Concerning the regularity of the paths of a supermartingale, the following result holds true.

**Lemma 94** *Suppose $X$ is a supermartingale such that $t \to EX_t$ is right-continuous. Then $X$ has a modification with all paths cadlag, i.e., there exists a process $Y$ with cadlag paths such that $X_t = Y_t$ $P$-a.s. for all $t \in \mathbb{R}_+$.*

So for a martingale, a submartingale, or a supermartingale with right-continuous expectation function, we can assume that it has cadlag paths. From now on we make the general assumption that all martingales, sub-martingales, and supermartingales are cadlag unless stated otherwise.

**Lemma 95** *Let $M$ be a martingale and consider a convex function $g : \mathbb{R} \to \mathbb{R}$ such that $X = g(M)$ is integrable. Then $X$ is a submartingale.*

*If $g$ is also nondecreasing, then the assertion remains true for submartingales $M$.*

**Proof.** Let $M$ be a martingale. Then by Jensen's inequality we obtain for $s \geq t$

$$X_t = g(M_t) = g(E[M_s|\mathcal{F}_t]) \leq E[g(M_s)|\mathcal{F}_t] = E[X_s|\mathcal{F}_t],$$

which shows that $X$ is a submartingale.

If $M$ is a submartingale and $g$ is nondecreasing, then

$$g(M_t) \leq g(E[M_s|\mathcal{F}_t])$$

shows that the conclusion remains valid. ∎

The last lemma is often applied with functions $g(x) = |x|^p, p \geq 1$. So, if $M$ is a square integrable martingale, then $X = M^2$ defines a submartingale.

One key result in martingale theory is the following convergence theorem (cf. [68], p. 72).

**Theorem 96** *Let $X$ be a supermartingale (martingale). Suppose that*

$$\sup_{t\in\mathbb{R}_+} E|X_t| < \infty,$$

*a condition that is equivalent to $\lim_{t\to\infty} EX_t^- < \infty$. Then the random variable $X_\infty = \lim_{t\to\infty} X_t$ exists and is integrable.*

*If the supermartingale (martingale) $X$ is uniformly integrable, $X_\infty$ exists and closes $X$ on the right in that for all $t \in \mathbb{R}_+$*

$$X_t \geq E[X_\infty|\mathcal{F}_t] \ (\text{respectively } X_t = E[X_\infty|\mathcal{F}_t]).$$

As a consequence we get the following characterization of the convergence of martingales.

**Theorem 97** *Suppose $M$ is a martingale. Then the following conditions are equivalent:*
*(i) $M$ is uniformly integrable.*
*(ii) There exists a random variable $M_\infty$ such that $M_t$ converges to $M_\infty$ in $L^1$ : $\lim_{t\to\infty} E|M_t - M_\infty| = 0$.*
*(iii) $M_t$ converges $P$-a.s. to an integrable random variable $M_\infty$, which closes $M$ on the right: $M_t = E[M_\infty|\mathcal{F}_t]$.*

**Example 52** If in Example 51 we assume that $Y$ is $\mathcal{F}_\infty$-measurable, then we can conclude that the martingale $M_t = E[Y|\mathcal{F}_t]$ converges $P$-a.s. and in $L^1$ to $Y$.

In Example 50 (i) we see that Brownian motion $(X_t)$ is not uniformly integrable as for any $c > 1$ we can find a $t > 0$ such that $P(|X_t| > c) \geq \epsilon$ for some $\epsilon, 0 < \epsilon < 1$. In this case we can conclude that $X_t$ does not converge to any random variable for $t \to \infty$ neither $P$-a.s. nor in $L^1$.

Next we consider conditions under which the (super-)martingale property also extends from fixed time points $s, t$ to stopping times $\sigma, \tau$.

**Theorem 98** (*Optional Sampling Theorem*). *Let $X$ be a supermartingale and let $\sigma$ and $\tau$ be two stopping times such that $\sigma \leq \tau$. Suppose either that $\tau$ is bounded or that $(X_t)$ is uniformly integrable. Then $X_\sigma$ and $X_\tau$ are integrable and*

$$X_\sigma \geq E[X_\tau|\mathcal{F}_\sigma]$$

*with equality if $X$ is a martingale.*

An often used consequence of Theorem 98 is the following: If $X$ is a uniformly integrable martingale, then setting $\sigma = 0$ we obtain $EX_0 = EX_\tau$ for all stopping times $\tau$ (all quantities are related to the same filtration $\mathbb{F}$). A kind of converse is the following proposition.

**Proposition 99** *Suppose $X$ is an adapted cadlag process such that for any bounded stopping time $\tau$ the random variable $X_\tau$ is integrable and $EX_0 = EX_\tau$. Then $X$ is a martingale.*

A further consequence of the Optional Sampling Theorem is that a stopped (super-) martingale remains a (super-) martingale.

**Corollary 100** *Let $X$ be a right-continuous supermartingale (martingale) and $\tau$ a stopping time. Then the stopped process $X^\tau = (X_{t \wedge \tau})$ is a super-martingale (martingale). If either $X$ is uniformly integrable or $I(\tau < \infty)X_\tau$ is integrable and $\lim_{t \to \infty} \int_{\{\tau > t\}} |X_t|\, dP = 0$, then $X^\tau$ is uniformly integrable.*

Martingales are often constructed in that an increasing process is subtracted from a submartingale (cf. Example 50 (ii), p. 229). This fact emanates from the celebrated *Doob–Meyer decomposition*, which is a cornerstone in modern probability theory.

**Theorem 101 (*Doob–Meyer decomposition*).** *Let the process $X$ be right-continuous and adapted. Then $X$ is a uniformly integrable submartingale if and only if it has a decomposition*

$$X = A + M,$$

*where $A$ is a right-continuous predictable nondecreasing and integrable process with $A_0 = 0$ and $M$ is a uniformly integrable martingale. The decomposition is unique within indistinguishable processes.*

**Remark 25** 1. Several proofs of this and more general results, not restricted to uniformly integrable processes, are known (cf. [68], p. 198 and [112], p. 412). Some of these also refer to local martingales, which are not needed for the applications we have presented and which are therefore not introduced here.

2. The process $A$ in the theorem above is often called compensator.

3. In the case of discrete time such a decomposition is easily constructed in the following way. Let $(X_n), n \in \mathbb{N}_0$, be a submartingale with respect to a filtration $(\mathcal{F}_n), n \in \mathbb{N}_0$. Then we define

$$X_n = A_n + M_n,$$

where

$$
\begin{aligned}
A_n &= A_{n-1} + E[X_n | \mathcal{F}_{n-1}] - X_{n-1}, n \in \mathbb{N},\ A_0 = 0, \\
M_n &= X_n - A_n, n \in \mathbb{N}_0.
\end{aligned}
$$

The process $M$ is a martingale and $A$ is nondecreasing and predictable in that $A_n$ is $\mathcal{F}_{n-1}$-measurable for $n \in \mathbb{N}$. This decomposition is unique, since for a second decomposition $X_n = \widetilde{A}_n + \widetilde{M}_n$ with the same properties we must have $M_n - \widetilde{M}_n = A_n - \widetilde{A}_n$, which is a predictable martingale. Therefore,

$$0 = E[A_n - \widetilde{A}_n | \mathcal{F}_{n-1}] = A_n - \widetilde{A}_n, n \in \mathbb{N}$$

and $A_0 = \widetilde{A}_0 = 0$.

The continuous time result needs much more care and uses several lemmas, one of which is interesting in its own right and will be presented here.

**Lemma 102** *A process $M$ is a predictable martingale of integrable variation, i.e., $E[\int_0^\infty |dM_s|] < \infty$, if and only if $M_t = M_0$ for all $t \in \mathbb{R}_+$.*

We will now use the Doob–Meyer decomposition to introduce two types of (co-)variation processes. For this we recall that $\mathcal{M}$ ($\mathcal{M}_0$) denotes the class of cadlag martingales (with $M_0 = 0$) and denote by $\mathcal{M}^2(\mathcal{M}_0^2)$ the set of martingales in $\mathcal{M}(\mathcal{M}_0)$, which are bounded in $L^2$, i.e., $\sup_{t \in \mathbb{R}_+} EM_t^2 < \infty$.

**Definition 40** *For $M \in \mathcal{M}^2$ the unique compensator of $M^2$ in the Doob –Meyer decomposition, denoted $\langle M, M \rangle$ or $\langle M \rangle$, is called the predictable variation process. For $M_1, M_2 \in \mathcal{M}^2$ the process*

$$\langle M_1, M_2 \rangle = \frac{1}{4}(\langle M_1 + M_2 \rangle - \langle M_1 - M_2 \rangle)$$

*is called the predictable covariation process of $M_1$ and $M_2$.*

**Proposition 103** *Suppose that $M_1, M_2 \in \mathcal{M}^2$. Then $A = \langle M_1, M_2 \rangle$ is the unique predictable cadlag process with $A_0 = 0$ such that $M_1 M_2 - A \in \mathcal{M}$.*

**Proof.** The assertion follows from the Doob–Meyer decomposition and

$$
\begin{aligned}
M_1 M_2 - \langle M_1, M_2 \rangle &= \frac{1}{4}\left((M_1 + M_2)^2 - (M_1 - M_2)^2\right) - \langle M_1, M_2 \rangle \\
&= \frac{1}{4}\left((M_1 + M_2)^2 - \langle M_1 + M_2 \rangle\right) \\
&\quad - \frac{1}{4}\left((M_1 - M_2)^2 - \langle M_1 - M_2 \rangle\right). \qquad \blacksquare
\end{aligned}
$$

To understand what predictable variation means, we give a heuristic explanation. Recall that for a martingale $M$ we have for all $0 < h < t$

$$E[M_t - M_{t-h} | \mathcal{F}_{t-h}] = 0,$$

or in heuristic form:

$$E[dM_t | \mathcal{F}_{t-}] = 0.$$

Since $M^2 - \langle M \rangle$ is a martingale and $\langle M \rangle$ is predictable, we obtain

$$E[dM_t^2|\mathcal{F}_{t-}] = E[d\langle M\rangle_t|\mathcal{F}_{t-}] = d\langle M\rangle_t.$$

Furthermore,

$$
\begin{aligned}
dM_t^2 &= M_t^2 - M_{t-}^2 \\
&= (M_{t-} + dM_t)^2 - M_{t-}^2 \\
&= (dM_t)^2 + 2M_{t-}dM_t,
\end{aligned}
$$

yielding

$$
\begin{aligned}
d\langle M\rangle_t &= E[(dM_t)^2|\mathcal{F}_{t-}] + 2M_{t-}E[dM_t|\mathcal{F}_{t-}] = E[(dM_t)^2|\mathcal{F}_{t-}] \\
&= \mathrm{Var}[dM_t|\mathcal{F}_{t-}].
\end{aligned}
$$

This indicates (and it can be proved) that $\langle M \rangle_t$ is the stochastic limit of the form

$$\sum_{i=1}^{n} \mathrm{Var}[M_{t_i} - M_{t_{i-1}}|\mathcal{F}_{t_{i-1}}]$$

as $n \to \infty$ and the span of the partition $0 = t_0 < t_1 < ... < t_n = t$ tends to $0$.

**Definition 41** *Two martingales $M, L \in \mathcal{M}^2$ are called orthogonal if their product is a martingale: $ML \in \mathcal{M}$.*

For two martingales $M, L$ of $\mathcal{M}^2$ that are orthogonal we must have $\langle M, L \rangle = 0$. If we equip $\mathcal{M}^2$ with the scalar product

$$(M, L)_{\mathcal{M}^2} = E[M_\infty L_\infty]$$

inducing the norm $\| M \| = (EM_\infty^2)^{1/2}$, then $\mathcal{M}^2$ becomes a Hilbert space. Because of $ML - \langle M, L \rangle \in \mathcal{M}$ and $\langle M, L \rangle_0 = 0$, it follows that

$$(M, L)_{\mathcal{M}^2} = E[M_\infty L_\infty] = E\langle M, L \rangle_\infty + EM_0L_0.$$

So two orthogonal martingales $M, L$ of $\mathcal{M}_0^2$ are also orthogonal in the Hilbert space $\mathcal{M}^2$ (cf. Elliott [75], p. 88).

The set of continuous martingales in $\mathcal{M}_0^2$, denoted $\mathcal{M}_0^{2,c}$, is a complete subspace of $\mathcal{M}_0^2$ and $\mathcal{M}_0^{2,d}$ is the space orthogonal to $\mathcal{M}_0^{2,c}$. The martingales in $\mathcal{M}_0^{2,d}$ are called *purely discontinuous*. As an immediate consequence we obtain that any martingale $M \in \mathcal{M}_0^2$ has a unique decomposition $M = M^c + M^d$, where $M^c \in \mathcal{M}_0^{2,c}$ and $M^d \in \mathcal{M}_0^{2,d}$.

A process strongly connected to predictable variation is the so-called square bracket process introduced in the following definition.

**Definition 42** *Suppose $M \in \mathcal{M}_0^2$ and $M = M^c + M^d$ is the unique decomposition with $M^c \in \mathcal{M}_0^{2,c}$ and $M^d \in \mathcal{M}_0^{2,d}$. The increasing cadlag process $[M]$ with*

$$[M]_t = \langle M^c \rangle_t + \sum_{s \leq t} \triangle M_s^2$$

*is called the quadratic variation of $M$, where $\triangle M_t = M_t - M_{t-}$ denotes the jump of $M$ at time $t > 0$ ($\triangle X_0 = X_0$). For martingales $M, L \in \mathcal{M}_0^2$ we define the quadratic covariation $[M, L]$ by*

$$[M, L] = \frac{1}{4}\left( [M + L] - [M - L] \right).$$

The following proposition helps to understand the name quadratic covariation.

**Proposition 104** *Suppose $M, L \in \mathcal{M}_0^2$.*
*1. Let $(t_i^n)$ be a sequence of partitions $0 = t_0^n < t_1^n < ... < t_n^n = t$ such that the span $\sup_i (t_{i+1}^n - t_i^n)$ tends to $0$ as $n \to \infty$. Then*

$$\sum_i (M_{t_{i+1}} - M_{t_i})(L_{t_{i+1}} - L_{t_i})$$

*converges $P$-a.s. and in $L^1$ to $[M, L]_t$ for all $t > 0$.*
*2. $ML - [M, L]$ is a martingale.*

## A.6  Semimartingales

A decomposition of a stochastic process into a (predictable) drift part and a martingale, as presented for submartingales in the Doob–Meyer decomposition, also holds true for more general processes. We start with the motivating example of a sequence $(X_n), n \in \mathbb{N}_0$, of integrable random variables adapted to the filtration $(\mathcal{F}_n)$. This sequence admits a decomposition

$$X_n = X_0 + \sum_{i=1}^{n} f_i + M_n$$

with a predictable sequence $f = (f_n), n \in \mathbb{N}$, (i.e., $f_n$ is $\mathcal{F}_{n-1}$-measurable) and a martingale $M = (M_n), n \in \mathbb{N}_0$, $M_0 = 0$. We can take

$$
\begin{aligned}
f_n &= E[X_n - X_{n-1}|\mathcal{F}_{n-1}], \\
M_n &= \sum_{i=1}^{n}(X_i - E[X_i|\mathcal{F}_{i-1}]).
\end{aligned}
$$

This decomposition is unique because a second decomposition of this type, say with a sequence $\widetilde{f}$ and a martingale $\widetilde{M}$, would imply that

$$M_n - \widetilde{M}_n = \sum_{i=1}^{n}(\widetilde{f}_i - f_i)$$

defines a predictable martingale, i.e., $E[M_n - \widetilde{M}_n|\mathcal{F}_{n-1}] = M_n - \widetilde{M}_n = M_0 - \widetilde{M}_0 = 0$, which shows the uniqueness.

   Unlike the time-discrete case, corresponding decompositions cannot be found for all integrable processes in continuous time. The role of increasing processes in the Doob–Meyer decomposition will now be taken by processes of bounded variation.

**Definition 43** *For a cadlag function $g : \mathbb{R}_+ \to \mathbb{R}$ the variation is defined as*

$$V_g(t) = \lim_{n \to \infty} \sum_{k=1}^{n} |g(tk/n) - g(t(k-1)/n)|.$$

*The function $g$ is said to have finite variation if $V_g(t) < \infty$ for all $t \in \mathbb{R}_+$. The class of cadlag processes $A$ with finite variation starting in $A_0 = 0$ is denoted $\mathcal{V}$.*

   For any $A \in \mathcal{V}$ there is a decomposition $A_t = B_t - C_t$ with increasing processes $B, C \in \mathcal{V}$ and

$$B_t + C_t = V_A(t) = \int_0^t |dA_s|.$$

**Definition 44** *A process $Z$ is a semimartingale if it has a decomposition*

$$Z_t = Z_0 + A_t + M_t,$$

*where $A \in \mathcal{V}$ and $M \in \mathcal{M}_0$.*

   There is a rich theory based on semimartingales that relies on the remarkable property that semimartingales are stable under many sorts of operations, e.g., changes of time, of probability measures, and of filtrations preserve the semimartingale property, also products and convex functions of semimartingales are semimartingales (cf. Dellacherie and Meyer [68], pp. 212–252). The importance of semimartingales lies also in the fact that stochastic integrals

$$\int_0^t H_s dZ_s$$

of predictable processes $H$ with respect to a semimartingale $Z$ can be introduced replacing Stieltjes integrals. It is beyond the scope of this book to present the whole theory of semimartingales; we confine ourselves to the case that the process $A$ in the semimartingale decomposition is absolutely continuous (with respect to Lebesgue-measure). The class of such processes is rich enough to contain most processes interesting in applications and allows the development of a kind of "differential" calculus.

**Definition 45** *A semimartingale $Z$ with decomposition $Z_t = Z_0 + A_t + M_t$ is called smooth semimartingale (SSM) if $Z$ is integrable and $A$ has the form*

$$A_t = \int_0^t f_s ds,$$

*where $f$ is a progressive process and $A$ has locally integrable variation, i.e.,*

$$E \int_0^t |f_s| ds < \infty$$

*for all $t \in \mathbb{R}_+$. Short notation: $Z = (f, M)$.*

As submartingales can be considered as stochastic analog to increasing functions, smooth semimartingales can be seen as the stochastic counterpart to differentiable functions. Some of the above-mentioned operations will be considered in the following.

### A.6.1  Change of Time

Let $(\tau_t), t \in \mathbb{R}_+$, be a family of stopping times with respect to $\mathbb{F} = (\mathcal{F}_t)$ such that for all $\omega$, $\tau_t(\omega)$ is nondecreasing and right-continuous as a function of $t$. Then for an $\mathbb{F}$-semimartingale $Z$ we consider the transformed process $\widetilde{Z}_t = Z_{\tau_t}$, which is adapted to $\widetilde{\mathbb{F}} = (\widetilde{\mathcal{F}}_t)$, where $\widetilde{\mathcal{F}}_t = \mathcal{F}_{\tau_t}$.

**Theorem 105** *If $Z$ is an $\mathbb{F}$-semimartingale, then $\widetilde{Z}$ is an $\widetilde{\mathbb{F}}$-semimartingale.*

One example of such a change of time is stopping a process at some fixed stopping time $\tau$ :

$$\tau_t = t \wedge \tau.$$

If we consider an SSM $Z = (f, M)$, then the stopped process $Z^\tau = \widetilde{Z} = (\widetilde{f}, \widetilde{M})$ is again an SSM with

$$\widetilde{f}_t = I(\tau > t) f_t.$$

### A.6.2  Product Rule

It is known that the product of two semimartingales is a semimartingale (cf. [68], p. 219). However, this does not hold true in general for SSMs. As an example consider a martingale $M \in \mathcal{M}_0^2$ with a predictable variation process $\langle M \rangle$ that is not continuous. Then $Z = M$ is an SSM with $f = 0$, but $Z^2 = M^2$ has a decomposition

$$Z_t^2 = \langle M \rangle_t + R_t$$

with some martingale $R$, which shows that $Z^2$ is not an SSM. To establish conditions under which a product rule for SSMs holds true, we first recall the integration by parts formula for ordinary functions.

**Proposition 106** *Let $a$ and $b$ be cadlag functions on $\mathbb{R}_+$, which are of finite variation. Then for each $t \in \mathbb{R}_+$*

$$
\begin{aligned}
a(t)b(t) &= a(0)b(0) + \int_0^t a(s-)db(s) + \int_0^t b(s)da(s) \\
&= a(0)b(0) + \int_0^t a(s-)db(s) + \int_0^t b(s-)da(s) \\
&\quad + \sum_{0 < s \le t} \triangle a(s) \triangle b(s),
\end{aligned}
$$

*where $a(s-)$ is the left limit at $s$ and $\triangle a(s) = a(s) - a(s-)$.*

Replacing $a$ and $b$ by SSMs $Z$ and $Y$ in this integration by parts formula we need to give

$$\int_0^t Y_{s-} dZ_s$$

a meaning. The finite variation part can be defined as an ordinary (pathwise) Stieltjes integral. It remains to define $\int_0^t Y_{s-} dM_s$ where $M$ is a martingale possibly of unbounded variation. Because we do not want to develop the theory of stochastic integration, we only quote the following theorem stating conditions to be used in the product formula we aim at.

**Theorem 107** *Suppose $M \in \mathcal{M}_0^2$ and let $X$ be a predictable process such that*

$$E \int_0^\infty X_s^2 d\langle M \rangle_s < \infty.$$

*Then there exists a unique process $\int_0^t X_s dM_s \in \mathcal{M}_0^2$ with the characterizing property*

$$\left\langle \int_0^t X_s dM_s, L \right\rangle = \int_0^t X_s d\langle M, L \rangle_s$$

*for all $L \in \mathcal{M}_0^2$.*

For two SSMs $Z$ and $Y$ with martingale parts $M$ and $L$, respectively, $M, L \in \mathcal{M}_0^2$, we define the covariation $[Z, Y]$ by

$$
\begin{aligned}
[Z, Y]_t &= \langle M^c, L^c \rangle_t + \sum_{s \leq t} \triangle Z_s \triangle Y_s \\
&= \langle M^c, L^c \rangle_t + Z_0 Y_0 + \sum_{s \leq t} \triangle M_s \triangle L_s \\
&= Z_0 Y_0 + [M, L]_t.
\end{aligned}
$$

After these preparations the following product rule can be established.

**Theorem 108** *Let $Z = (f, M)$ and $Y = (g, L)$ be $\mathbb{F}$-SSMs with orthogonal martingales $M, L \in \mathcal{M}_0^2$, i.e., $ML \in \mathcal{M}_0$. Assume that*

$$
E \int_0^t (|Z_s g_s| + |Y_s f_s|) ds < \infty, \quad E|Z_0 Y_0| < \infty,
$$

$$
E \int_0^\infty Y_{s-}^2 d\langle M \rangle_s < \infty, \quad E \int_0^\infty Z_{s-}^2 d\langle L \rangle_s < \infty.
$$

*Then $ZY$ is an $\mathbb{F}$-SSM with representation*

$$
Z_t Y_t = Z_0 Y_0 + \int_0^t (Y_s f_s + Z_s g_s) ds + R_t,
$$

*where $R = (R_t)$ is a martingale in $\mathcal{M}_0$.*

**Proof.** To prove the product rule we use a form of integration by parts for semimartingales, which is an application of Ito's formula (see [75], p. 140):

$$
Z_t Y_t = \int_{(0,t]} Z_{s-} dY_s + \int_{(0,t]} Y_{s-} dZ_s + [Z, Y]_t.
$$

The definition of stochastic integrals implies

$$
\int_{(0,t]} Z_{s-} dY_s = \int_{(0,t]} Z_{s-} d \left( \int_0^s g_u du \right) + \int_{(0,t]} Z_{s-} dL_s.
$$

The second term of the sum is a martingale of $\mathcal{M}_0^2$ by virtue of

$$
E \int_0^\infty Z_{s-}^2 d\langle L \rangle_s < \infty.
$$

The first term of the sum is an ordinary Stieltjes integral. Since the paths of $Z$ have at most countably many jumps, it follows that

$$
\int_{(0,t]} Z_{s-} d \left( \int_0^s g_u du \right) = \int_0^t Z_s g_s ds.
$$

The second integral in the integration by parts formula is treated in the same way.

It remains to show that in $[Z, Y]_t = Z_0 Y_0 + [M, L]_t$ the second term of the sum is a martingale. From Proposition 104, p. 235, we know that $ML - [M, L]$ is a martingale. By virtue of the assumption that $ML \in \mathcal{M}_0$ the square bracket process $[M, L]$ must also have the martingale property. Altogether the product semimartingale has the representation

$$Z_t Y_t = Z_0 Y_0 + \int_0^t (Z_s g_s + Y_s f_s) ds + R_t,$$

where

$$R_t = \int_{(0,t]} Z_{s-} dL_s + \int_{(0,t]} Y_{s-} dM_s + [M, L]_t$$

is a martingale in $\mathcal{M}_0$. This completes the proof.  ∎

Sometimes the product rule is used for a product one factor of which is the one point process $I(\zeta \le t)$ with a stopping time $\zeta$. Because of the special structure of this factor less restrictive conditions are necessary to establish a product rule.

**Proposition 109** *Let $Z = (f, M)$ be an $\mathbb{F}$-SSM and $\zeta > 0$ a (totally inaccessible) $\mathbb{F}$-stopping time with*

$$Y_t = I(\zeta \le t) = \int_0^t g_s ds + L_s.$$

*Furthermore, it is assumed that for all $t \in \mathbb{R}_+$*

$$E \int_0^t |Z_s g_s| ds < \infty, \quad E \int_0^t |Z_{s-}| \, |dL_s| < \infty$$

*and $\triangle M_\zeta = 0$. Then $ZY$ is an SSM with representation*

$$Z_t Y_t = \int_0^t (Z_s g_s + Y_s f_s) ds + R_t,$$

*where $R \in \mathcal{M}_0$.*

**Proof.** The product $ZY$ can be represented in the form

$$Z_t Y_t = Z_t - Z_{t \wedge \zeta} + \int_0^t Z_s dY_s$$

with the pathwise defined Stieltjes integral

$$\int_0^t Z_s dY_s = \int_0^t Z_s q_s ds + \int_0^t Z_s dL_s.$$

The second term in this sum can be decomposed as

$$\int_0^t Z_s dL_s = \int_0^t Z_{s-} dL_s + \sum_{s \le t} \triangle M_s \triangle L_s.$$

The sum of jumps is 0, since $L$ is continuous outside $\{(t, \omega) : \zeta(\omega) = t\}$ and $\triangle M_\zeta = 0$. The martingale $L$ is of finite variation and the condition $E \int_0^t |Z_{s-}| \, |dL_s| < \infty$ implies that the integral of the predictable process $Z_{s-}$ with respect to $L$ is a martingale (cf. [112]).

To sum up we get

$$Z_t Y_t = \int_0^t (f_s - I(\zeta > s) f_s + Z_s q_s) ds + M_t - M_t^\zeta + \int_0^t Z_{s-} dL_s,$$

which proves the assertion.    ■

# Appendix B
## Renewal Processes

In this appendix we present some definitions and results from the theory of renewal processes, including renewal reward processes and regenerative processes. Key references are [1, 9, 46, 64, 145, 168].

The purpose of this appendix is not to give an all-inclusive presentation of the theory. Only definitions and results needed for establishing the results of Chapters 1–5 (in particular Chapter 4) is covered.

## B.1 Basic Theory of Renewal Processes

Let $T, T_j, j = 1, 2, \ldots$, be a sequence of nonnegative independent identically distributed (i.i.d.) random variables with distribution function $F$. To avoid trivialities, we assume that $P(T = 0) < 1$. From the nonnegativity of $T$, it follows that $ET$ exists, although it may be infinite, and we denote

$$\mu = ET = \int_0^\infty P(T > t)\, dt.$$

The variance of $T$ is denoted $\sigma^2$. Let

$$S_0 = 0, \quad S_j = \sum_{i=1}^j T_i, \quad j \in \mathbb{N}$$

and define

$$N_t = \sup\{j : S_j \leq t\},$$

or equivalently,

$$N_t = \sum_{j=1}^{\infty} I(S_j \leq t). \tag{B.1}$$

The processes $(N_t), t \in \mathbb{R}_+$, and $(S_j), j \in \mathbb{N}_0$, are both called a *renewal process*. We say that a renewal occurs at $t$ if $S_j = t$ for some $j \geq 1$. The random variable $N_t$ represents the number of renewals in $[0, t]$. Since the *interarrival times* $T_j$ are independent and identically distributed, it follows that after each renewal the process restarts.

Let $M(t) = EN_t$, $0 \leq t < \infty$. The function $M(t)$ is called the renewal function. It can be shown that $M(t)$ is finite for all $t$. From (B.1) we see that

$$M(t) = \sum_{j=1}^{\infty} F^{*j}(t), \tag{B.2}$$

where $F^{*j}$ denotes the $j$-fold convolution of $F$. If, for example, $F$ is a Gamma distribution with parameters 2 and $\lambda$, i.e., $F(t) = 1 - e^{-\lambda t} - \lambda t e^{-\lambda t}$, it can be shown that

$$M(t) = \frac{\lambda t}{2} - \frac{1 - e^{-2\lambda t}}{4}.$$

Refer to [1, 33, 34] for more general formulas for the renewal function of the Gamma distribution and expressions and bounds for other distributions. In Proposition 110 we show how $M$ can be determined (at least in theory) from $F$. It turns out that $M$ uniquely determines $F$.

**Proposition 110** *There is a one-to-one correspondence between the inter-arrival distribution $F$ and the renewal function $M$.*

**Proof.** We introduce the Laplace transform $L_B(s) = \int_0^{\infty} e^{-sx} dB(x)$, where $B : \mathbb{R}_+ \rightarrow \mathbb{R}_+$ is a nondecreasing and right-continuous function. By taking the Laplace transform $L$ on both sides of formula (B.2) we obtain

$$
\begin{aligned}
L_M(s) &= \sum_{j=1}^{\infty} L_{F^{*j}}(s) \\
&= \sum_{j=1}^{\infty} (L_F(s))^j \\
&= \frac{L_F(s)}{1 - L_F(s)}, \tag{B.3}
\end{aligned}
$$

or equivalently

$$L_F(s) = \frac{L_M(s)}{1 + L_M(s)}.$$

Hence $L_F$ is determined by $M$ and since the Laplace transform determines the distribution, it follows that $F$ also is determined by $M$. ∎

The function $M(t)$ satisfies the following integral equation:

$$M(t) = F(t) + \int_0^t M(t-x)dF(x),$$

i.e., $M = F + M * F$, where $*$ means convolution. This equation is referred to as the renewal equation, and is seen to hold by conditioning on the time of the first renewal. Upon doing so we obtain

$$
\begin{aligned}
M(t) &= \int_0^\infty E[N_t|T_1 = x]dF(x) \\
&= \int_0^t [1 + M(t-x)]dF(x) \\
&= F(t) + (M * F)(t),
\end{aligned}
$$

noting that if the first renewal occurs at time $x$, $x \leq t$, then from this point on the process restarts, and thus the expected number of renewals in $[0,t]$ is just 1 plus the expected number to arrive in a time $t - x$ from an equivalent renewal process. A more formal proof is the following;

$$
\begin{aligned}
M(t) &= EN_t = E\sum_{j=1}^\infty I(S_j \leq t) = F(t) + E\sum_{j=2}^\infty I(S_j \leq t) \\
&= F(t) + E\sum_{j=2}^\infty I(S_j - S_1 \leq t - S_1) \\
&= F(t) + \int_0^t E\sum_{j=2}^\infty I(S_j - S_1 \leq t - s)dF(s) \\
&= F(t) + \int_0^t M(t-s)dF(s).
\end{aligned}
$$

To generalize the renewal equation, we write

$$g(t) = h(t) + (g * F)(t), \tag{B.4}$$

where $h$ and $F$ are known and $g$ is an unknown function to be determined as a solution to the equation (B.4). The solution of this equation is given by the following result.

**Theorem 111** *If the function $g$ satisfies the equation (B.4) and $h$ is bounded on finite intervals, then*

$$g(t) = h(t) + (h * M)(t)$$

*is a solution to (B.4) and the unique solution which is bounded on finite intervals.*

**Proof.** A proof of this result is given in Asmussen [9], p. 113. A simpler proof can however be given in the case where the Laplace transform of $h$ and $g$ exists: Taking Laplace transforms in (B.4), yields

$$L_g(s) = L_h(s) + L_g(s)L_F(s),$$

and it follows that

$$
\begin{aligned}
L_g(s) &= \frac{L_h(s)}{1 - L_F(s)} \\
&= L_h(s)[1 + \frac{L_F(s)}{1 - L_F(s)}] \\
&= L_h(s) + L_h(s)L_M(s) \\
&= L_{h+h*M}(s),
\end{aligned}
$$

where the second last equality follows from (B.3). Since the Laplace transform uniquely determines the function, this gives the desired result.    ∎

Using the (strong) law of large numbers, many results related to renewal processes can be established, including the following.

**Theorem 112** *With probability one,*

$$\frac{N_t}{t} \to \frac{1}{\mu} \quad as \quad t \to \infty.$$

**Proof.** By definition of $N_t$, it follows that

$$S_{N_t} \le t \le S_{N_t+1}.$$

Hence,

$$\frac{S_{N_t}}{N_t} \le \frac{t}{N_t} \le \frac{S_{N_t+1}}{N_t}.$$

Now the strong law of large numbers states that with probability one, $S_j/j \to \mu$ as $j \to \infty$. As can be easily shown, $N_t \to \infty$ as $t \to \infty$, and thus

$$\frac{S_{N_t}}{N_t} \to \mu \quad as \quad t \to \infty \qquad (P\text{-a.s.}).$$

By the same argument, we also see that with probability one,

$$\frac{S_{N_t+1}}{N_t} = \frac{S_{N_t+1}}{N_t+1}\frac{N_t+1}{N_t} \to \mu \cdot 1 = \mu \quad as \quad t \to \infty.$$

The result follows.    ∎

We now formulate some limiting results, without proof, including the Elementary Renewal Theorem, the Key Renewal Theorem, Blackwell's Theorem, and the Central Limit Theorem for renewal processes. Refer to Alsmeyer [1], Asmussen [9], Daley and Vere-Jones [64], and Ross [145] for proofs; see also Birolini [46]. Some of the results require that the distribution $F$ is not periodic (lattice). We say that $F$ is periodic if there exists a constant $c$, $c > 0$, such that $T$ takes only values in $\{0, c, 2c, 3c, \ldots\}$.

**Theorem 113 (*Elementary Renewal Theorem*)**

$$\lim_{t \to \infty} \frac{M(t)}{t} = \frac{1}{\mu}.$$

**Theorem 114 (*Tightened Elementary Renewal Theorem*).** *Assume that $\sigma^2 = Var[T] < \infty$. If the distribution $F$ is not periodic, then*

$$\lim_{t \to \infty} \left[ M(t) - \frac{t}{\mu} \right] = \frac{\sigma^2 - \mu^2}{2\mu^2}.$$

**Theorem 115** *Assume that $\sigma^2 = Var[T] < \infty$. If the distribution $F$ is not periodic, then*

$$\lim_{t \to \infty} \frac{Var[N_t]}{t} = \frac{\sigma^2}{\mu^3}.$$

Before we state the Key Renewal Theorem, we need a definition. Let $g$ be a function defined on $\mathbb{R}_+$ and for $h > 0$ let

$$g_-^h(x) = \inf_{0 \le \delta \le h} g(x - \delta), \quad g_+^h(x) = \sup_{0 \le \delta \le h} g(x - \delta).$$

We say that $g$ is *directly Riemann integrable* if for any $h > 0$;

$$h \sum_{n=1}^{\infty} |g_-^h(nh)| \quad \text{and} \quad h \sum_{n=1}^{\infty} |g_+^h(nh)|$$

are finite, and

$$\lim_{h \to 0+} h \sum_{n=1}^{\infty} g_-^h(nh) = \lim_{h \to 0+} h \sum_{n=1}^{\infty} g_+^h(nh).$$

In particular, a nonnegative, nonincreasing and integrable function is directly Riemann integrable. See [64, 97] for some other sufficient conditions for a function to be directly Riemann integrable.

**Theorem 116 (*Key Renewal Theorem*).** *Assume that the distribution $F$ is not periodic and $g$ is a directly Riemann integrable function. Then*

$$\lim_{t \to \infty} \int_0^t g(t - s) \, dM(s) = \frac{1}{\mu} \int_0^{\infty} g(s) \, ds.$$

**Remark 26** An alternative formulation of the Key Renewal Theorem is the following: If $g$ is bounded and integrable with $g(t) \to 0$ as $t \to \infty$, then $\lim_{t\to\infty} \int_0^t g(t-s)\,dM(s) = (1/\mu) \int_0^\infty g(s)\,ds$ provided that $F$ is spread out. A distribution function is spread out if there exists an $n$ such that $F^{*n}$ has a nonzero absolutely continuous component with respect to Lebesgue measure, i.e., we can write $F^{*n} = G_1 + G_2$, where $G_1, G_2$ are nonnegative measures on $\mathbb{R}_+$, and $G_1$ has a density with respect to Lebesgue measure.

The Key Renewal Theorem is equivalent to Blackwell's Theorem below.

**Theorem 117 (Blackwell's Theorem).** *For a renewal process with a nonperiodic distribution $F$,*

$$\lim_{t\to\infty} [M(t) - M(t-s)] = \frac{s}{\mu}.$$

If $F$ has a density $f$, then $M$ has a density $m$, and

$$m(t) = \sum_{j=1}^\infty f^{*j}(t),$$

where $f^{*1} = f$ and

$$f^{*j}(t) = \int_0^t f^{*(j-1)}(t-s)f(s)ds, \ j = 2, 3, \ldots.$$

Under certain conditions the renewal density $m(t)$ converges to $1/\mu$ as $t \to \infty$.

**Theorem 118 (Renewal Density Theorem).** *Assume that $F$ has a density $f$ with $f(t)^p$ integrable for some $p > 1$, and $f(t) \to 0$ as $t \to \infty$. Then $M$ has a density $m$ such that*

$$\lim_{t\to\infty} m(t) = \frac{1}{\mu}.$$

**Remark 27** The conclusion of the theorem also holds true if $F$ has a density $f$, which is directly Riemann integrable, or if $F$ has finite mean and a bounded density $f$ satisfying $f(t) \to 0$ as $t \to \infty$.

**Theorem 119 (Central Limit Theorem).** *Assume that $\sigma^2 = Var[T] < \infty$. Then $N_t$, suitably standardized, tends to a normal distribution as $t \to \infty$, i.e.,*

$$\lim_{t\to\infty} P(\frac{N_t - t/\mu}{\sqrt{t\sigma^2/\mu^3}} \le x) = \frac{1}{\sqrt{2\pi}} \int_{-\infty}^x e^{-\frac{1}{2}u^2}\,du.$$

Next we formulate the limiting distribution of the forward and backward recurrence times $\alpha_t$ and $\beta_t$, defined by

$$\alpha_t = S_{N_t+1} - t,$$
$$\beta_t = t - S_{N_t}.$$

The recurrence times $\alpha_t$ and $\beta_t$ are the time intervals from $t$ forward to the next renewal point and backward to the last renewal point (or to the time origin), respectively. Let $F_{\alpha_t}$ and $F_{\beta_t}$ denote the distribution functions of $\alpha_t$ and $\beta_t$, respectively. The following result is a consequence of the Key Renewal Theorem.

**Theorem 120** *Assume that the distribution $F$ is not periodic. Then the asymptotic distribution of the forward and backward recurrence times are given by*

$$\lim_{t\to\infty} F_{\alpha_t}(x) = \lim_{t\to\infty} F_{\beta_t}(x) = \frac{\int_0^x \bar{F}(s)\,ds}{\mu}.$$

This asymptotic distribution of $\alpha_t$ and $\beta_t$ is called the *equilibrium distribution.*

A simple formula exists for the mean forward recurrence time; we have

$$ES_{N_t+1} = \mu(1 + M(t)). \tag{B.5}$$

Formula B.5 is a special case of Wald's equation (see, e.g., Ross [145]), and follows by writing

$$
\begin{aligned}
ES_{N_t+1} &= E\sum_{k\geq 1} S_k I(N_t + 1 = k) = E\sum_{k\geq 1}\sum_{j=1}^{k} T_j I(N_t + 1 = k)\\
&= E\sum_{j\geq 1} T_j I(N_t + 1 \geq j) = E\sum_{j\geq 1} T_j I(S_{j-1} \leq t)\\
&= \sum_{j\geq 1} ET_j EI(S_{j-1} \leq t) = \mu\sum_{j\geq 0} F^{*j}(t) = \mu(1 + M(t)).
\end{aligned}
$$

Finally in this section we prove a result used in the proof of Theorem 42, p. 114.

**Proposition 121** *Let $g$ be a real-valued function which is bounded on finite intervals. Assume that*

$$\lim_{t\to\infty} g(t) = g.$$

*Then*

$$\lim_{t\to\infty} \frac{1}{t}\int_0^t g(s)\,dM(s) = \frac{g}{\mu}.$$

**Proof.** To prove this result we use a standard $\epsilon$ argument. Given $\epsilon > 0$, there exists a $t_0$ such that $|g(t) - g| < \epsilon$ for $t \geq t_0$. Hence for $t > t_0$ we have

$$\frac{1}{t} \int_0^t |g(s) - g| dM(s)$$

$$\leq \frac{1}{t} \int_0^{t_0} |g(s) - g| dM(s) + \frac{1}{t} \int_{t_0}^t \epsilon \, dM(s).$$

Since $t_0$ is fixed, this gives by applying the Elementary Renewal Theorem,

$$\limsup_{t \to \infty} \frac{1}{t} \int_0^t |g(s) - g| dM(s) \leq \frac{\epsilon}{\mu}.$$

The desired conclusion follows.                                    ∎

## B.2    Renewal Reward Processes

Let $(T, Y), (T_1, Y_1), (T_2, Y_2), \ldots$, be a sequence of independent and identically distributed pairs of random variables, with $T, T_j \geq 0$. We interpret $Y_j$ as the "reward" ("cost") associated with the $j$th interarrival time $T_j$. The random variable $Y_j$ may depend on $T_j$. Let $Z_t$ denote the total reward earned by time $t$. We see that if the reward is earned at the time of the renewal,

$$Z_t = \sum_{j=1}^{N_t} Y_j.$$

The limiting value of the average return is established using the law of large numbers and is given by the following result (cf. [145]).

**Theorem 122** *If $E|Y|$ is finite, then*

(i)  *With probability 1*

$$\frac{Z_t}{t} \to \frac{EY}{ET} \qquad as \ t \to \infty,$$

(ii)

$$\frac{EZ_t}{t} \to \frac{EY}{ET} \qquad as \ t \to \infty.$$

**Remark 28** The conclusions of Theorem 122 also hold true if $Y \geq 0$, $EY = \infty$ and $ET < \infty$.

Many results from renewal theory can be generalized to renewal reward processes. For example Blackwell's Theorem holds:

$$\lim_{t \to \infty} [Z_t - Z_{t-s}] = \frac{sEY}{ET}.$$

The following theorem, which is a reformulation of Theorem 3.2, p. 136, in [9], generalizes the Central Limit Theorem for renewal processes, Theorem 119.

**Theorem 123** *Suppose* $\mathrm{Var}[Y] < \infty$ *and* $\mathrm{Var}[T] < \infty$. *Then as* $t \to \infty$

$$\sqrt{t}\left[\frac{Z_t}{t} - \frac{EY}{ET}\right] \xrightarrow{D} \mathrm{N}\left(0, \frac{\tau^2}{ET}\right),$$

*where*

$$
\begin{aligned}
\tau^2 &= \mathrm{Var}\left[Y - \frac{EY}{ET}T\right] \\
&= \mathrm{Var}[Y] + \left(\frac{EY}{ET}\right)^2 \mathrm{Var}[T] - 2\frac{EY}{ET}\,\mathrm{Cov}[Y, T].
\end{aligned}
$$

## B.3  Regenerative Processes

The stochastic process $(X_t)$ is called regenerative if there exists a renewal process $(T_j)$ such that for $k \in \mathbb{N}$, $(X_t)_{t \geq 0} \overset{D}{=} (X_{t+S_k})_{t \geq 0}$, and

$$((X_{t+S_k})_{t \geq 0}, (T_j), j > k) \text{ and } ((X_t)_{0 \leq t \leq S_k}, T_1, T_2, ..., T_k)$$

are stochastically independent. Thus the continuation of the process beyond $S_k$ is a probabilistic replica of the whole process starting at 0. The random times $S_k$ are said to be regenerative points for the process $(X_t)$ and the time interval $[S_{k-1}, S_k)$ is called the $k$th cycle of the process.

In the following assume that the state space of $(X_t)$ equals $\mathbb{N}_0 = \{0, 1, 2, ...\}$. Let

$$P_k(t) = P(X_t = k), \quad k \in \mathbb{N}_0.$$

The following result taken from Ross [145] is stated without proof.

**Theorem 124** *If the distribution of* $T_1$ *has an absolutely continuous component and* $ET_1 < \infty$, *then*

$$\lim_{t \to \infty} P_k(t) = \frac{E \int_0^{T_1} I(X_t = k)\, dt}{ET_1}, \quad k \in \mathbb{N}_0.$$

**Remark 29** We see that if $\lim_{t \to \infty} P_k(t) = P_k$ exists, then

$$\lim_{t \to \infty} \frac{1}{t} E \int_0^t I(X_s = k) \, ds = P_k.$$

The quantity $(1/t)E \int_0^t I(X_s = k) \, ds$ represents the expected portion of time the process is in state $k$ in $[0, t]$. Since

$$\frac{1}{t} E \int_0^t I(X_s = k) \, ds = \frac{1}{t} \int_0^t EI(X_s = k) \, ds = \frac{1}{t} \int_0^t P_k(s) \, ds,$$

this quantity is also equal to the average probability that the process is in state $k$.

## B.4   Modified (Delayed) Processes

Consider a renewal process $(S_j)$ as defined in Section B.1, but assume now that the first interarrival time $T_1$ has a distribution $\tilde{F}$, that is not necessarily identical to $F$. The process is referred to as a *modified renewal process* (or a *delayed renewal process*). Similarly, we define a *modified (delayed) renewal reward process* and a *modified (delayed) regenerative process*. For the modified renewal reward process the distribution of the pair $(Y_1, T_1)$ is not necessarily the same as the pairs $(Y_i, T_i)$, $i = 2, 3, \ldots$.

It can be shown that all the asymptotic results presented in the previous sections of this appendix still hold true for the modified processes. If we take the first distribution to be equal to the asymptotic distribution of the recurrence times, given by Theorem 120, p. 249, the renewal process becomes stationary in the sense that the distribution of the forward recurrence time $\alpha_t$ does not depend on $t$. Furthermore,

$$M(t + h) - M(t) = h/ET.$$

# References

[1] Alsmeyer, G. (1991) *Erneuerungstheorie*. Teubner Skripten zur Mathematischen Stochastik. B.G. Teubner, Stuttgart.

[2] Andersen, P. K., Borgan, Ø., Gill, R. and Keiding, N. (1992) *Statistical Models Based on Counting Processes*. Springer, New York.

[3] Arjas, E. (1993) Information and reliability: A Bayesian perspective. In: Barlow, R., Clarotti, C. and Spizzichino, F. (eds.): *Reliability and Decision Making*. Chapman & Hall, London, pp. 115–135.

[4] Arjas, E. (1989) Survival models and martingale dynamics. *Scand. J. Statist.* **16**, 177–225.

[5] Arjas, E. (1981) A stochastic process approach to multivariate reliability systems: Notions based on conditional stochastic order. *Mathematics of Operations Research* **6**, 263–276.

[6] Arjas, E. (1981) The failure and hazard processes in multivariate reliability systems. *Mathematics of Operations Research* **6**, 551–562.

[7] Arjas, E. and Norros, I. (1989) Change of life distribution via hazard transformation: An inequality with application to minimal repair. *Mathematics of Operations Research* **14**, 355–361.

[8] Asmussen, S. (1989) Risk theory in a Markovian environment. *Scand. Actuarial J.*, 69–100.

[9] Asmussen, S. (1987) *Applied Probability and Queues*. Wiley, New York.

[10] Asmussen, S. (1984) Approximations for the probability of ruin within finite time. *Scand. Actuarial J.*, 31–57.

[11] Asmussen, S., Frey, A., Rolski, T. and Schmidt, V. (1995) Does Markov-modulation increase risk? *Astin Bulletin* **25**, 49–66.

[12] Athreya, K. and Kurtz, T. (1973) A generalization of Dynkin's identity and some applications. *Ann. Prob.* **1**, 570–579.

[13] Aven, T. (1996) Availability analysis of monotone systems. In: S. Özekici (ed.): *Reliability and Maintenance of Complex Systems*. NATO ASI Series F, Springer, Berlin, pp. 206–223.

[14] Aven, T. (1996) Optimal replacement of monotone repairable systems. In: S. Özekici (ed.): *Reliability and Maintenance of Complex Systems*. NATO ASI Series F, Springer, Berlin, pp. 224–238.

[15] Aven, T. (1996) Condition based replacement times - a counting process approach. *Reliability Engineering and System Safety*. Special issue on Maintenance and Reliability **51**, 275–292.

[16] Aven, T. (1992) *Reliability and Risk Analysis*. Elsevier Applied Science, London.

[17] Aven, T. (1990) Availability evaluation of flow networks with varying throughput-demand and deferred repairs. *IEEE Trans. Reliability* **38**, 499–505.

[18] Aven, T. (1987) A counting process approach to replacement models. *Optimization* **18**, 285–296.

[19] Aven, T. (1985) A theorem for determining the compensator of a counting process. *Scand. J. Statist.* **12**, 69–72.

[20] Aven, T. (1985) Reliability evaluation of multistate systems of multistate components. *IEEE Trans. Reliability* **34**, 473–479.

[21] Aven, T. (1983) Optimal replacement under a minimal repair strategy – A general failure model. *Adv. Appl. Prob.* **15**, 198–211.

[22] Aven, T. and Bergman, B. (1986) Optimal replacement times, a general set-up. *J. Appl. Prob.* **23**, 432–442.

[23] Aven, T. and Dekker, R. (1997) A useful framework for optimal replacement models. *Reliability Engineering and System Safety* **58**, 61–67.

[24] Aven, T. and Haukås, H. (1997) Asymptotic Poisson distribution for the number of system failures of a monotone system. *Reliability Engineering and System Safety* **58**, 43–53.

[25] Aven, T. and Haukås, H. (1997) A note on the steady state availability of monotone systems. *Reliability Engineering and System Safety* **59**, 269–276.

[26] Aven, T. and Jensen, U. (1998) A general minimal repair model. Research report, University of Ulm.

[27] Aven, T. and Jensen, U. (1998) Information based hazard rates for ruin times of risk processes. Research Report, University of Ulm.

[28] Aven, T. and Jensen, U. (1999) A note on the asymptotic Poisson limit for the distribution of the number of system failures of a monotone system. Research Report University of Oslo.

[29] Aven, T. and Jensen, U. (1997) Asymptotic distribution of the downtime of a monotone system. *Mathematical Methods of Operations Research* . Special issue on Stochastic Models of Reliability, **45**, 355–375.

[30] Aven, T. and Opdal, K. (1996) On the steady state unavailability of standby systems. *Reliability Engineering and System Safety* **52**, 171–175.

[31] Aven, T. and Østebø, R. (1986) Two new component importance measures for a flow network system. *Reliability Engineering* **14**, 75–80.

[32] Barlow, R. and Hunter, L. (1960) Optimum preventive maintenance policies. *Operations Res.* **8**, 90–100.

[33] Barlow, R. and Proschan, F. (1965) *Mathematical Theory of Reliability*. Wiley, New York.

[34] Barlow, R. and Proschan, F. (1975) *Statistical Theory of Reliability and Life Testing*. Holt, Rinehart and Winston, New York.

[35] Basu, A. (1988) Multivariate exponential distributions and their applications in reliability. In: Krishnaiah, P. R. and Rao, C. R. (eds.): *Handbook of Statistics 7. Quality Control and Reliability*. North-Holland, Amsterdam, pp. 99–111.

[36] Baxter, L. A. (1981) Availability measures for a two-state system. *J. Appl. Prob.* **18**, 227–235.

[37] Beichelt, F. (1993) A unifying treatment of replacement policies with minimal repair. *Nav. Res. Log. Q.* **40**, 51–67.

[38] Beichelt, F. and Franken, F. (1984) *Zuverlässigkeit und Instandhaltung.* Carl Hanser Verlag, München.

[39] Berg, M. (1996) Economics oriented maintenance analysis and the marginal cost approach. In: Özekici, S. (ed.): *Reliability and Maintenance of Complex Systems.* NATO ASI Series F, Springer, Berlin, pp. 189–205.

[40] Bergman, B. (1978) Optimal replacement under a general failure model. *Adv. Appl. Prob.* **10**, 431–451.

[41] Bergman, B. (1985) On reliability theory and its applications. *Scand. J. Statist.* **12**, 1–41.

[42] Bergman, B. and Klefsjö, B. (1994) *Quality.* Studentlitteratur, Lund.

[43] Bertsekas, D. (1995) *Dynamic Programming and Optimal Control.* Vol. 1 and 2. Athena Scientific, Belmont.

[44] Billingsley, P. (1979) *Probability and Measure.* Wiley, New York.

[45] Birnbaum, Z. W. (1969) On the importance of different components in a multicomponent system. In: Krishnaiah, P. R. (ed.) *Multivariate Analysis II*, Academic Press, pp. 581–592.

[46] Birolini, A. (1994) *Quality and Reliability of Technical Systems.* Springer, Berlin.

[47] Birolini, A. (1985) *On the use of Stochastic Processes in Modeling Reliability Problems.* Lecture notes in Economics and Mathematical Systems **252**, Springer, Berlin.

[48] Block, H. W. and Savits, T. H. (1997) Burn-In. *Statistical Science* **12**, 1–19.

[49] Block, H. W. and Savits, T. H. (1994) Comparison of maintenance policies. In: Shaked, M. and Shanthikumar, G. (eds.): *Stochastic Orders and their Applications.* Academic Press, Boston, pp. 463–484.

[50] Block, H. W., Borges, W. and Savits, T. H. (1985) Age-dependent minimal repair. *J. Appl. Prob.* **22**, 370–385.

[51] Boland, P. and Proschan, F. (1994) Stochastic order in system reliability theory. In: Shaked, M. and Shanthikumar, G. (eds.): *Stochastic Orders and their Applications.* Academic Press, Boston, pp. 485–508.

[52] Brémaud, P. (1981) *Point Processes and Queues. Martingale Dynamics.* Springer, New York.

[53] Brown, M. and Proschan, F. (1983) Imperfect repair. *J. Appl. Prob.* **20**, 851–859.

[54] Butler, D. A. (1979) A complete importance ranking for components of binary coherent systems, with extensions to multi-state systems. *Nav. Res. Log. Q.* **26**, 565–578.

[55] Chow, Y., Robbins, H., Siegmund, D. (1971) *Great Expectations, the Theory of Optimal Stopping.* Houghton Mifflin, Boston.

[56] Chung, K. L. (1974) *A Course in Probability Theory.* 2nd ed. Academic Press, London.

[57] Çinlar, E. (1975) Superposition of point processes. In: Lewis, P. (ed.) *Stochastic Point Processes.* Wiley, New York, pp. 549–606.

[58] Çinlar, E. (1984) Markov and semi-Markov models of deterioration. In: Abdel-Hameed, M., Çinlar, E. and Quinn, J. (eds.): *Reliability Theory and Models.* Academic Press, Orlando, pp. 3–41.

[59] Çinlar, E. and Özekici, S. (1987) Reliability of complex devices in random environments. *Probability in the Enginering and Informational Sciences* **1**, 97–115.

[60] Çinlar, E., Shaked, M. and Shanthikumar, G. (1989) On lifetimes influenced by a common environment. *Stoch. Proc. Appl.* **33**, 347–359.

[61] Constantini, C. and Spizzichino, F. (1997) Explicit solution of an optimal stopping problem: The burn-in of conditionally exponential components. *J. Appl. Prob.* **34**, 267–282.

[62] Csenki, A. (1994) Cumulative operational time analysis of finite semi-Markov reliability models. *Reliability Engineering and System Safety* **44**, 17–25.

[63] Csenki, A. (1995) An integral equation approach to the interval reliability of systems modelled by finite semi-Markov processes. *Reliability Engineering and System Safety* **47**, 37–45.

[64] Daley, D. J. and Vere-Jones, D. (1988) *An Introduction to the Theory of Point Processes.* Springer, Berlin.

[65] Davis, M. H. A. (1993) *Markov Models and Optimization.* Chapman & Hall, London.

[66] Delbaen, F. and Haezendonck, J. (1985) Inversed martingales in risk theory. *Insurance: Mathematics and Economics* **4**, 201–206.

[67] Dellacherie, C. and Meyer, P. A. (1978) *Probabilities and Potential A*. North-Holland, Amsterdam.

[68] Dellacherie, C. and Meyer, P. A. (1980) *Probabilities and Potential B*. North-Holland, Amsterdam.

[69] Dekker, R. (1996) A framework for single-parameter maintenance activities and its use in optimisation, priority setting and combining. In: Özekici, S. (ed.): *Reliability and Maintenance of Complex Systems*. NATO ASI Series F, Springer, Berlin, pp. 170–188.

[70] Dekker, R. and Groenendijk, W. (1995) Availability assessment methods and their application in practice. *Microelectron. Reliab.* **35**, 1257–1274.

[71] Doksum, K. (1991) Degradation rate models for failure time and survival data. *CWI Quarterly* **4**, 195–203.

[72] Dominé, M. (1996) First passage time distribution of a Wiener process with drift concerning two elastic barriers. *J. Appl. Prob.* **33**, 164–175.

[73] Donatiello, L. and Iyer, B. R. (1987) Closed-form solution for system availability distribution. *IEEE Trans. Reliability* **36**, 45–47.

[74] Dynkin, E. B. (1965) *Markov Processes*. Springer, Berlin.

[75] Elliott, R. (1982) *Stochastic Calculus and Applications*. Springer, New York.

[76] Feller, W. (1968) *An Introduction to Probability and its Applications*. Wiley, New York.

[77] Freund, J. E. (1961) A bivariate extension of the exponential distribution. *J. Amer. Stat. Ass.* **56**, 971–977.

[78] Funaki, K. and Yoshimoto, K. (1994) Distribution of total uptime during a given time interval. *IEEE Trans. Reliability* **43**, 489–492.

[79] Gaede, K.-W. (1977) *Zuverlässigkeit, Mathematische Modelle*. Carl Hanser Verlag, München.

[80] Gåsemyr, J. and Aven, T. (1999) Asymptotic distributions for the downtimes of monotone systems. *J. Appl. Prob.*, to appear.

[81] Gaver, D. P. (1963) Time to failure and availability of paralleled systems with repair. *IEEE Trans. Reliability* **12**, 30–38.

[82] Gertsbakh, I. B. (1989) *Statistical Reliability Theory*. Marcel-Dekker, New York.

[83] Gertsbakh, I. B. (1984) Asymptotic methods in reliability: A review. *Adv. Appl. Prob.* **16**, 147–175.

[84] Gnedenko, B. V. and Ushakov, I. A. (1995), edited by Falk, J. A. *Probabilistic Reliability Engineering.* Wiley, Chichester.

[85] Grandell, J. (1991) *Aspects of Risk Theory.* Springer, New York.

[86] Grandell, J. (1991) Finite time ruin probabilities and martingales. *Informatica* **2**, 3–32.

[87] Griffith, W. S. (1980) Multistate reliability models. *J. Appl. Prob.* **15**, 735–744.

[88] Grigelionis, B. (1993) Two-sided Lundberg inequalities in a Markovian environment. *Liet. Matem. Rink.* **1**, 30–41.

[89] Grimmelt, G. R. and Stirzaker, D. R. (1992) *Probability and Random Processes.* 2nd ed. Oxford Science Publication, Oxford.

[90] Haukås, H. and Aven, T. (1997) A general formula for the downtime of a parallel system. *J. Appl. Prob.* **33**, 772–785.

[91] Haukås, H. and Aven, T. (1996) Formulae for the downtime distribution of a system observed in a time interval. *Reliability Engineering and System Safety* **52**, 19–26.

[92] Heinrich, G. and Jensen, U. (1995) Parameter estimation for a bivariate lifetime distribution in reliability with multivariate extensions. *Metrika* **42**, 49–65.

[93] Heinrich, G. and Jensen, U. (1996) Bivariate lifetime distributions and optimal replacement. *Mathematical Methods of Operations Research* **44**, 31–47.

[94] Heinrich, G. and Jensen, U. (1992) Optimal replacement rules based on different information levels. *Nav. Res. Log. Q.* **39**, 937–955.

[95] Henley, E.J. and Kumamoto, H. (1981) *Reliability Engineering and Risk Assessment.* Prentice Hall, New Jersey.

[96] Herberts, T. and Jensen, U. (1998) Optimal stopping in a burn-in model. Research report, University of Ulm.

[97] Hinderer, H. (1987) Remarks on directly Riemann integrable functions. *Mathematische Nachrichten* **130**, 225–230.

[98] Hokstad, P. (1997) The failure intensity process and the formulation of reliability and maintenance models. *Reliability Engineering and System Safety* **58**, 69–82.

[99] Høyland, A. and Rausand, M. (1994) *System Reliability Theory*, Wiley, New York.

[100] Hughes, R. P. (1987) Fault tree truncation error bounds. *Reliability Engineering* **1**, 37–46.

[101] Hutchinson, T. P. and Lai, C. D. (1990) *Continuous Bivariate Distributions, Emphasising Applications*. Rumbsby Scientific Publishing, Adelaide.

[102] Jacod, J. (1975) Multivatiate point processes: predictable projection, Radon-Nikodym derivatives, representation of martingales. *Z. für Wahrscheinlichkeitstheorie und Verw. Gebiete* **31**, 235–253.

[103] Jasmon, G.B. and Kai, S. (1985) A new technique in minimal path and cut set evaluation. *IEEE Trans. Reliability* **34**, 136–143.

[104] Jensen, U. and Hsu, G. (1993) Optimal stopping by means of point process observations with applications in reliability. *Mathematics of Operations Research* **18**, 645–657.

[105] Jensen, U. (1996) Stochastic models of reliability and maintenance: an overview. In: S. Özekici (ed.): *Reliability and Maintenance of Complex Systems*. NATO ASI Series F, Springer, Berlin, pp. 3–36.

[106] Jensen, U. (1997) An optimal stopping problem in risk theory. *Scand Actuarial J.* 149–159.

[107] Jensen, U. (1990) A general replacement model. *ZOR-Methods and Models of Operations Research* **34**, 423–439.

[108] Jensen, U. (1990) An example concerning the convergence of conditional expectations. *Statistics* **21**, 609–611.

[109] Jensen, U. (1989) Monotone stopping rules for stochastic processes in a semimartingale representation with applications. *Optimization* **20**, 837–852.

[110] Kalashnikov, V.V. (1989) Analytical and simulation estimates of reliability for regenerative models. *Syst. Anal. Model. Simul.* **6**, 833–851.

[111] Kallianpur, G. (1980) *Stochastic Filtering Theory*. Springer, New York.

[112] Kallenberg, O. (1997) *Foundations of Modern Probability*. Springer, New York.

[113] Kaplan, N. (1981) Another look at the two-lift problem. *J. Appl. Prob.* **18**, 697–706.

[114] Karr, A. F. (1986) *Point Processes and their Statistical Inference.* Marcel Dekker, New York.

[115] Keilson, J. (1966) A limit theorem for passage times in ergodic regenerative processes. *Ann. Math. Stat.* **37**, 866–870.

[116] Keilson, J. (1979) *Markov Chain Models – Rarity and Exponentiality.* Springer, Berlin.

[117] Keilson, J. (1987) Robustness and exponentiality in redundant repairable systems. *Annals of Operations Research* **9**, 439–447.

[118] Kijima, M. (1989) Some results for repairable systems. *J. Appl. Prob.* **26**, 89–102.

[119] Koch, G. (1986) A dynamical approach to reliability theory. *Proc. Int. School of Phys. "Enrico Fermi," XCIV.* North-Holland, Amsterdam, pp. 215–240.

[120] Kovalenko, I. N. (1994) Rare events in queueing systems – a survey. *Queueing Systems* **16**, 1–49.

[121] Kovalenko, I. N., Kuznetsov, N. Y., and Pegg, P. A. (1997) *Mathematical Theory of Reliability of Time Dependent Systems with Practical Applications.* Wiley, New York.

[122] Kovalenko, I. N., Kuznetsov, N. Y., and Shurenkov, V. M. (1996) *Models of Random Processes.* CRC Press, London.

[123] Kozlov, V. V. (1978) A limit theorem for a queueing system. *Theory of Probability and its Application* **23**, 182–187.

[124] Kuo, W. and Kuo, Y. (1983): Facing the headaches of early failures: a state-of-the-art review of burn-in decisions. *Proceedings of the IEEE* **71**, 1257–1266.

[125] Lam, T. and Lehoczky, J. (1991) Superposition of renewal processes. *Adv. Appl. Prob.* **23**, 64–85.

[126] Last, G. and Brandt, A. (1995) *Marked Point Processes on the Real Line - The Dynamic Approach.* Springer, New York.

[127] Last, G. and Szekli, R. (1998) Stochastic comparison of repairable systems. *J. Appl. Prob.* **35**, 348–370.

[128] Last, G. and Szekli, R. (1998) Time and Palm stationarity of repairable systems. *Stoch. Proc. Appl.,* to appear.

[129] Leemis, L. M. and Beneke, M. (1990) Burn-in models and methods: a review. *IIE Transactions* **22**, 172–180.

262    References

[130] Lehmann, A. (1998) Boundary crossing probabilities of Poisson counting processes with general boundaries. In: Kahle, W., Collani, E., Franz, J., and Jensen, U. (eds.): *Advances in Stochastic Models for Reliability, Quality and Safety*. Birkhäuser, Boston, pp. 153–166.

[131] Marcus, R. and Blumenthal, S. (1974) A sequential screening procedure. *Technometrics* **16**, 229–234.

[132] Marshall, A. W. and Olkin, I. (1967) A multivariate exponential distribution. *J. Amer. Stat. Ass.* **62**, 30–44.

[133] Meilijson, I. (1981) Estimation of the lifetime distribution of the parts from the autopsy statistics of the machine. *J. Appl. Prob.* **18**, 829–838.

[134] Métivier, M. (1982) *Semimartingales, a Course on Stochastic Processes*. De Gruyter, Berlin.

[135] Natvig, B. (1990) On information-based minimal repair and the reduction in remaining system lifetime due to the failure of a specific module. *J. Appl. Prob.* **27**, 365–375.

[136] Natvig, B. (1988) Reliability: Importance of components. In: Johnson, N. and Kotz, S. (eds.): *Encyclopedia of Statistical Sciences*, vol. 8, Wiley, New York, pp. 17–20.

[137] Natvig, B. (1994) Multistate coherent systems. In: Johnson, N. and Kotz, S. (eds.): *Encyclopedia of Statistical Sciences*, vol. 5. Wiley, New York.

[138] Osaki, S. (1985) *Stochastic System Reliability Modeling*. World Scientific, Philadelphia.

[139] Phelps, R. (1983) Optimal policy for minimal repair. *J. Opl. Res.* **34**, 425–427.

[140] Pierskalla, W. and Voelker, J. (1976) A survey of maintenance models: The control and surveillance of deteriorating systems. *Nav. Res. Log. Q.* **23**, 353–388.

[141] Puterman, M. L. (1994) *Markov Decision Processes: Discrete Stochastic Dynamic Programming*. Wiley, New York.

[142] Rai, S. and Agrawal, D. P. (1990) Distributed Computing network reliability. 2nd ed. IEEE Computer Soc. Press, Los Alamitos, California.

[143] Rogers, C. and Williams, D. (1994) *Diffusions, Markov Processes and Martingales*, Vol. 1, 2nd ed. Wiley, Chichester.

[144] Rolski, T., Schmidli, H., Schmidt, V. and Teugels, J. (1999) *Stochastic Processes for Insurance and Finance*. Wiley, Chichester.

[145] Ross, S. M. (1970) *Applied Probability Models with Optimization Applications*. Holden-Day, San Francisco.

[146] Ross, S. M. (1975) On the calculation of asymptotic system reliability characteristics. In: Barlow R. E., Fussel, J. B. and Singpurwalla, N. D. (eds.) *Fault Tree Analysis*. Society for Industrial and Applied Mathematics, SIAM, Philadelphia, PA.

[147] Schöttl, A. (1997) Optimal stopping of a risk reserve process with interest and cost rates. *J. Appl. Prob.* **35**, 115–123.

[148] Serfozo, R. (1980) High-level exceedances of regenerative and semi-stationary processes. *J. Appl. Prob.* **17**, 423–431.

[149] Shaked, M. and Shanthikumar, G. (1993) *Stochastic Orders and their Applications*. Academic Press, Boston.

[150] Shaked, M. and Shanthikumar, G. (1991) Dynamic multivariate aging notions in reliability theory. *Stoch. Proc. Appl.* **38**, 85–97.

[151] Shaked, M. and Shanthikumar, G. (1990) Reliability and maintainability. In: Heyman, D. and Sobel, M. (eds.): *Stochastic Models, vol. 2*. North-Holland, Amsterdam, pp. 653–713.

[152] Shaked, M. and Shanthikumar, G. (1986) Multivariate imperfect repair. *Oper. Res.* **34**, 437–448.

[153] Shaked, M. and Szekli, R. (1995) Comparison of replacement policies via point processes. *Adv. Appl. Prob.* **27**, 1079–1103.

[154] Shaked, M. and Zhu, H. (1992) Some results on block replacement policies and renewal theory. *J. Appl. Prob.* **29**, 932–946.

[155] Sherif, Y. and Smith, M. (1981) Optimal maintenance models for systems subject to failure. A review. *Nav. Res. Log. Q.* **28**, 47–74.

[156] Smeitink, E., Dijk, N., and Haverkort, B. R. (1992) Product forms for availabilty models. *Applied Stochastic Models and Data Analysis* **8**, 283–302.

[157] Smith, M. (1998) Insensitivity of the *k* out of *n* system. *Probability in the Engineering and Informational Sciences*, to appear.

[158] Smith, M. (1997) *On the availability of failure prone systems*. PhD thesis Erasmus University, Rotterdam.

[159] Smith, M., Aven, T., Dekker, R. and van der Duyn Schouten, F.A. (1997) A survey on the interval availability of failure prone systems. In: Proceedings ESREL'97 conference, Lisbon, 17–20 June, 1997, pp. 1727–1737.

[160] Solovyev, A.D. (1971) Asymptotic behavior of the time to the first occurrence of a rare event. *Engineering Cybernetics* **9** (6), 1038–1048.

[161] Spizzichino, F. (1991) Sequential burn-in procedures. *J. Statist. Plann. Inference* **29**, 187–197.

[162] Srinivasan, S.K. and Subramanian, R. (1980) *Probabilistic Analysis of Redundant Systems.* Lecture Notes in Economic and Mathematical Systems 175, Springer, Berlin.

[163] Stadje, W. and Zuckerman, D. (1991) Optimal maintenance strategies for repairable systems with general degreee of repair. *J. Appl. Prob.* **28**, 384–396.

[164] Szász, D. (1977) A problem of two lifts. *The Annals of Probability* **5**, 550–559.

[165] Szász, D. (1975) On the convergence of sums of point processes with integer marks. In: Lewis, P. (ed.) *Stochastic Point Processes.*, Wiley, New York, pp. 607–615.

[166] Takács, L. (1957) On certain sojourn time problems in the theory of stochastic processes. *Acta Math. Acad. Sci. Hungar.* **8**, 169–191.

[167] Thompson, W. A. (1988) *Point Process Models with Applications to Safety and Reliability.* Chapman and Hall, New York.

[168] Tijms, H. C. (1994) *Stochastic Modelling and Analysis: A Computational Approach.* Wiley, New York.

[169] Ushakov, I. A. (ed.) (1994) *Handbook of Reliability Engineering.* Wiley, Chichester.

[170] Valdez-Flores, C. and Feldman, R. (1989) A survey of preventive maintenance models for stochastically deteriorating single-unit systems. *Nav. Res. Log. Q.* **36**, 419–446.

[171] Van der Duyn Schouten, F. A. (1983) *Markov Decision Processes with Continuous Time Parameter.* Math. Centre Tracts 164, Amsterdam.

[172] Van Heijden, M. and Schornagel, A. (1988) Interval uneffectiveness distribution for a $k$-out-of-$n$ multistate reliability system with repair. *European Journal of Operational Research* **36**, 66–77.

[173] Van Schuppen, J. (1977) Filtering, prediction and smoothing observations, a martingale approach. *SIAM J. Appl. Math.* **32**, 552–570.

[174] Voina, A. (1982) Asymptotic analysis of systems with a continuous
component. *Kibernetika* **18**, 516–524.

[175] Wendt, H. (1998) A model describing damage processes and resulting
first passage times. Research Report University of Magdeburg.

[176] Williams, D. (1991) *Probability with Martingales*. Cambridge University Press, Cambridge.

[177] Yashin, A. and Arjas, E. (1988) A note on random intensities and
conditional survival functions. *J. Appl. Prob.* **25**, 630–635.

[178] Yearout, R. D., Reddy, P., and Grosh, D. L. (1986) Standby redundancy in reliability − a review. *IEEE Trans. Reliability* **35**, 285–292.

# Index

# Applications of Mathematics

*(continued from page ii)*